A Date with the Two Cerne Giants

*Reinvestigating an Iconic British Hill Figure
(The National Trust Excavations 2020)*

Edited by Michael J. Allen

Published in association with the National Trust
and Allen Environmental Archaeology

Windgather Press is an imprint of Oxbow Books

Published in the United Kingdom in 2024 by
OXBOW BOOKS
81 St Clements, Oxford OX4 1AW

and in the United States by
OXBOW BOOKS
1950 Lawrence Road, Havertown, PA 19083

© Windgather Press and the authors 2024

Paperback Edition: ISBN 978-1-914427-37-4
Digital Edition: ISBN 978-1-914427-38-1 (epub)

A CIP record for this book is available from the British Library

All rights reserved. No part of this book may be reproduced or transmitted in any form or by any means, electronic or mechanical including photocopying, recording or by any information storage and retrieval system, without permission from the publisher in writing.

Printed in Malta by Melita Press

For a complete list of Windgather titles, please contact:

United Kingdom	United States of America
OXBOW BOOKS	OXBOW BOOKS
Telephone (0)1226 734350	Telephone (610) 853-9131, Fax (610) 853-9146
Email: oxbow@oxbowbooks.com	Email: queries@casemateacademic.com
www.oxbowbooks.com	www.casemateacademic.com/oxbow

Oxbow Books is part of the Casemate group

Front cover: The Giant Jigsaw, created by Declan Ingram
Back cover: Sheep grazing the Giant © Peter Stanier 2004

Contents

List of figures and tables — vi
Acknowledgements — x
Contributors — xii
Foreword: Our Cerne Giant by *Kate Adie* — xvi

Part 1: The Cerne Giant: excavation and dating the Giant — 1

1. Place, person and context: an introduction to the Cerne Giant — 3
   Michael J. Allen

2. The new research: dating the Giant; reconnaissance, aims and methods — 50
   Michael J. Allen

3. Research results: fieldwork, dating and analysis — 61
   Michael J. Allen

   High resolution photogrammetric survey (3-D ground surface model) (*Michael J. Allen*) — 62
   Geophysical surveys (*Andrew David, Tony Clark, Alister Bartlett, Paul Linford, Megan Clements and Paul Cheetham*) — 68
   Excavation results (*Michael J. Allen and Martin Papworth*) — 77
   Auger surveys (*Michael J. Allen*) — 103
   Optically stimulated luminescence dating (*Phillip Toms, Jamie Wood and Michael J. Allen*) — 108
   The land-use history of a hillside: land mollusc evidence (*Michael J. Allen*) — 113
   Discussion and conclusions: putting the Giant in his place in the landscape (*Michael J. Allen*) — 120

4. The Giant and the early medieval history of Cerne — 122
   Barbara Yorke

5. Hide and seek on a Dorset hillside — 136
   Brian Edwards

6. Know your Giant — 150
   Brian Edwards

Contents

7.	The Giant's story: the archaeological results considered *Michael J. Allen*	157
	Main conclusions	157
	The date of the Giant	158
	The sleeping Giant	162
	At least two Giants	163
	Outline and form	163
	Recording scouring and maintenance activities	173
	Conclusion	174
	The Giant timeline *(Brian Edwards and Michael J. Allen)*	175
	Acknowledgements	180

Part 2: The Giant in context — 181

8.	The Saxon Abbey of Cerne: an introduction to the Abbey and recent archaeological research *Michael J. Allen*	183
9.	The tenth-century Cerne Abbey: Benedictine ecclesiastical reform and land management *Katherine Barker*	190
10.	The Cerne Giant: an antiquity on trial 1996; a summary *Katherine Barker*	195
11.	Why did we think the Giant was ancient? *Timothy Darvill*	203
12.	Giant assumptions: locating chalk figures within prehistory *Susan Greaney*	208
13.	Images of the Giant *Sarah Fry*	215
14.	A research agenda for the Giant *Michael J. Allen*	222

Part 3: Giant considerations: wider reflections on the results — 229
Context and contrasts

15.	The Long Man of Wilmington: a progress report on a giant conundrum *Martin Bell and Chris Butler*	231

16.	I will survive: the continuing story of the Uffington White Horse *David Miles and Simon Palmer*	247
17.	Two chalk giants: Wilmington and Cerne revisited *Rodney Castleden*	258

Essays and reflections

18.	Implications of the hill figure dates *Ronald Hutton*	271
19.	Heroes, kings and giants at assembly places *Stuart Brookes*	275
20.	Wiltshire's chalk equine hill figures: what's the problem? *Garry Gibbons*	284
21.	Hill figures in the landscape: contexts, survival and function *Tom Williamson*	297
22.	Hill figures: retrospective and a national research agenda *Michael J. Allen and Win Scutt*	307

Appendices 321

Appendix 1: Description of Giant Hill and chalk grassland vegetation; loose insert in National Trust Management Plan November 1974 — 323

Appendix 2: Placing Cerne Abbas 'On the Map'; Stuart Piggott's 1946 BBC radio broadcast (21 June 1946) on the theme of the Giant (*Jan Lewis*) — 325

Appendix 3: National Trust Management Plan November 1974; Appendix 2, The Cerne Giant: Schedule of Works — 329

Appendix 4: Location of OSL sample 1 — 331

Bibliography — 334

Index — 354

List of figures and tables

List of figures

Figure 1.1.	Location of Cerne Abbas and the Giant in relation to the topography and archaeological sites	9
Figure 1.2.	Cerne Abbas in relation to the Giant and the Trendle	13
Figure 1.3.	Aerial view of Giant Hill and Cerne Abbas and location of the former abbey	14
Figure 1.4.	Plan of the abbey site showing extant earthworks	15
Figure 1.5.	Sir Flinders Petrie's 1920s survey of the Giant	16
Figure 1.6.	Photograph of the Giant with the Trendle on the hilltop	17
Figure 1.7.	Antiquarian representations of the changes in the Giant's genitalia	19
Figure 1.8.	Giant with cloak and severed head	21
Figure 1.9.	Symbols and numbers between the Giant's legs (Hutchins 1774)	23
Figure 1.10.	Giant and Giantess 1997	24
Figure 1.11.	The Giant on Giant Hill	25
Figure 1.12.	The earliest illustration of the Giant (Society of Antiquaries minute book 1763)	27
Figure 1.13.	The Cerne Giant drawing from the *Gentleman's Magazine* 1764	31
Figure 1.14.	Cerne Giant emasculated with symbols between his legs by Hutchins (1774)	32
Figure 1.15.	The main illustrations of the Giant, seventeenth to twentieth century	33
Figure 1.16.	The Dorset Ooser with the Wessex Morris Men	36
Figure 1.17.	(left) Rodney Castleden undertaking resistivity survey; (right) geophysical plot	42
Figure 1.18.	Alister Bartlett and Mike Allen with the survey of the Giant on the wall	43
Figure 2.1.	The left leg showing rainwater scouring, September 2019	52
Figure 2.2.	The Giant's left foot with rills and chalk accumulation, September 2019	53
Figure 2.3.	The Giant's left foot showing areas of soil accumulation	54
Figure 2.4.	Plan of the Giant showing the four 2020 trenches and auger hole	56
Figure 3.1.	Colour vertical image of the drone survey in 2020 of the Giant	63
Figure 3.2.	The processed and rendered image of the Giant and Trendle	66
Figure 3.3.	Detail of the Giant in the rendered image	67
Figure 3.4.	Vertical vs perpendicular views to give a true plan image	68
Figure 3.5.	New (2020) outline rectified image of the Giant	68

Figure 3.6.	Location of the 1979 earth resistance survey	70
Figure 3.7.	Plots of the earth resistance survey	72
Figure 3.8.	Location of the three 2023 geophysical survey areas	74
Figure 3.9.	Electrical imaging profiles across the top of the legs and down the phallus	75
Figure 3.10.	Geophysics survey Area C: GPR and resistance plots	76
Figure 3.11.	Plan of the Giant showing the location of the four 2020 trenches and auger hole	78
Figure 3.12.	Plans of the location of the four 2020 trenches	79
Figure 3.13.	Excavation of the four small trenches on the Giant	79
Figure 3.14.	(left) Section drawings of trenches B–D, north sections; (right) Section drawings of trenches A, B and D, south sections	83
Figure 3.15.	Trench B; north-facing section	95
Figure 3.16.	Trenches B and D; sections of chalk outline fills with proposed infill dates	97
Figure 3.17.	The wooden stake *in situ* in Trench B	97
Figure 3.18.	Auger profiles (2024) across the top of the legs and phallus	105
Figure 3.19.	Interpretation of the development outline and accumulating and colluvium	106
Figure 3.20.	Phil Toms monitoring the OSL sample, Trench B	109
Figure 3.21.	The two main sampled sections showing snail and OSL samples	112
Figure 3.22.	The land snail histograms from Trench B and C	118
Figure 5.1.	Detail from early Edwardian postcard with highlighted penis and navel	139
Figure 5.2.	a) The 1763 sketch of the Giant from the Society of Antiquaries minute book; b) The Giant (after Benjamin Pryce 1768)	142
Figure 5.3.	The Giant by Samuel Hieronymus Grimm in 1790	143
Figure 5.4.	The Cerne Giant drawn in 1764 and again in 1774	144
Figure 5.5.	The Cerne Giant drawn in 1897	145
Figure 6.1.	The Cerne Giant 'intact' on a rechargeable milk bottle	150
Figure 6.2.	William III as Hercules from a Dutch broadside of 1689	153
Figure 6.3.	The dates of the 'repairing of the Giant' from the churchwardens' accounts 1694	154
Figure 6.4.	William III when Prince of Orange *c.* 1667	155
Figure 7.1.	Chalk outline infill, Trench A	165
Figure 7.2.	Detail of photograph of the Giant by Rev E.V. Tanner *c.* 1920s–1940s	165
Figure 7.3.	Aerial photographs of the Giant with trenched and chalk-filled outline 1925 and 1971	167
Figure 7.4.	Ravilious' paintings of the Cerne Giant and Long Man of Wilmington	168
Figure 7.5.	Photograph of the Giant, September 2023	169
Figure 7.6.	Aerial photograph with a white Giant 1935	171
Figure 8.1.	Plan of the Cerne abbey site showing extant earthworks	186

Figure 8.2.	Plan of Sherborne Abbey	187
Figure 8.3.	Cerne Abbey: location of the GPR survey and 2023 excavation trenches	189
Figure 10.1.	Detail of the Giant from a 1920s aerial photograph	196
Figure 10.2.	Giantess 'grid' 1997	199
Figure 10.3.	Skewering the plastic sheeting to create the Giantess in 1997	200
Figure 10.4.	Aerial photograph of the Giant and Giantess 1997	201
Figure 10.5.	Still from computer-generated animation of the Giant	201
Figure 13.1.	Cover of *One for the Record: The authorised biography of Galahad*	216
Figure 13.2.	Giant sporting moustache to support Movember	217
Figure 13.3.	Drawing of the Giant with white mask	219
Figure 13.4.	Giant faces up to Covid with his mask – Dorset Council's reminder	220
Figure 13.5.	National Trust volunteers polishing up the Giant's parts	220
Figure 15.1.	Wilmington Long Man in its South Downs escarpment setting	232
Figure 15.2.	Wilmington: excavating Trench 2 with a circle of pagan worshippers below	233
Figure 15.3.	Wilmington: location of trenches across the Long Man	234
Figure 15.4.	The 2002 trench at the base of the slope below the Long Man	236
Figure 15.5.	Section of the 2002 trench at the base of the slope below the Long Man	237
Figure 15.6.	Section drawing 2002 trench showing the buried soil and OSL sample	238
Figure 15.7.	Excavated faces showing the shallow soil and lack of features cutting chalk	239
Figure 16.1.	The Uffintgon White Horse above the Manger and Dragon Hill	248
Figure 16.2.	Location of the Uffington monuments	250
Figure 16.3.	Variations in the Uffington horse 'Beak'	253
Figure 16.4.	The scouring of the Uffington Horse in 1857	255
Figure 16.5.	Comparison of the horse in 1989 with 2022	256
Figure 17.1.	The Long Man: the 1776 Burrell sketch, and the Long Man before 1873–74	262
Figure 17.2.	Two images of the burning of Protestant martyrs (in Lewes in 1557)	263
Figure 17.3.	The rack, as depicted in Foxe's *Book of Martyrs*	264
Figure 17.4.	Changes to the phallus: as today and before 1908	265
Figure 17.5.	Figurines and images of Gaulish warriors	266
Figure 17.6.	The 1996 site meeting of National Trust archaeologists at the Cerne giant	267
Figure 17.7.	Cerne Giant; how he might have appeared in the seventh century	270
Figure 19.1.	Map showing the boundaries of the Domesday shires	276
Figure 19.2.	The figures of Wandlebury Camp identified by Lethbridge (1957)	281
Figure 20.1.	Wiltshire's extant and lost chalk horses	285
Figure 20.2.	Devizes chalk horse, topographic survey	292

Figure 20.3.	Devizes chalk horse, partial outline and dimensions	292
Figure 20.4.	Photographs of the Marlborough chalk horse *c.* 1860, and *c.* 1866	294
Figure 20.5.	Marlborough chalk horse plan survey and contour survey	295
Figure 21.1.	Whiteleaf Cross, viewed from the air	300
Figure 21.2.	Whiteleaf Cross, as illustrated by Francis Wise in 1742	301
Figure 21.3.	Bledlow Cross in the early twentieth century	302
Figure 21.4.	The relationship of the Chiltern hill figures to the natural topography	304
Figure A4.1.	Location of OSL sample 1 in Trench C	331

List of tables

Table 1.1.	Dimensions of the Giant	41
Table 3.1.	Context categories for colluvium and chalk fill in the four trenches	85
Table 3.2.	Wooden stake measurements	88
Table 3.3.	The location of the OSL samples	108
Table 3.4.	OSL aliquot type, grain fraction and ages	110
Table 3.5.	The OSL results	111
Table 3.6.	Land snails from the Cerne Giant Trenches B and C	111
Table 13.1.	Selected images of the Giant referenced in the chapter	217
Table 20.1.	History of Devizes White Horse	291
Table 22.1.	Hill figures in England: description and designation	309
Table A4.1.	Fractions of the sieved 2 kg bulk sample	332
Table A4.2.	The OSL 1 sample location	333

Acknowledgements

The 2020 fieldwork at the Cerne Giant was directed by Martin Papworth with Mike Allen who undertook the geoarchaeology and archaeological science, with fieldwork assisted by former National Trust staff (Nancy Grace) and National Trust volunteers (Beth Darlington, Carol Lewis, Peter Moore and Fay Pendell). The OSL sampling was undertaken by Phil Toms (University of Gloucestershire) and the palaeo-environmental sampling was assisted by Julie Gardiner. The 2023 geophysical survey was led by Paul Cheetham, organised by Mike Allen and assisted by nine volunteers, in particular by Dave Stewart, Anne Brown, Sarah Fry and Steve Griffin. The 2024 augering and belt survey was undertaken by Mike Allen and Sarah Fry. At Historic England we would like to thank Helen Woodhouse, and Keith Miller and Hugh Beamish (former Inspector and Assistant Inspector of Ancient Monuments respectively, Historic England) for their support and encouragement, and Keith Miller for discussing the work on site with us (2020), and subsequently, Sasha Chapman and Dan Bashford (Inspector and Assistant Inspector of Ancient Monuments respectively, Historic England) for advice and assistance with the section 42 licence for targeted geophysics (2023) and SMC for additional augering (2024). Providing advice and assisting with permissions to auger on the SSSI land adjacent to the Giant on Giant Hill was David Charman, lead advisor West Dorset Team, Natural England and Ellie MacConnachie, Natural England.

The geoarchaeological, palaeo-environmental work and drone photogrammetry was made possible by funding from the National Trust and supported by Hannah Jefferson, General Manager for North and West Dorset National Trust Property Portfolio and her team. It was the centenary events surrounding the celebration of the acquisition of the Giant by National Trust in 1920 that created the opportunity to at last fund this work.

Publication of this book was funded by the National Trust (Hannah Jefferson and Christopher Tinker, Publisher, Curatorial Content), with a grant towards illustrations (Chapters 1, 3 and 8) from CBA Wessex. The research and writing of the final archaeological report, and the compilation of this book was generously supplied in time and limited finance by Mike Allen (Allen Environmental Archaeology), which enabled the satisfactory completion of this project, and the production of this book.

We particularly thank Phil Toms and the University of Gloucestershire for the analysis of more OSL dates than were funded, but also for continued discussion of deposit taphonomy in relation to the results. Special thanks have to go to Brian Edwards for constant information (and guidance), discussion, and historical information about the Giant, and to Martin Bell for discussing his work at the Long Man of Wilmington (see also *Epilogue* in archive reports). We also offer our thanks to Garry Gibbons, curator at Uffington's Tom Brown Museum for information on maintenance of chalk figure sites and in particular Peter Hawes (formerly Monks Wood National Nature Reserve) for detailed ecological information of chalk grassland recolonisation.

In completing the post-excavation research and writing this book we thank in particular Becky Loughead (Librarian, Society of Antiquaries of London), Sarah Cottom (AC Archaeology) and Peter Bellamy (Terrain Archaeology) for grey literature (evaluation and fieldwork) reports, Michael Clarke (National Trust area ranger) for information on recent scouring and rechalking, Claire Pinder (Dorset County Council), Francesca Ratcliffe, and Cerne Historical Society especially Gordon Bishop and John Charman. We also thank David Hinton, Martin Brown, James Clark, Chris Copson, several contributors and eight anonymous referees amongst others for advice on texts. In particular Brian Edwards was a constant fount of knowledge, inspiration and support.

The Wessex Morris Men via the auspices of David Chiplen, Dorset Ooser Beast Handler, and Jane Tearle, provided the image of the Dorset Ooser. Downland Partnership, Eduardo Vaquero and Robin Ault (Downland Exact Survey) for rendering of the photogrammetry and Keith Challis (National Trust) for digital manipulation of these data. We also thank the National Trust business support staff at Tisbury (Eleanor Daniel, Nicola Ravenhill and Rachel Rider) for archive searching and preparing scans and jpeg images for this book. We also thank Clare Harrup, Peter Stanier, Stu Nicholson (Galahad) and Jeremy Geller for providing and allowing reproduction of their images. Published in association with the National Trust's Cultural Heritage Publishing programme, the National Trust gratefully acknowledges a generous bequest from the late Mr and Mrs Kenneth Levy that has supported the cost of preparing this book. At Oxbow we thank Jess Hawxwell (Managing Editor), Eduard Cojocaru (former Editorial Administrator), Dec Ingram (Designer) and Julie Gardiner (Publisher) for assistance.

Contributors

Author/Editor

MICHAEL J. ALLEN
Geoarchaeologist/environmental archaeologist, Allen Environmental Archaeology and visiting research fellow, Bournemouth University, Allen Environmental Archaeology, Redroof, Green Road, Codford, Wiltshire, BA12 0NW (www.themolluscs.com), and Department of Archaeology & Anthropology, Bournemouth University, Christchurch House, Talbot Campus, Fern Barrow, Poole, BH12 5BB
Email: aea.escargots@gmail.com

Key Contributors

KATE ADIE, CBE
Author and broadcaster, Cerne Abbas

KATHERINE BARKER
Visiting research fellow, University of Bournemouth
Email: katherinebarker@lanprobi.org.uk

ALISTER BARTLETT†
Formerly Geophysics Section, Ancient Monuments Laboratory, then Department of the Environment (pre-adoption of English Heritage/Historic England)

MARTIN BELL
Geoarchaeologist and archaeologist, emeritus professor of archaeology, Department of Archaeology, University of Reading, Whiteknights, Reading, RG6 6AB
Email: m.g.bell@reading.ac.uk

CHRIS BUTLER
Managing director CBAS Ltd, commercial archaeologist, CBAS Ltd, Unit 12, Mays Farm, Selmeston, Polegate, East Sussex, BN26 6TS
Email: chris.butler@cbasltd.co.uk

STUART BROOKES
Lecturer in British archaeology, UCL Institute of Archaeology, 31–4 Gordon Square, London, WC1H 0PY, and Editor *Antiquaries Journal*
Email: s.brookes@ucl.ac.uk

Rodney Castleden
Author, archaeologist, geographer, former head of humanities at Roedean School, Rookery Cottage, Blatchington Hill, Seaford, East Sussex
Email: rodneycastleden35@gmail.com

Tony Clark†
Formerly Head of Ancient Monuments Laboratory (and Geophysics section), then Department of the Environmental (pre-adoption of English Heritage/Historic England)

Megan Clements
Geophysicist, Historic England, Fort Cumberland, Fort Cumberland Road, Eastney, Portsmouth, PO4 9LD
Email: Megan.Clements@HistoricEngland.org.uk

Paul Cheetham
Archaeological scientist and visiting research fellow, formerly Department of Archaeology and Anthropology, Bournemouth University, Christchurch House, Talbot Campus, Fern Barrow, Poole, BH12 5BB
Email: PCheetham@bournemouth.ac.uk

Timothy Darvill
Prehistorian, archaeologist, professor of archaeology, Department of Archaeology and Anthropology, Bournemouth University, Christchurch House, Talbot Campus, Fern Barrow, Poole, BH12 5BB
Email: tdarvill@bournemouth.ac.uk

Andrew David
Formerly Geophysics Section, Ancient Monuments Laboratory, English Heritage
Email: andrew.david1@hotmail.co.uk

Brian Edwards
Historiographer and public historian, visiting research fellow, University of West of England, Regional History Centre, University of West of England, Coldharbour Lane, Stoke Gifford, Bristol, BS16 1QY
Email: public.history@gmail.com

Sarah Fry
Associate fellow, Institute of Historical Research, University of London, Senate House, Malet St, London WC1E 7HU
Email: sarah1.fry@btinternet.com

Garry Gibbons
Former curator, Tom Brown's School Museum, Uffington
Email: garry.gibbons@btinternet.com

xiv   *Contributors*

NANCY GRACE
Former National Trust archaeologist
Email: nance.grace@btinternet.com

SUSAN GREANEY
Prehistorian, archaeologist, university lecturer, Department of Archaeology and History, University of Exeter, Laver Building, North Park Road, Exeter, Devon, EX4 4QE
Email: S.Greaney@exeter.ac.uk

RONALD HUTTON
Professor of history, Department of History (Historical Studies), University of Bristol, Beacon House, Queens Road, Bristol, BS8 1QU, and Gresham professor of divinity, Gresham College, London
Email: R.Hutton@bristol.ac.uk

JAN LEWIS
Visiting research fellow, Faculty of Media and Communication, Bournemouth University, Christchurch House, Talbot Campus, Fern Barrow, Poole, BH12 5BB
Email: JLewis@bournemouth.ac.uk

PAUL LINFORD
Geophysics principal, Historic England, Fort Cumberland, Fort Cumberland Rd, Eastney, Portsmouth, PO4 9LD
Email: Paul.Linford@historicengland.org.uk

DAVID MILES
Former chief archaeologist English Heritage, former director Oxford Archaeology
Email: david.miles66@btinternet.com

SIMON PALMER
Former archaeologist at Oxford Archaeology, co-director Uffington White Horse Project
Email: nomispalmer@gmail.com

MARTIN PAPWORTH
Dating Cerne Giant project director, archaeologist south west region, National Trust, SW Region, Wiltshire Office, Tisbury Hub, Place Farm Courtyard, Court Street, Tisbury, Wiltshire, SP3 6LW
Email: Martin.Papworth@nationaltrust.org.uk

WIN SCUTT
Senior Properties Curator (West), English Heritage, 1st Floor Fermentation North, Finzels Reach, Hawkins Lane, Bristol, BS1 6JQ
Email: Win.Scutt@english-heritage.org.uk

PHILLIP TOMS
Professor of physical geography, Luminescence dating laboratory, University of Gloucestershire, Swindon Road, Cheltenham, Gloucestershire, GL50 4AZ
Email: ptoms@glos.ac.uk

TOM WILLIAMSON
Emeritus professor of landscape history, School of History, University of East Anglia, Norwich Research Park, Norwich, Norfolk, NR4 7TJ
Email: T.Williamson@uea.ac.uk

JAMIE WOOD
Research fellow, Luminescence dating laboratory, University of Gloucestershire, Swindon Road, Cheltenham, Gloucestershire, GL50 4AZ
Email: jwood1@glos.ac.uk

BARBARA YORKE
Emeritus professor of early medieval history, School of History and Archaeology, University of Winchester, Sparkford Road, Winchester, Hampshire, SO22 4NR
Email: barbara.yorke@winchester.ac.uk

Project publication contributors

CATHIE BARNETT
Environmental archaeologist, technical director archaeology and heritage & environmental specialist services lead, Stantec, The Blade, Abbey Square, Reading, United Kingdom, RG1 3BE
Email: Catherine.Barnett@stantec.com

TONY DAVIES; DOWNLAND PARTNERSHIP LTD.
Survey and measurement, Unit 6, Roundway Hill Business Centre, Devizes, Wiltshire, SN10 2LT
Email: info@exactsurvey.com

Foreword: Our Cerne Giant

Kate Adie

Any giant commands attention. Our giant in Cerne Abbas certainly does. Throughout history, such figures – real and mythical – have fascinated people. And on a hillside with a commanding view over the Dorset countryside, our Giant, the largest in Britain, draws tourists, acts as an unofficial navigation point for low-flying aircraft, and is the source of endless tales and traditions – mostly involving young maidens …

It is also a creature of fascination to historians and archaeologists. And as this book reveals, scientific advances are cutting through centuries of speculation about his origin and his age. The result is both surprising – and fascinating – and involves long-dead snails!

The villagers have always speculated about his age too, as we direct tourists towards the Giant Hill, and point out that the hike up the hill is not a mere stroll. As for how long it's been there … There has been debate about his age – ranging from the Bronze Age, to the Roman occupation – is he Hercules? – to the seventeenth century. The village hosts several ancient sites – a wellspring connected in legend to St Augustine, a medieval St Catherine's Chapel, a tithe barn and timber-framed houses. Farmers and local gardeners regularly turn up Roman coins and artefacts. Underneath a grassy area called Belvoir lie the crumbled remains of a huge Benedictine Abbey, courtesy of Henry VIII in his Dissolution of the monasteries. But unusually, for a busy, often prosperous village, with several centuries of pilgrims trekking to the Abbey in medieval times, on a trail that went straight past the Giant Hill, there's no early description, not a mention to be found of our Giant. The first known reference is in the village's churchwardens' accounts, in 1694, which state that 3 shillings were paid 'for repairing the Giant'. For the moment, his outline on the green hill is a broad line of chalk, which makes him visible for miles. The chalk fades quickly and the sheep nibble at him. His gaze falls on the local school, which every few years, adds a cohort of children to help the National Trust attend to his outline (17 tonnes of chalk are needed to pack into the earth). Clearly, over the centuries, he needed attention – hence the 1694 expense. But no one, local or traveller, seems to have mentioned him until then.

So until recently, everything about him has been speculation – or a good guess. Even his outline is not settled – we're all used to seeing the sharp chalked figure. But for centuries it was a brown outline on the grassy slope, with a trench to mark the body and the details – details that may have varied through the ages – possibly a severed head dangling from his hand, a cloak hanging from his arm and, possibly even mysterious numbers and letters between his thighs. All subject to speculation, depending on how you imagine your giant should be.

Speculation which ensures that, over the years, we have had a steady stream of historians, archaeologists, writers and artists passing through the village, as well as tourists and curious visitors. So the Giant is a source of local pride; his presence engenders a genuine fascination with the ground on which we live – most gardeners have turned up the odd Roman tile or coin, and several fireplaces have huge stone lintels clearly acquired from the Abbey. And the curious grass humps and hillocks at the base of the Giant Hill on the west side may yet hold more secrets.

In 2023, the village hosted a team from Sheffield University investigating a possible discovery of the exact whereabouts of the Benedictine Abbey. The excavations had a permanent bunch of sightseers, eager to view what was found 10 feet down. No one expected to find a complete skeleton. And there he was, on day four, possibly a medieval abbot. Part of a tiled floor emerged, large fragments of carved stone, and buckets of animal bones. All this gives the village a sense of history all around; that people have lived here for millennia, building, planting, cooking, praying – and seeing strangers – Romans, pilgrims, travellers, artists, tourists. The Giant is the most curious part of our history and in the last century he gained more professional attention, with science helping to pin down some details.

However, the latest investigation led by Martin Papworth with Mike Allen and the National Trust has revealed that there are likely to be not one but *three* giants. And that he was here before William the Conqueror arrived … This is the story of geoarchaeology, archaeology, ancient snail shells and the modern curse of Covid, all coming together to reveal not so much a different giant, but a tale of repeated efforts to maintain a figure on the hill over a thousand years.

Yet again, he commands our attention.

Part 1

THE CERNE GIANT: EXCAVATION AND DATING THE GIANT

Painting of Cerne Giant on the slopes of Giant Hill. © Clare Harrup 2023

CHAPTER ONE

Place, person and context: an introduction to the Cerne Giant

Michael J. Allen

The Giant is unique and in my view one of our most important properties, justifying much trouble and attention (National Trust Warden undated *c.* 1974)

A research design (2019), excavation (2020) and post-excavation analyses (2022) all aimed primarily to date the Giant (and provide a palaeo-environmental context for him). This book does that (Part 1), but goes well beyond by providing a history of the Giant and his maintenance, and of his contemporaneous landscape (Part 2), before placing him in the world of other geoglyphs: chalk hill figures (Part 3).

In 2019 Martin Papworth, National Trust archaeologist, South West Region, conceived the idea of finally providing a scientific date for the Giant. Hitherto, his age had only been one of speculation and debate amongst antiquarians, archaeologists and historians for centuries. This was an aspiration he'd articulated as long ago as 1996 (Bergamar 1997, 27). Consequently, following a reconnaissance site visit (Allen 2019) and the development of project design and method statement (Allen 2020a), four small clinical incisions were made into the Giant in March 2020 with the aim of principally obtaining OSL (optically stimulated luminescence) dates. This also provided an opportunity to look at his accompanying land-use and environmental history. Martin's original aim was to present the results to the Cerne Abbas village and the nation as a whole in July 2020 to mark the centenary of the National Trust's acquisition of the Giant; a gift from the Pitt Rivers family on 20 July 1920. COVID-19 obviously obviated that desire and announcement. Although publication of the results of four small trenches, in a book rather than a simple paper as originally planned, is considerably more work, it does allow much greater consideration of the results and the landscape in which the Giant stands and provides a more rounded archaeology that he deserves.

This book, therefore, provides the background and history of the Giant, the excavation results, the dates, and discusses his identity. This enables us firstly to place the Giant into his relevant local archaeological/historical context and retrospectively to consider the dates previously considered for the Giant. The first published research agenda for the Cerne Giant (Chapter 14) is based on our new data and review of previous research. This provides the basis for wider considerations such as comparison of the Giant with other excavated hill figures and other geoglyphs, and sets out some concluding comments about hill figures generally, outlining a national research agenda. Little of this would have been possible in a short journal paper – and we hope that in total this does justice to the research undertaken and to the Cerne Giant. It also provides the opportunity to expand upon wider themes relating to the Giant and to hill figures in general. Consequently, Part 1 deals with the excavation, analysis and consideration of the results; Part 2 provides seven essays placing the Giant into his chronological and archaeological context, ending with the first research design for the monument. The final part, Part 3, explores aspects of hill figures as archaeological monuments, again finishing with an outline research design for hill figures as a whole. Liaison was maintained with the Cerne Historical Society during the excavation and post-excavation research and they were invited to contribute to this book but declined.

National Trust centennial Cerne project

The Cerne Giant is an iconic enigmatic figure standing alone on the Dorset hillside. Among many of the mysteries that engulf him is that of his age. There has been a strong desire for centuries amongst antiquarians, archaeologists and historians, as well as the public at large, to know the true date of the Giant; although for some, it must be admitted, there was a wish for this mystery to be preserved. Amongst many past notable enquirers have been William Stukeley and Rev John Hutchins in the eighteenth century; John Sydenham and Dr Wake Smart in the nineteenth century; Sir Flinders Petrie and Stuart Piggott in the twentieth century. Despite attention, discussion and even survey stretching back several hundred years, during which many articles and books had been written about him, and television programmes dedicated to him, before 2020 his age had only been debated and not established. At the start of the twenty-first century, the best that could be said was 'the origin of the Giant is unknown and hotly debated' (Papworth & Keighley 2004; see also Darvill *et al.* 1999).

Of all the questions most commonly asked of him (when was he carved on to the hillside, who by and who was he?), only dating him was a question that could, potentially, be achieved scientifically, clinically and with any certitude. The desire to answer this question was no less strong than with the archaeologists in the National Trust, the organisation who had owned and tended him for a century. It was apposite then that Martin Papworth, National Trust archaeologist for the South West, set out to do this, and at such a time when the result could also celebrate the centenary gift of the Giant by Alexander and George Pitt Rivers, to the National Trust. The ambition (in 2019) was to undertake excavation, sampling and dating in time to announce the date of the Giant (ideally in the village) on 20 July 2020, exactly 100 years after the Giant was gifted to the National Trust.

Both the Uffington White Horse, Oxfordshire and the Long Man of Wilmington, East Sussex were successfully dated by OSL dating of colluvium or soils associated with the chalk figures. The Uffington White Horse was dated to somewhere between 1400–600 BC (Miles *et al.* 2003; Rees-Jones & Tite 2003), and a buried soil associated with the Long Man of Wilmington produced an OSL date of AD 1420±620, and bricks that originally outlined him gave an OSL date of AD 1545±30 (Bell & Butler 2014). Previous research designs had already proposed to date the Cerne Giant in a similar fashion (Oxford Archaeology Unit 1998).

OSL has an advantage in that it is based in characteristics of the sediments, rather than any cultural debris or detritus (charcoal or charred items for radiocarbon dating, pottery or other diagnostic artefacts), which we'd expect to be absent or sparse from hill figures essentially devoid of domestic and settlement activity. Instead OSL provides a measure of time since quartz grains in soils and sediments were deposited and shielded from further light (Aitken 1985). It is dating based on the principal that exposure to light releases trapped electrons in crystals (e.g., quartz), and the quantity remaining trapped since last exposure to light provides a measure of the age of sediments (Grün 2001).

Interest in the date and identity of the Giant had escalated as a result of the *Cerne Giant: an antiquity on trial*, a concept conceived and the event organised by Katherine Barker (see Chapter 10). A one-day conference/trial debating the origin of the Giant was held in the Cerne village hall in March 1996 (Barker, Chapter 10; Darvill *et al.* 1999). Inspired by this, and following from a meeting on site in October 1996, the Oxford Archaeological Unit had, at the invitation of the National Trust, produced a detailed archaeological research design by December 1998. Unfortunately at this time funds could not be found to realise this. Although 'Further research, including investigating of the potential buried soils, and to develop an understanding of the construction, and possibly the date of the figure by careful excavation are currently being discussed (1997)' (Keithley *et al.* 1999, 25), a second opportunity arose in June 2002 when a film company offered to fund excavation and date the Giant for a programme entitled *Landscape Mysteries*. They had contacted Mike Allen, who proposed a trench across the Giant's left foot as the best location from which to take OSL samples for dating (National Trust Archaeology, Cerne Abbas correspondence files 21 June 2002). The lead in time was too short for all the approvals and consents to be agreed, and the National Trust Property Manager for West Dorset concluded the discussion by writing 'We discussed this today and my thoughts were. This could disappoint a lot of people, do we really need to do it? It would lead to changed expectations' (National Trust Archaeology, Cerne Abbas correspondence file 3 July 2002). And for some he is just important in himself: 'For his own folk he is something intimate and natural. His exact age does not concern them. He is just 'old' and was always there' (Harvey Darton 1935, quoted by Vale & Vale 2000, 14).

Undefeated, an attempt was made to date the Giant for the programme via proxy, by identifying the presence of colluvium (hillwash) at the foot of slope that had accumulated as a result of the initial construction; and then examining the colluvium for included artefacts (cf. Bell 1983; Allen 1988; 1991). This failed as, surprisingly, no colluvium could be detected during the augering programme (see previous fieldwork, probablistic augering 2002, below).

Here, the research ambitions for dating the Giant rested for another 17 years. In 2019, plans were made to date the Giant to celebrate the centenary of the gifting of the Giant to the National Trust. A research design was devised in conjunction with the National Trust (September 2019 to February 2020), and when completed (Allen 2020a), pleasingly it unknowingly contained a similar approach and number of similar elements to that produced by Oxford Archaeology 22 years earlier, despite its existence being unknown to me. Martin Papworth's request for funding was welcomed, and the Trust's West Dorset General Manager Hannah Jefferson allocated money for photogrammetry, excavation, OSL dating and palaeo-environmental analysis. The ambition remained to announce the date of the Giant on 20 July 2020, exactly 100 years after the Giant was gifted to the National Trust, and excavation needed to take place early that year to achieve this.

Clinical incisions into the Giant would also offer the opportunity to examine the history of rechalking, and the environmental and land-use history contemporary with the Giant. Application to the Secretary of State for Scheduled Monument Consent was successful, and five days of excavation and sampling were undertaken from 16–20 March 2020. The plan was to excavate four small clinical incisions into the Giant with the clear primary aim:

- To enable sampling to provide a *terminus ante quem* date for the construction of the Cerne Abbas Giant, via optically stimulated luminescence (OSL) dating.

And with subsidiary aims defined as:

- To examine the geoarchaeology of the chalk figure (and its associated deposits);
- To attempt to examine the palaeo-environment contemporary with the construction of the Giant (and its hillside land-use history) (Allen 2020b, 3).

On the last day of the excavation (20 March 2020) the OSL samples were taken by Prof Phil Toms (University of Gloucestershire), and environmental samples by Mike Allen, and later that same afternoon the then prime minster (Boris Johnson) announced closure of cafés, pubs, bars, restaurants and hotels, which was the start of the COVID-19 pandemic and widespread, worldwide lockdown (with full stay at home lockdown the following weekend). Most post-excavation analyses were delayed (University of Gloucestershire laboratories were locked with OSL samples inside; project leader Martin Papworth was furloughed by the National Trust), however, the final analysis phase (funded by National Trust) then ensued later that year, largely comprising the completion of the OSL dating and palaeo-environmental record. The publication project was organised and led in 2023 for the National Trust by Mike Allen in his own time.

Publishing the Giant project

The National Trust's idea of dating the Giant for the centennial anniversary of the ownership was an inspired one. The announcement of the OSL date was unfortunately thwarted by the COVID-19 measures announced on the last day of the excavation; samples were locked in university laboratories and key staff (e.g., Martin Papworth), were furloughed. However, the results ultimately were significant and unexpected, and deserved wider publication than that originally proposed in the project design; all the academic and project advisors unanimously suggested the results, and discussion of them, merited a national audience. The National Trust, in their generosity and naivety had, however, given the results to other archaeologists, allowing them to publish the dates with a seminal discussion of the archaeological and chronological relevance (Morcom & Gittos 2024) prior to the National Trust's own project publication. Consequently, national journal editors were then wary of publishing the project results as the key element (and the single principle aim of the project – dating the Giant) had already been published (Morcom & Gittos 2024). The largely drafted journal paper now had no obvious home or outlet, and on its own was too short for a stand-alone publication such as a monograph or book. The solution, however, was to allow much wider discussion of the Giant himself, and his social and chronological context, and then to use this as a platform and vehicle to invite some wide comparisons with, and discussion of, other hill figures and the hill figure phenomenon. In 2023 the project director was too busy with other National Trust commitments, several other post-excavation programmes and excavation publications, to undertake this prior to retirement. The project was reluctantly taken on by the editor, and through the final writing programme (May–December 2023) Martin Papworth became busier with other projects so withdrew from many of his own proposed writing commitments of the excavation report and for this book. Nevertheless, the publication in this book allows a much wider consideration of the Giant than has been achieved anywhere previously, with a diverse range of invited key authors it outlines comparisons with the two other OSL-dated hill figures (the Uffington White Horse and the Long Man of Wilmington), and the production for the first time of research agendas for both the Cerne Giant (Chapter 14), and hill figures generally (Chapter 22).

Although the announcement of the dates on the centennial anniversary of the National Trust's ownership of the monument was thwarted, this publication satisfyingly coincides with the centenary of its incorporation into the national listings as a Scheduled Monument (The Giant no. 1003202) on 15 October 1924.

The chalk figure

Hill figures are unique to England and are seen on the chalk scarp slopes mainly across southern England from Dorset to Kent, with a few examples in Yorkshire. Many take the form of horses (see Table 22.1; and Gibbons, Chapter 20) or military badges, but notably among them there are two human figures: the Long Man of Wilmington in East Sussex, and probably the most spectacular of them all, the Cerne Abbas Giant, Dorset. He is the subject of the research outlined here, and the inspiration for the essays and reflections we present in Part 3. Throughout this book we refer to the Giant's left and right feet/arms and trenches located in his right or left feet etc., not the right and left (as you look at them) feet.

The 55 m (180 ft) high chalk-cut white figure of the Cerne Abbas Giant has dominated the chalkland hillside (ST 6665 0168) overlooking the Dorset landscape for centuries in his full naked glory, arm outstretched and club in one hand. He lies (or stands) on the south-western scarp slope of Giant Hill (formerly known as Trendle Hill until the mid-nineteenth to early twentieth century) at 182 m OD, ½ km north of the village of Cerne Abbas (Fig. 1.1). The Giant stands boldly and proudly on the hillside overlooking the River Cerne and its wide valley with the village of Cerne Abbas and its abbey beyond his outstretched left hand to the south. For centuries archaeologists, antiquarians, villagers and visitors have debated his age and origin among themselves, in pubs, and in the pages of magazines, books and learned journals. The Giant, quite rightly, is a focus of interest in the village, and indeed in that part of Dorset. He is also typically the subject of myth, legends, folklore and japes, some of which we will touch upon later.

Because of his notoriety and antiquarian and archaeological interest (see Castleden 1996; Darvill *et al.* 1999), he has been a continued fascination for archaeologists (Petrie 1926; Crawford 1929; Piggott 1932; 1938; Darton 1935: Daniel 1976; Bettey 1981) and antiquarians (e.g., Society of Antiquities 1763 and *Gentleman's Magazine* 1764; Morgan Evans 1998; 1999; Edwards 2020; Edwards, Chapter 6). He has been the subject of speculation especially regarding his age, identity and meaning, yet there has been little archaeological fieldwork there, and no formal excavation. Despite this long public and academic interest, until this project (see high precision photogrammetry, Chapter 3) he had not been fully surveyed since Flinders Petrie's investigation of the 1920s (Petrie 1926) and this replaced the measured drawing of 1762/4. Apart from probably the first scientific geophysical survey of the whole figure (Clark 1980; 1983; David *et al.*, Chapter 3), most others have tended to be conducted of specific areas (Castleden 1993; 1996; Gale 1999; see Keithley *et al.* 1999) such as those by Rodney Castleden in 1989 and 1995 focusing on the phallus and the space under the left arm. Surprisingly, apart from limited test augering in the Cerne landscape in 2002 by Mike Allen (see *previous archaeological work*, below), no formal official invasive field investigation had ever been conducted on the Giant, unlike the Long Man of Wilmington (Holden 1971; Bell & Butler 2014; Chapter 15) and

1. Place, person and context: an introduction to the Cerne Giant

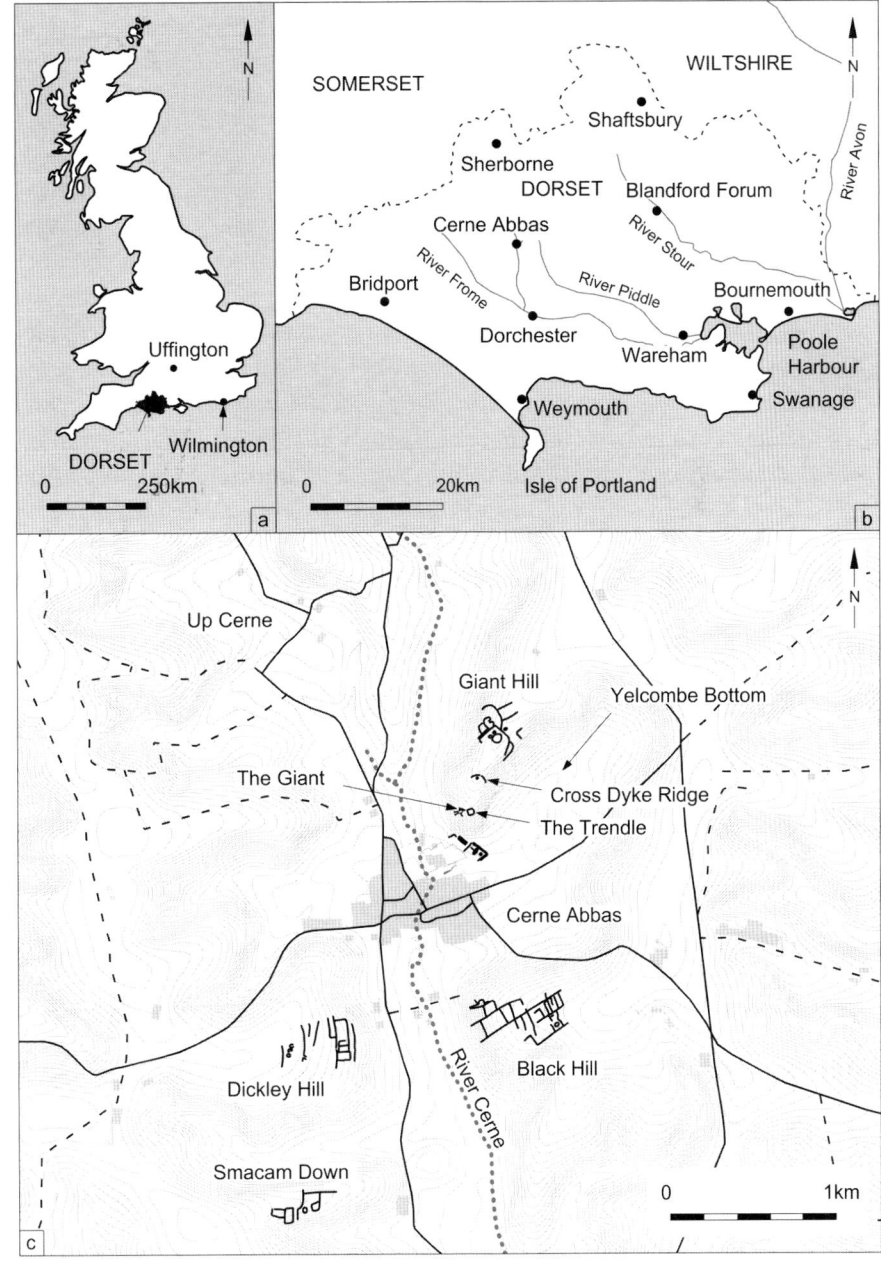

FIGURE 1.1. Location of Cerne Abbas. (top) Cerne Abbas in relation to the Dorset drainage; (bottom) Cerne Abbas in relation to the River Cerne, Yelcombe valley and main prehistoric field systems and archaeological sites. Image: Justin Russell

the Uffington White Horse (Miles *et al.* 2003; Chapter 16), both of which have been dated by those fieldwork campaigns.

The age of the Giant has, therefore, long been a matter of discussion, even among the general public at large, with no consensus, and with suggestions ranging from prehistoric to later historic. For centuries many antiquarians and archaeologists (e.g., Stukeley 1764; Piggott 1938) had allied him with the

Roman god Hercules (see Castleden 1999; Appendix 2), and Piggott stated, 'I feel it almost inevitable that the Giant of Cerne must be Romano-British, and may possibly date from years immediately following 191' (1938, 327). In more recent times, however, he has generally been considered to be either later prehistoric (Late Bronze Age/Iron Age) or historic, specifically seventeenth to eighteenth century, but rarely medieval or early medieval/Saxon save perhaps by Smart (1872). The strongly phallic nature of the Giant with his club has long been assumed to indicate that he was early i.e., prehistoric (Bell & Butler 2014; see Darvill, Chapter 11; Greaney, Chapter 12). The concept that he was Romano-British has generally lost favour. In 1999 these opposing views (prehistoric vs post-medieval) were exacerbated in the Trial set up in the Cerne village hall on 23 March 1996, where a number of archaeologists and historians presented their case for the date of the Giant. The resulting publication divided the presentations into prehistoric (the case for an ancient Giant, led by Prof Tim Darvill), and historic (the case of a post-medieval Giant, led by Prof Ronald Hutton) (Darvill *et al.* 1999). Of the prehistoric protagonists Darvill proffers a later prehistoric (Late Bronze Age to Iron Age date, and Newman and Putnam indicate an Iron Age (Celtic or Durotriges) association. Countering this, Hutton, Vale, Bettey and Morgan Evans generally suggested a seventeenth-century origin, with some throwing out the old adage that he was a ridicule of Cromwell. Needless to say, the trial did not start, nor end, with any decisive consensus (Barker, Chapter 10). See *Consideration of the date of the Giant* below for further discussion.

Much has been written about the Giant, including several books, notably Castleden (1996) and Legg (1986; 1990) and the trial (Darvill *et al.* 1999), and he takes a prominent position in many books on hill figures (e.g., Marples 1949; 1981; Bergamar 1968; 1986; 1997; Newman 1987; Goodman 1998; Castleden 2000), popular magazines (e.g., Copson 1988; Legg 1992) and academic papers (e.g., Smart 1872; Petrie 1926; Piggott 1932; 1938; Daniel 1976; Grinsell 1980a; Bettey 1981; Lloyd 1982; Clark 1983; Willcox 1988; Morgan Evans 1998), as well as amusing the committee of the Society of Antiquaries in the eighteenth century (Stukeley 1764). While we don't review or reproduce the debate and records of the Giant in detail, it is necessary at least to digest and revisit some the history of the Giant, the records of the figure and changing attributes, and review some of the previous survey work (e.g., Clark 1983; Gale 1999; Castleden 1996; 2000) to place our new evidence in the context of the Giant as a whole.

Although the original aim of presenting the results on the centennial anniversary of the National Trusts ownership was thwarted due to the COVID-19 pandemic and lockdown, this book provides the first publication of the excavation and of those dates considered together with their implications for our understanding of the Giant and other geoglyphs. The Giant is not alone in England. Although there is only one other true human figure (or geoglyph; rock figure), there are a number of other chalk hill figures, predominantly horses, in the country and along with considering the Giant in this wider context, this volume also considers aspects of some of these other archaeological monuments.

Setting and location

The hill figure known as the Cerne Giant (ST 66649 01679) is inscribed on a west-facing steep downland on the side of Giant Hill north of the village of Cerne Abbas, about 12 miles (19 km) south of Sherborne and 8 miles (12.5 km) north of Dorchester. Cerne Abbas lies within the block of Dorset chalkland north of Dorchester in quintessentially English rolling downland. The flora typical of chalk downland (e.g., *Hippocrepis comosa* or horse vetch, with Chalk Hill Blue and Adonis Blue butterflies), but species are starting to suffer severely (and thus declining) from 'agricultural improvement' of the hill i.e., fertiliser and spray (National Trust 1974a). A description of the grassland flora of Giant Hill was given in foolscap typescript insertion into the National Trust 1974 management plan and the flora from this is given in Appendix 1. The area can be seen to be defined by three north–south flowing rivers cutting through the Down to Dorchester: Sydling Water to the west, and the River Cerne flowing from Minterne Magna, though Cerne Abbas both flowing into the Frome just to the north-west of Dorchester, and the Piddle to the east flowing into Poole Harbour a few hundred metres north of the Frome (Fig. 1.1). Cerne Abbas itself lies nestled on the eastern side of the valley in a large embayment and the junction of the Cerne Valley and Yelcombe Bottom dry valley which heads eastwards into the down.

The giant stands on the south-western scarp (west-facing) slope of Giant Hill at 192 m OD, ½ km north of the village of Cerne Abbas and about 60 m above the valley floor of the Cerne Valley (Chartrand 1999). He is situated on a promontory just south of the confluence of the River Cerne and the western branch of the river (a stream) to Up Cerne, and lies north of Cerne Abbas village and the Yelcombe dry valley. He overlooks the 500 m wide Cerne valley and faces the chalk scarp of the opposite valley (Weam Common Hill in particular). The hillside of Giant Hill is Chalk over Sandstone with a modicum of floodplain alluvium in the valley at the foot of the scarp, and Clay-with-Flints on the ridge above and *c.* 500 m north of the Giant (Bird 1995). The Giant on the scarp face is etched into New Pit Chalk Formation (Middle Chalk), with Lewes Nodular Chalk Formation (Upper Chalk) above on the top of the hill and surmounted by Clay-with-Flints. Below the Giant is Holywell Nodular Chalk (Middle Chalk) and Zig Zag Chalk Formations (Lower Chalk) overlying the Shaftesbury Sand Member and Cann Sand Member (Upper Greensand), into which the Cerne Stream is cut, and onto which limited Holocene overbank floodplain alluvium is mapped. The white Chalk Formations provide the striking outline against the green, and summer browning, downland grass.

The steep 45° scarp slope supports shallow (<300 mm) humic rendzinas of the Icknield Association, with typical calcareous brown earths of the Bromyard

Association in the main valley, and stagnogleyic paleo-argillic brown earths of the Batcombe Association on the ridge forming in the Clay-with-Flints (Findlay *et al.* 1983; 1984). Bronze Age barrows lie on the spur, and above his head is the rectangular (almost square) undated, but possibly Iron Age or Romano-British enclosure of the Trendle (or 'frying pan'). The village itself is the home of the Saxon Cerne Abbey. Today he is fenced-off to protect the figure from damage by animal grazing and human footfall and the site is owned (since 1920) and managed by the National Trust (Keithley *et al.* 1999).

The Cerne valley is about ½ km wide with the River Cerne hugging its eastern edge at the foot of the Down at Cerne. Today the river is heavily canalised and modified, not least at the new flood defences north of Kettle Bridge, but also south of the bridge. It now flows, in part, in an artificial channel (a former mill race) slightly above the floodplain to the mill a couple of hundred metres south. The base of the scarp has two spring lines; one just below the Giant at the junction of the New Pit Chalk/Holywell Nodular Chalk Formations and the Zig Zag Chalk Formation and the second at the base of the slope at the junction of the Zig Zag Chalk Formation and the Upper Greensand (the Shaftesbury Sand Member) (Charman 2020).

The former River Silley was, in part, a canal built by the monks to provide the abbey with water. Water was diverted from the Cerne at Minton Parva, and following the topography at the side of the valley it ran around the foot of Giant Hill (across North Mead). It passes to the east of North Barn (now known as Silley Court Barn) downstream from Kettle Bridge, and then flowed into the former north-east–south-west orientated mill pond. This is not to be confused with the town pond at the top (north) end of Abbey Street (or the former abbey fishponds). From the former mill pond it is supposed to have then flowed westwards to Abbey Mill on the east bank of the River Cerne and to have returned to the Cerne, possibly via the former fishponds of the abbey. The extant spring (St Augustine's well) below the abbey, in the floor of Yelcombe Bottom, flows from the former abbey site into the town pond at the mouth of Yelcombe Bottom and was incorporated into Cerne Abbey probably from early medieval times. A winterbourne commonly rises in Yelcombe Bottom in the winter months and the valley floor is often moist to wet during the rest of the year (Charman 2020).

Archaeology (Figure 1.2)

There are, surprisingly, no Neolithic moments on this down although stray finds include a Neolithic axe. However, a few extant obvious Bronze Age barrows lie on the ridge of Giant Hill including a scheduled bowl barrow along with a cross ridge dyke. The 100 m long dyke almost forms a semicircle on the chalk ridge, surviving about 5 m wide and 1 m high with a gap or entrance in the middle through which modern trackway passes. The shallow ditch on its northern side is only about 0.5 m deep and 4 m wide. Iron Age (Durotriges) settlement is well known in the area (Putnam 1999; Papworth 2011), though some may have their origins in the Bronze Age. Late Bronze Age and Iron Age

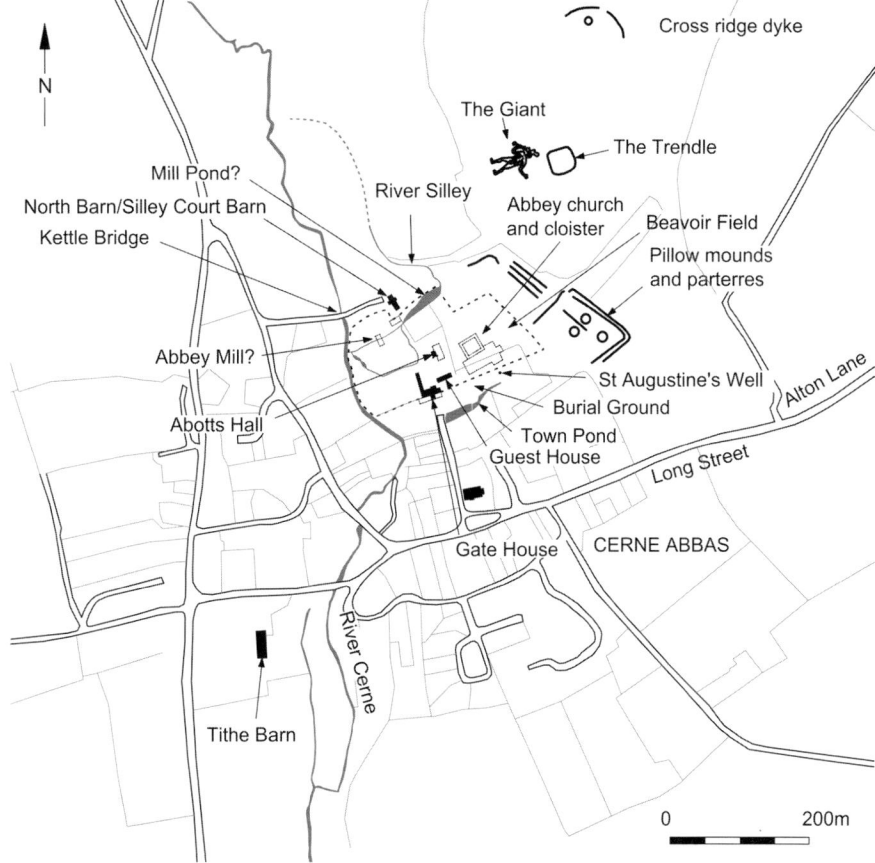

FIGURE 1.2. Cerne Abbas showing the relationship of the Giant to the Trendle and cross ridge dyke on Giant Hill, and key buildings and features relating to the abbey (see Fig. 1.4) in the town (based on Putnam 1999, fig. 31). Image: Justin Russell

occupation evidence including settlement sites (enclosures) set amongst field systems defined by prehistoric lynchets exist on Giant Hill to the north of the Giant (RCHME 1974, 82–5), those excavated by the Central Excavation Unit on Black Hill south of Cerne (Bond 1982; Woodward & Cox 1984; Woodward *et al.* 1988), and on the opposite side of the valley on Dickley Hill facing Black Hill (Fig 1.1c), and also at Smacam Down just to the south. These suggest widespread occupation and farming from the earlier Bronze Age to the Romano-British period. Though undated, it is possible that the rectangular, unexplored earthwork of The Trendle (or 'Frying Pan'/'Giant's Frying Pan') (RCHME 1974, xxxiv; Darvill 1999a, fig. 9; Fig. 1.3) enclosing an area of 23 m × 19 m just above the Giant's head, belongs to the Iron Age, or the Romano-British period. It may represent a Late Iron Age or Romano-British temple as Piggott suggested (1938, 328) and others have concurred (e.g., Castleden 1996, 100–1); Late Iron Age and Romano-British pottery has been found from animal burrows in the bank and ditch. The earthwork may enclose a masonry structure (see Figs 1.3, 1.6 & 3.2) like those which can be seen at Maiden Castle, Dorset (Wheeler 1943; Woodward 1992), and Brean Down, Somerset (ApSimon 1964), or wooden examples at Hayling Island, Hampshire (Downey *et al.* 1979; King & Soffe 1994), or that the enclosure surrounded a chapel linked to Cerne Abbey on the

FIGURE 1.3. Aerial view of Giant Hill and Cerne Abbas with the location of the former abbey background (circled). © Historic England

hill like St Michael's on Glastonbury Tor and St Catherine's above Abbotsbury. By the Saxon or early medieval period, the abbey had been established in Cerne. Tradition has it that it was established by St Augustine in AD 597 baptising newly converted Christians in a fountain that issued from the ground on his command (VCH 1908, 53; from a thirteenth-century account of William of Malmesbury). The abbey was refounded by Æthelmær (the Stout) in AD 987, and it probably incorporated, or was founded conceptually around, the spring of St Augustine's well. The abbey was located on the north-eastern edge of the village between the foot of the Down and the thirteenth-century church of St Mary, and is the object of current fieldwork research by Hugh Willmott and Helen Gittos. By the height of the medieval period the abbey included a large church and cloister, Abbots Hall, Abbey Mill, Brewhouse, mill pond and fishponds, enclosed by a perimeter wall with a Northern and Southern Gatehouse (Fig. 1.4). A small chapel next to St Augustine's well and the burial ground lay next to the town pond and immediately beyond the southern perimeter wall. The abbey and recent research is described in more detail in Chapter 8.

FIGURE 1.4. Plan of the abbey site showing extant earthworks, the Scheduled Monument area, and location of extant remnants of the abbey buildings (solid) and conjectural plan of other abbey buildings and abbey features (dashed), grey extant buildings. Speculative abbey boundary after Dorset County Council. Image © P. Bellamy, Terrain Archaeology 2023

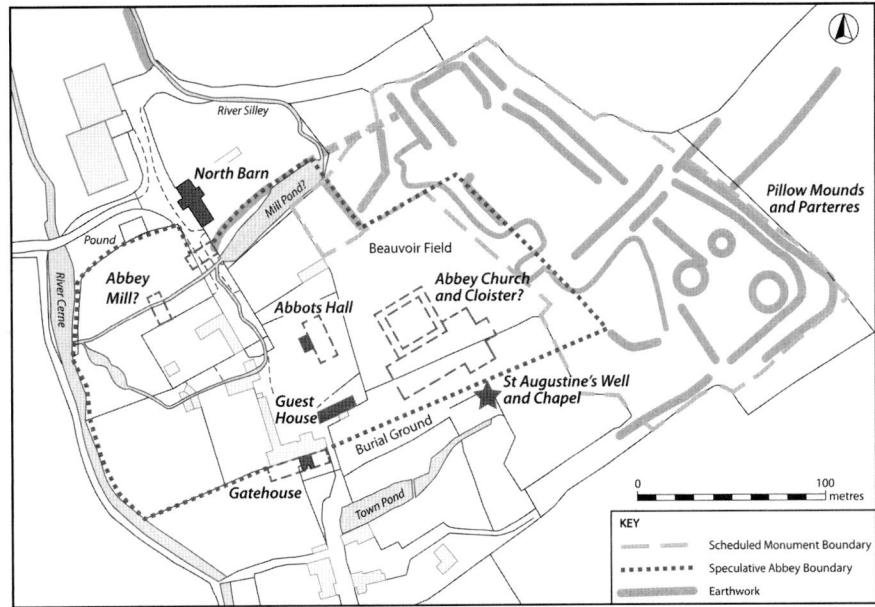

The town has Saxon origins; the earlier core lies on the edge of the River Cerne, while east of this lies the fourteenth-century planned town centred on Long Street and bounded by the abbey precinct to the north. A number of fine medieval buildings still survive and a medieval fifteenth-century ham stone preaching (or market cross) exists in the burial ground (Mortimer 2020), and agricultural ridge and furrow cultivation on its outskirts and certainly on Alton Lane (Freeman 1998). A substantial Saxon ditch and some pottery was found in evaluations at Simsay off Alton Lane (east of the town and south of the abbey site) with charred free-threshing cereal grains pears/beans and chaff (Robinson & Valentin 2004). Saxon/early medieval animal bones included sheep/goat, cattle, pig, deer (red and roe) and dog (Higbee in Robinson & Valentin 2004). Butchery on both deer species suggested high status since it was restricted to the aristocracy. Earthworks including a small part of the abbey and medieval and post-medieval pillow mounds and garden features to the east of Beauvoir Field were surveyed by Hazel Riley and Robert Wilson-North (1999; Wilson-North & Riley 2003). The old town, the abbey site and the Giant were designated a Conservation Area in 1971 (Penn 1980, 30–1).

Introducing the Giant: the anatomy of the Giant (Figures 1.5–1.7)

The Giant is a 55 m (180 ft; head to toe) tall, bare man, holding a 37 m (121 ft) long knobbly club with a pommel, or shaped end, raised above his head in his right hand, with his left arm outstretched to the south towards Cerne (and the abbey). His legs are slightly flexed and both feet face right as if he is strolling across the hillside. The raised club almost looks like the Giant is going to strike but for the surprised, un-fierce and un-angered facial features. He is unlike the Long Man of Wilmington in Sussex, whose feet point downwards and leftward

16 *Michael J. Allen*

and give the impression of a stationary, static, standing man. The Giant's most striking feature is his erect phallus and testes which now, after incorporating a former navel, is 7 m long (formerly it was 4.8 m).

Viewing the Giant

The Giant must have been marked out from a pre-determined plan or drawing, probably gridded in a similar way to how Katherine Barker's team mapped out the Giantess (Barker 1997; see Fig. 1.10 below). Consequently, when viewed from the ground he is foreshortened and squat. The Giant, therefore, only shows his true proportions when seen from the air. This does not mean he was

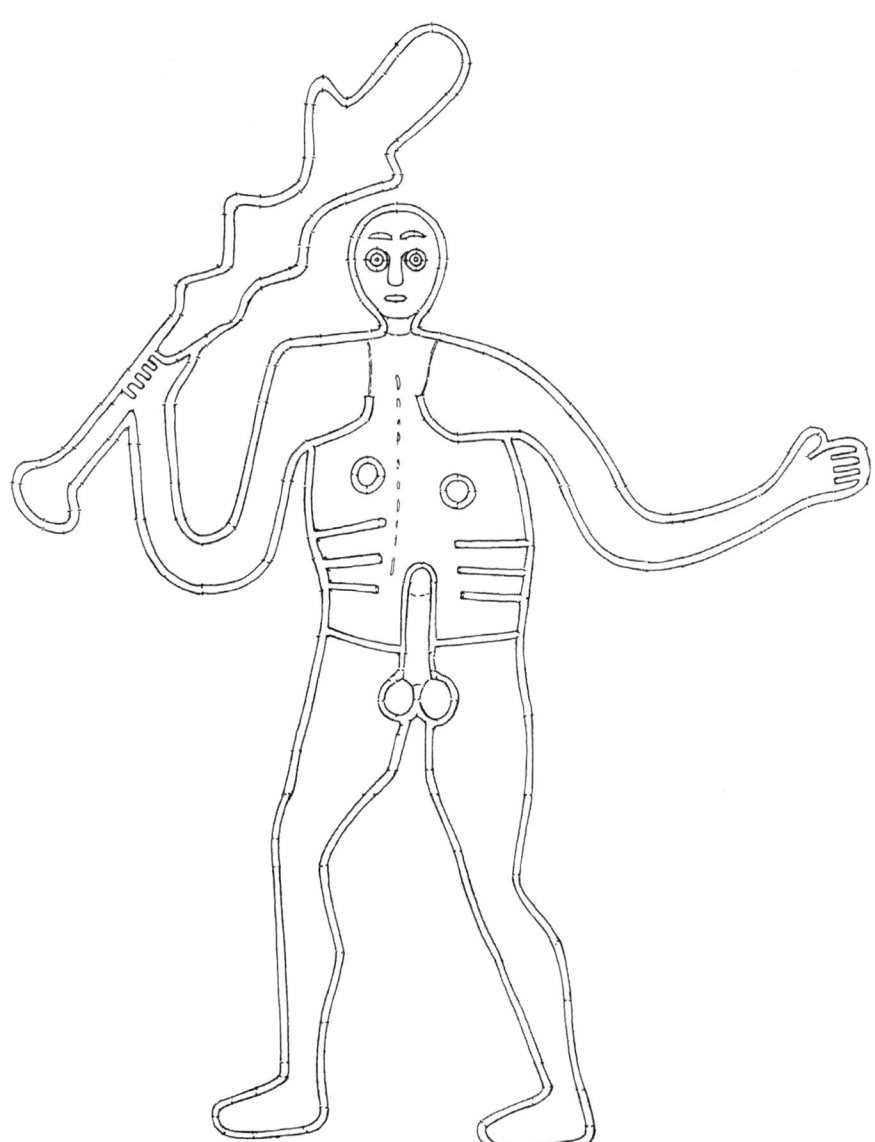

FIGURE 1.5. Sir Flinders Petrie's 1920s survey of the Giant. Note the dots representing his survey points recorded by offset and triangulation (from Petrie 1926, fig. III)

1. Place, person and context: an introduction to the Cerne Giant 17

designed to be viewed from the air, just that his design was translated from a 'paper' plan to the ground without full appreciation of the effects of the topography, perspective, and the steep slope located higher than the observer. From the Giant, the Cerne Valley and the scarp slope of the opposite valley side is in plain sight, and from which the Giant can be seen in better proportion albeit at a distance of over 1½ km. The view northwards from the Giant is largely into the valley towards Up Cerne, but the village of Cerne Abbas is hidden, and the former abbey church is out of sight. Significantly, the abbey would not have been visible from the Giant, nor it from the abbey grounds.

Outline

The Giant is a cut chalk figure; the soil on this scarp was probably only ever 20–30 cm thick. The turves were cut and the soil removed to expose, and slightly cut into, the chalk creating a white outline about 40 cm wide in the first instance. This would expose a clean white chalk line; but the bare (weathered) chalk was, if not originally certainly subsequently, covered with chalk rubble to near the turf surface. Through the actions of removing, scouring and replenishing the chalk, combined with soil erosion and accumulation, deep troughs have been created

FIGURE 1.6. Photograph of the Giant (alone) with the Trendle on the hilltop above his head and club. © National Trust Images/ Ray Gaffney

which have been reported as being 2 ft deep since at least the eighteenth century (Hutchins 1774, 292) and confirmed in the 1920s (Petrie 1926, 9). Subsequently, these deeper troughs have been filled with a succession of chalk rubble and kibbled chalk leaving, for the most part, a white chalk figure 'painted' on the hillside, though he has not always looked as well-defined. (Note that 'kibbled chalk' is typically stream-rounded and washed chalk, however, the kibbled chalk here refers to the specification of the industrially sieved and graded chalk after quarrying at Shillingstone Quarry.) More details are provided in Chapter 3.

Face and body

His face and body are also adorned. He is hairless and hat- or helmet-less, with an oval (or teardrop-shaped) earless head tapering towards an absent chin, and Newman describes this rounded head as 'babylike' (1987, 76). Though it is possible his head was once a complete oval as Castleden comments that 'the face may originally have been a complete oval with a line marking the chin' (1996, 27). Petrie had also commented on the presence of a groove (1926, 10), and indeed a slight depression is just detectable in the grass today (pers. obs. March 2020), which also suggests this possibility (see Chapter 3, *High resolution photogrammetric survey*). His face comprises strongly defined features with simple eyebrows, circles for eyes and short horizontal line for a mouth. The nose is a sculpted physical relief feature rather than cut soil and exposed chalk. The National Trust had noticed that it had become denuded in the 1970s and was remodelled and restored by them on 6 April 1993 (Keithley *et al.* 1999, 24; Edwards, Chapter 6); although there are varying suggestions that the nose was never as raised, or didn't even exist. With no pre-restoration survey or plans this cannot be resolved with ease. However, of all the features of the face, it is perhaps the eyebrows that are the more surprising; a detail infrequently seen on crude and simple images of the human face, where all the other features of face and body are more, or less, commonly depicted. His neck-less head sits directly on his shoulders and torso. His right shoulder, like his right nipple, is higher than his left.

His hands each have four fingers and a thumb, rudimentary pectorals are indicated ironically incorrectly *above* two lopsided circle nipples, and he has three ribs on each side.

The penis has been the subject of much discussion and publication. Perhaps the most notable element of this discourse relates to the disappearance of the Giant's navel and its incorporation into, and the consequent enlargement of, the phallus. The original, and earliest surviving, measured illustration (in the *Gentleman's Magazine* of 1694) shows the navel as separate to, and above, the penis. Leslie Grinsell argued it still appears in late nineteenth-century postcards or photographs (1980a). In 1926, Petrie's image shows a longer penis incorporating the navel, which has been interpreted as resulting from scouring in an as yet unidentified earlier period (Pitman 1978; Darvill *et al.* 1999, 133; Grinsell 1980a; and see Edwards, Chapter 6). However, their separate identity was still recognised after this. In 1978 Rodney Legg noted that the navel 'has since been grassed over' (*Dorset County Magazine*, April 1978), but it was Gerald Pitman

1. Place, person and context: an introduction to the Cerne Giant 19

(a teacher from Sherborne and founder of the community museum) who pointed out in the same year that the navel had been incorporated into the penis extending it by 1.8 m (6 ft) (Pitman 1978; Legg 1990, 7), and this was reported in more detail with comparative illustrations (Fig. 1.7) by Leslie Grinsell (1980a). There is, however, still some ambiguity as to whether this was just the result of the Giant becoming grassed over and being rechalked differently in error,

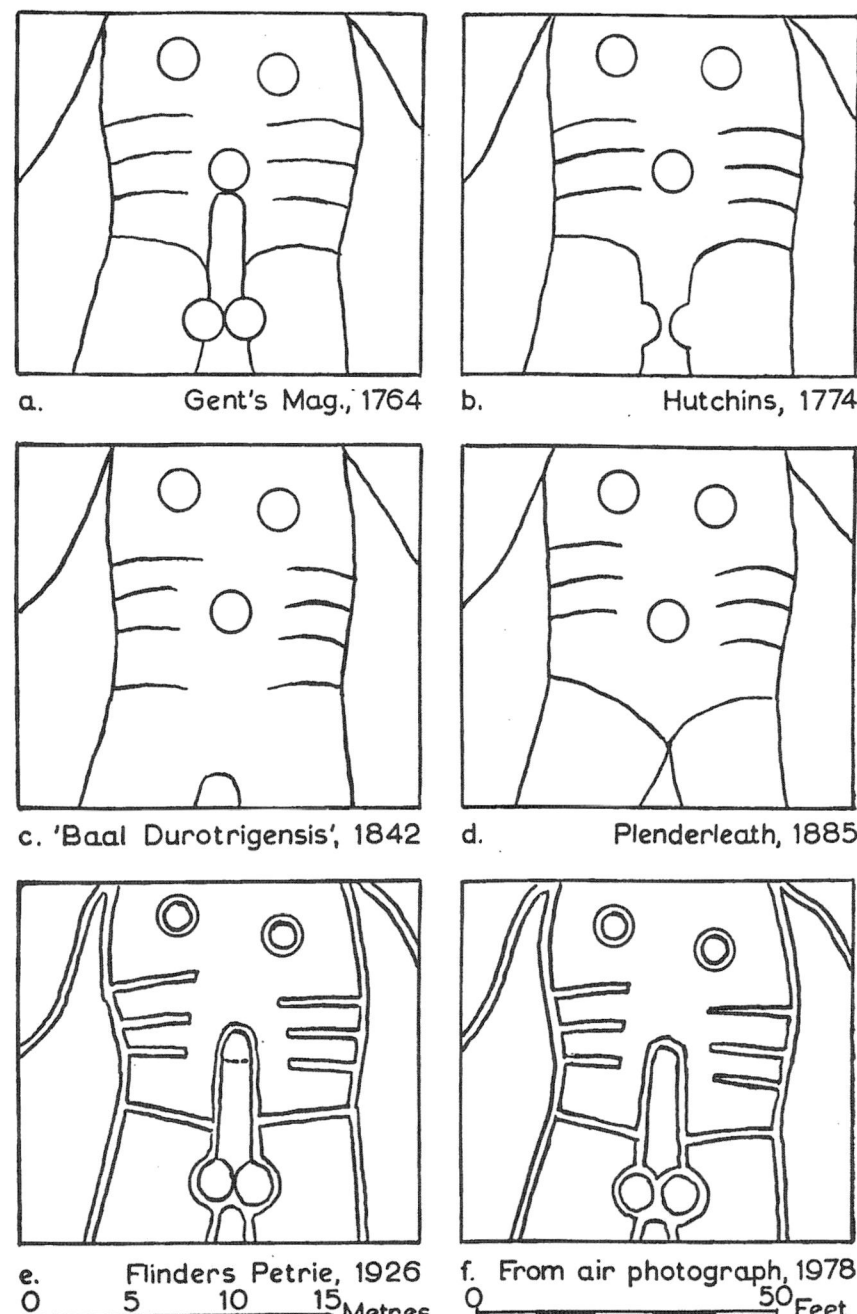

FIGURE 1.7. Antiquarian representations of the changing social and physical observation of the Giant's genitalia (after Grinsell 1980a, fig. 1), thanks to *Antiquity* (R. Witcher) © Antiquity Publications Ltd 1980

or whether this was in part a mischievous and purposeful modification during rechalking. Rodney Castleden confirms the presence of a separate navel by resistivity survey, but his illustrations show them as not only two separate items, but separated by a clear gap (Castleden 1996, 177; 1999, 45, fig. 23; Fig. 17.4), contrary to the illustration from the *Gentleman's Magazine* (1764, 335) where the navel is balancing on the tip of the penis. Amongst all this discourse, there is the possibility too, that the testicles were originally two complete circles, like the nipples, eyes, and the head (Castlelden 1996, 27).

Lion skin, cloak and severed head

The outstretched left arm suggests this may not just be a welcoming, pointing or signalling gesture but that he was holding something. Two things are commonly referenced: a skin (or cloak) draped over the forearm (Lethbridge 1957, 1; cited by Castleden 1996, 34) with its head, or possibly a decapitated head (Fig. 1.8), either being gripped by the hair, or in a bag, dangling below the outstretched left hand. Various geoarchaeological surveys have been conducted to test these ideas. A resistivity survey of the whole figure by Tony Clark and his team in 1979 for a Yorkshire television programme *Arthur C. Clarke's Mysterious World* aired in summer 1979 (Legg 1990, 7; David *et al.*, Chapter 3) seemed to indicate some anomalies (Clark 1983), which Tony, who, believing the Giant was Hercules (often depicted carrying a lion skin), reported that 'a pattern of indications in exactly the expected position', were found. But in reality this seems to be more a blur of high resistivity, than any clear outlined garment that others illustrate (see David *et al.*, Chapter 3). These were ground-truthed and confirmed with hand augering by Alister Bartlett, though what form that confirmation was, is not given (Clark 1983), except that there was a curving area below the arm. These unpublished auger results are analysed in summary in Chapter 3 (and form the study of new ongoing research). More importantly, and contrary to many who have reported his findings, Tony went on to say 'the interpretation must be regarded as tentative until more thorough auger tests or small-scale excavations have been carried out', though he did admit that the Giant's outstretched arm is better explained with something draped over it, and in his opinion it 'improves the compositional balance of the figure'. This survey was undertaken by Tony Clark, Andrew David and Alister Bartlett in their own time, with official approval (A. David pers. comm.), and is published for the first time here (see Chapter 3, David, Clark†, Bartlett†, Linford & Clements). Further resistivity survey by Rodney Castleden in 1989 and 1990 and then again in 1995 (see below) seemed to confirm Clark's areas of high resistivity. Significantly, however, Rodney Castleden claimed the lion skin to have a curvier, wavier clear line demarking some areas of higher readings, and indicating perhaps a cloak even with three possible vertical lines amongst the geophysical noise, which he interpreted as the outlines of folds in a cloak (Castleden 1993; 1995a; 1996, 158–7 and figures). In retrospect, however, this 2-probe resistivity survey may be flawed (see *Previous archaeological work*, below).

FIGURE 1.8. Giant with cloak and severed head (after Castleden 1999, fig. 24)

1. Place, person and context: an introduction to the Cerne Giant 21

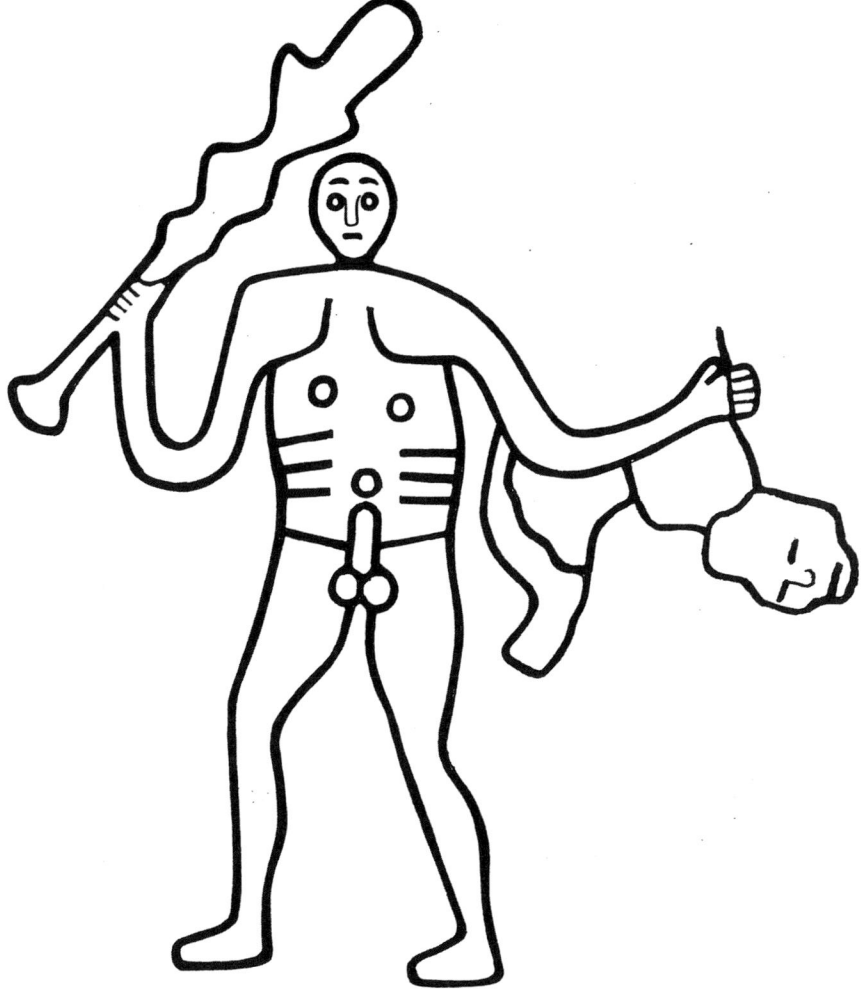

In 1996 John Gale again surveyed nearly the entire figure but with both resistivity and magnetometry (1999, fig. 33). Apart from clearly detecting the former coffin-shaped fence line particularly on the Giant's right hand side, the Giant's then recently remodelled nose containing metal chicken-wire, and the 1979 wooden construction track put in place during the National Trust's restoration (Gale 1999, fig. 34; see Keithley *et al.* 1999, fig. 12), the magnetometer survey with a fluxgate gradiometer did not bring anything new to light. Similarly, although the resistivity survey didn't pick up much, and did not resolve any of the issues relating to items beneath the Giant's left arm (*op cit.*, fig. 35), it did record some, but not all, of the outline of the Giant. Gale himself admitted that the short programme of geophysical prospection only 'had limited success' (*op cit.*, 62).

A head, whether attached to a lion's pelt or a decapitated victim's severed head held by the Giant, has been suggested below the Giant's left hand. There is indeed a low knoll (pers. obs.; Castleden 1996, 162); it is about 8 m across and 0.3–0.35 m high (pers. obs. 2020). Geophysical survey results, even after re-reprocessing did not clarify this (*op cit.*, 162), however, computer processing (in 1995) of Papworth's close interval topographic survey by Castleden (*op cit.*, 163, and figure) was potentially illuminating. Although the knoll could just be the severely weathered base of a former rubble dump from earlier rechalking, or even a former large ants' nest, Castleden's computer modelling of Papworth's contour survey of the small 0.4 m high mound below the Giant's hand, allowed him to claim it was a head complete with 'a mop of hair' and facial features: closed eyes and a mouth (Castleden 1996, 163–5). Limited hand augering as a part of this (2020) fieldwork could not resolve this satisfactorily, but did indicate loose mixed soil with chalk pieces and a large ants' nest. Tony Clark's team conducted 'auger borings' to test for a feature (lion skin or cloak) draped over the left arm. Although he admitted that augering did confirm the presence of 'the feature', it is not certain whether this was the cloak or severed head, and Tony Clark did go on to say that this 'interpretation must be regarded as tentative until more thorough auger tests or small-scale excavations have been carried out'. The full analysis of his auger survey (by Alister Bartlett) is now presented in Chapter 3. Certainly test augering associated with this project (2020) did confirm increased soil thickness and the deposits of the 'severed head'; see Chapter 3.

Whether the cloak/severed head are a part of the original design is ambiguous, they are clearly not an outline like the rest of the figure but could be a combination cut fill and/or just enhanced soil such as the raised nose. Whether the cloak is real rather than pedological (soil) or geological variation, and a difference created by the multifarious constructions also remains uncertain. Certainly, whether it represents a lion skin, or a cloak draped over the arm is not fully resolved. It does seem at odds, however, that is not in outline like the rest of the figure (see the Cerne Giantess for instance, Barker 1997; Barker & Darvill 1999, fig. 52). This discussion is not, however, the aim of this fieldwork project, nor is it for this discourse to engage with this matter in any great detail. We do, however, discuss this in Chapter 7 in the light of our new 2020 photogrammetry survey, the full auger data from 1979, and the 2023 geophysical surveys (Chapter 3).

Symbols

There has been a lot made of the letters and symbols between the Giant's shins. The 1764 depiction of the Giant (see Fig. 1.14, below) contained no symbols between his lower legs, though three numbers were reported to be present between his shins. In 1774 Hutchins' republication of that image in his *History and Antiquities of Dorset* shows three numbers (probably 798) with three symbols or letters (Fig. 1.9). They are only depicted here, and never with any certitude since. Although this is the only antiquarian depiction of them, all others until Castleden (1996, 173–6) were reproductions of the Hutchins drawing. A letter to the *Royal Magazine* in September 1763 mentions a date/ three figures between the legs (Anon 1763a; Morgan Evans 1999, 109), but does

FIGURE 1.9. Hutchins' 1774 versions of the symbols and numbers between the Giant's legs (see Fig. 1.14).

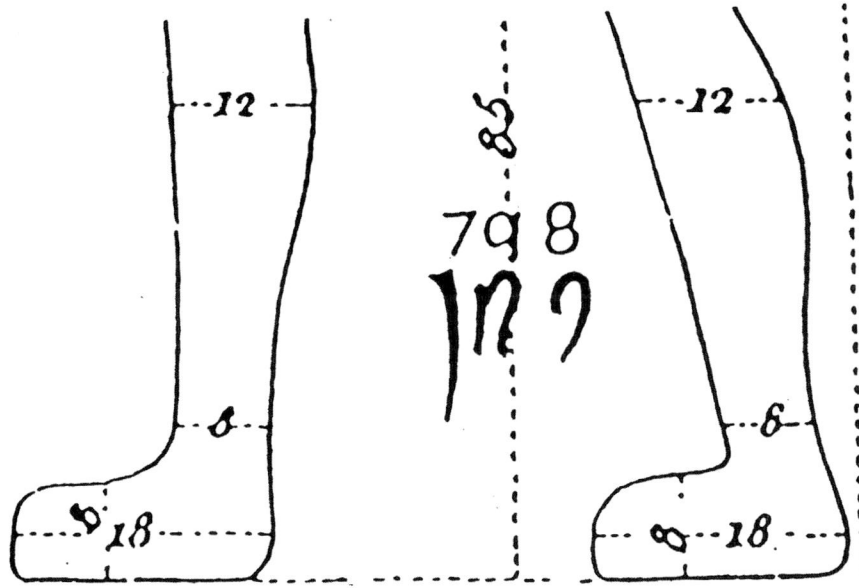

not show them. These figures and letters are not obvious in any topographical or geophysical survey, and I could not see them in field in 2020. Nevertheless Rodney Castleden specifically undertook resistivity survey of a 9 m square in April and July 1995 and did not reveal them either; he reported that 'It seemed as if the numbers and letters had vanished without trace' (Castleden 1996, 174). He did, however, report that some mid-twentieth-century photographs apparently showed rough linear depressions where three letters should be, although even more detailed investigations of the data was inconclusive. His resurvey in July 1995 and re-examination of the data may have recovered some variations in the ground, which may equate to the letters Hutchins records (*op. cit.* 174–5). Rodney traced, illustrated, and discussed these (*op. cit.* 173–6), but as he admits the results are inconclusive. It is clear that if these numbers and symbols existed they were not marked in the field in the same way as the Giant's outline as we know it. These figures have been discussed by others and may represent ANO (anno, i.e., year) or IAO (Jehovah), (Marples 1981, 164) or I or JHD (*Jehovah/Jesus hoc destruxit* – god has overthrown this idol/destroyed this: March 1901; Castleden 1996, 22) or I/J H/N D (Castleden 1996, 174–5) and readers are directed to those discourses for further information (e.g., Marples 1981, 163–5; Newman 1987, 81; Castleden 1996, 21–2, 173–6: Darvill 1999a, 9), but also see the 2020 photogrammetric and 2023 geophysical surveys (Chapter 3).

The Giant's companions

The Giant stands alone on the hillside, as does the Long Man of Wilmington and most other hill figures. However, Rodney Legg suggested in the hot summer July of 1976 about 36 m (120 ft) to the left (north) of the Giant's outline he

FIGURE 1.10. The Giant's companion; the Giantess 1997. Image © Francesca Radcliffe 1997

could discern the outline of a very large squat terrier dog about 46 m (150 ft) long in the parched grass (Newman 1987, 94–7; Legg 1992; Castleden 1996, 170–2); a figure he'd also claimed to have seen nine years previously in January 1967. The terrier, however, is of disproportionate size to the Giant, and unlike the Giant does not seem to have been foreshortened by the steep scarp slope. Despite numerous visits over several decades in different conditions neither I, Rodney Castleden (*op cit.*, 171–2), nor others have managed to see it; not that that means he does not exist, but that if he does his construction, scale and perspective differ from the Giant suggesting, if he does exist, they are not contemporary with each other.

The only other fleeting companions (other than animals and humans) was the Cerne Giantess, created next to him by Katherine Barker and students of Bournemouth University in 1997 (Barker, Chapter 10; Barker 1997; Fig. 1.10) and Homer Simpson in 2007 (Chapter 13, *Dorset Echo* 2007). In 1980 the Devon artist Kenneth Evans-Loude proposed placing a 70 m high figure of Marilyn Monroe in the unmistakable striking pose from the *Seven Year Itch* (which perhaps might have explained the Giant's excitement), and although the landowner approved, both an Arts Council grant and planning permission were refused (Castleden 1996, 37), resulting in *The Times* reporting the issue as 'some like it not'. In 1992 a mature art student, Mrs Rachel Edwards of Yetminster, proposed the creation of a Giantess using bodies (Edwards 1992).

The Giant today

Today the Giant lies in a rectangular fenced enclosure with The Trendle (Fig. 1.11), which the National Trust installed in 1977, replacing the former coffin-shaped enclosure of metal railings installed by Pitt Rivers in 1886 after

FIGURE 1.11. The Giant on Giant Hill. Peter Harlow 2021 Creative Commons Attribution

he'd erected similar railings at Kits Coty, Kent. Although reported to protect the Giant in association with the beacon for the Jubilee celebrations of June 1887 on Giant Hill, and elsewhere reported to be fenced off in 1889 with iron railings, the fencing was not associated with Victoria's jubilee celebrations, as the railings were installed in 1886, not 1887 as commonly reported or even 1889 (see *Timeline*, Edwards & Allen, Chapter 7). The Giant had been given to the National Trust on 20 July 1920 and four years later was endowed by Sir Henry Hoare of Stourhead to provide for its upkeep and maintenance until the capital sum was exhausted.

Maintenance and renewal

The Giant has to be maintained regularly to prevent the chalk outline becoming overgrown with moss, weeds and eventually engulfed in the downland grass. In 1807 Cooke reported that he was repaired from time to time by the townspeople (1807, 92–3). This maintenance took the form of just rewhitening the Giant by either topping up the chalk, or the larger more industrial process of removing old chalk and renewing. In some places a trench seems to have been cut almost

0.6 m deep and filled with chalk, heavily rammed down and compressed on its surface (Castleden 1996, 31; pers. comm.), in others the chalk is much shallower (>20 cm) and when it rains the surface of the weathered chalk geology is almost exposed in steeper downhill straights of the Giant.

Although the earliest record of maintenance is the cost of repair recorded in the Cerne Abbas churchwardens' account in November 1694, the first record of the frequency of maintenance is mentioned by Dr Richard Pococke dated 1754 'The lord of the manor gives some thing once in 7 or 8 years to have the lines clear'd and kept open' (Pococke 1754; Hutchins 1774, 292; Shippey 2016, 11). His scouring had been carried out since at least the eighteenth century, and Hutchins records that it was done every seven years (1774, 219) by people of the town, which, if maintained, would have resulted in 20 scourings by the time the National Trust took over the ownership, management and curation of the Giant. Cleaning was done in 1694, 1868 and 1887 (in honour of Queen Victoria's jubilee), and the coffin-shaped fence was erected around him in July 1886. Further scourings occurred in 1905 (see summary below), and 1924, but by 1934 the Giant was overgrown again (Keithley *et al.* 1999). Regular cleaning, scouring and rechalking of the outline lead to digging out the accumulated chalk, which left clearly defined trenches first described by Rev John Hutchins (1774, 292) as 'the outlines are two feet broad and as many deep'. This depth of trench seems to have been maintained into the twentieth century. A note from Thomas Hardy written at Max Gate (21 September 1925) to Alda and Henry Hoare at Stourhead describes the Giant as being defined by trenches 'fairly deep and all that can be done to make his shape clear is to keep the trenches cleaned out & spread white chalk over the bottom of them. This will remain white many years if raked over and weeded now and then' (National Trust archive; WRO 383/954/97; Grinsell 1980b). Flinders Petrie (1926) states the trenches were 2 ft deep during his survey of *c.* 1919.

Despite these major scouring events, the plan dimensions of the Giant had hardly changed from 1764 to when it was presented to the National Trust by Alex and George Pitt Rivers (son of General Pitt Rivers), and when it was surveyed again by Petrie (1926). It is clear, therefore, that these scourings were undertaken with care as this didn't significantly alter the figure, although as we will see later, several parts of the outline have become significantly wider, and drifted downslope in the past 50 years or so.

Other scourings are recorded by the National Trust in 1920 and 1924, and then with Messers Beard & Co. of Swindon who were subcontracted to do this on an almost industrial scale in 1956, and again over three months in 1979. These included the installation of a wooden trackway to haul up carts of chalk in 1979 (see Keithley *et al.* 1999, fig. 12). Detailed specifications were drawn up for the scouring (in 1979) and stated that the work should 'cut out all loose chalk from the parts of the Giant to be indicated on the site. This cutting *down to hard chalk is to be carried out*' (National Trust 1974b; Appendix 3). Indicating fairly extensive removal and renewal of deposits from most of the Giant. Thereafter, more routine basic maintenance (weeding, cleaning, tidying) was organised and supervised by the Trust's wardens (Ian Davey 1979–81; William Keighley[1]

FIGURE 1.12. The earliest illustration of the Giant; sketch from the Society of Antiquaries minute book 1763

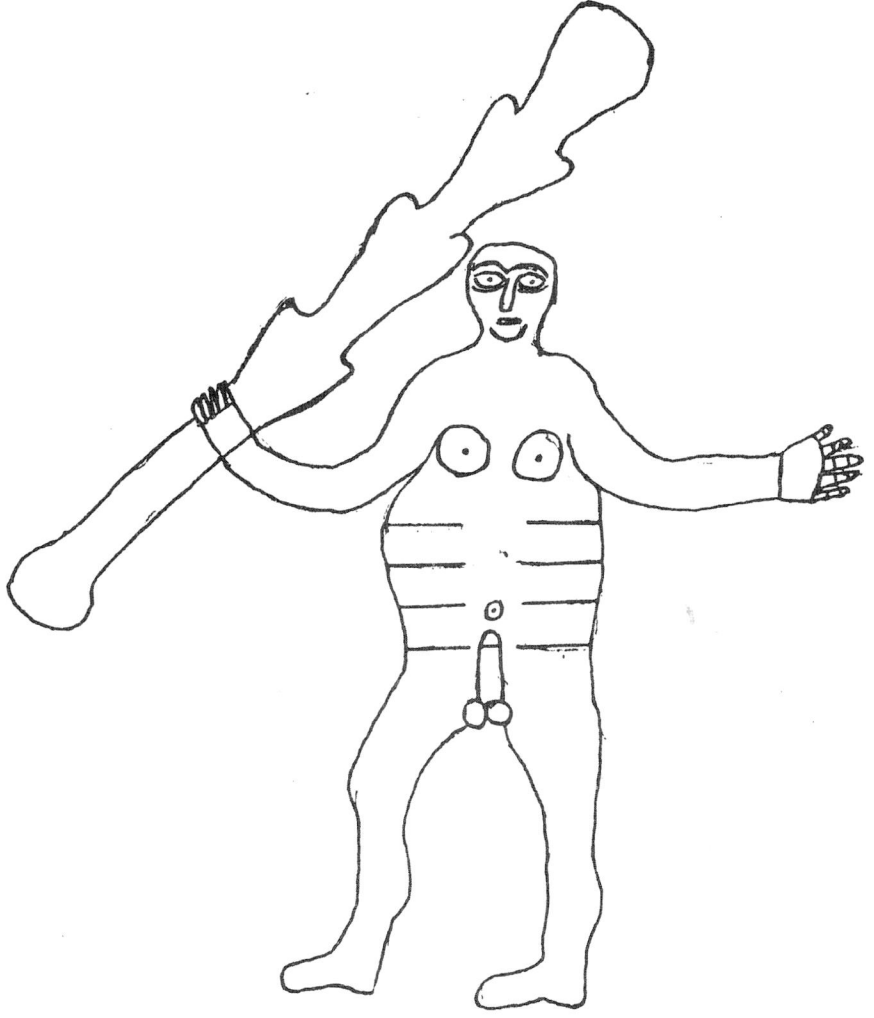

1981–2002 and then Michael Clarke from 2002 to present). Rewhitening, rather than scouring, was undertaken in 1983, sponsored by Heineken and undertaken with volunteers and required over 4 tonnes of chalk (Keithley *et al.* 1999; Edwards pers. research). Minor more cosmetic cleaning and refurbishment was undertaken with National Trust volunteers in 1995 and 2008 and most recently in August and September 2019, just before our research excavations in March 2020, using 17 tonnes of chalk, which was brought to site in numerus loads with a Land Rover and trailer. Michael Clarke reports that full scouring generally comprises nine days of work cleaning the trenches 18 inches (46 cm) wide and replenishing the chalk just 4–5 inches (10–13 cm) deep takes 40 tonnes of new chalk. Since 1979 this has been done with 60 (National Trust) volunteers representing 160 volunteer work days supervised by the National Trust warden.

Some of these larger scale scourings of the trenches and replenishing with tonnes of chalk involved trackways, carts, pulleys and hoists, and more recently

4×4 vehicles to bring in quantities of chalk. The chalk for works in from 1979 and subsequently was kibbled chalk (sieved and graded) from a quarry at Shillingstone. Before that date, chalk was presumably acquired from more local quarries and may have been raw chalk allowed to weather and breakdown. Between the major scouring and rechalking events the National Trust rangers bring a trailer of chalk to rewhiten the figure every few years (Michael Clarke pers. comm.) and minor works like these are not discernible in the section drawings, nor formally recorded in the National Trust archives.

Primary list of major scouring, cleaning, rewhitening and refurbishment (but see Giant Timeline, Chapter 7)

- 2019: National Trust – rechalking and refurbishment (the top dirty chalk was scraped off 'before hammering in 17 tonnes of new chalk by hand'– National Trust website https://www.nationaltrust.org.uk/cerne-giant/projects/cerne-giant-given-new-lease-of-life)
- 2008: National Trust – rechalking and refurbishment
- 2003: National Trust – partial rewhitening, sponsored by Heineken
- 1995: National Trust – rechalking and refurbishment
- 1983: Cleaning, if not scouring, sponsored by Heineken
- 1979: Contractor scouring (Keithley *et al.* 1999, 20), with kibbled chalk quarried from Shillingstone (Keithley *et al.* 1999, 21–2) carried out again by E.W. Beard Ltd., Swindon
- 1974: Six days of cleaning (*Western Daily Press*, Wednesday 1 May 1974, 4, col. 4)
- 1956: Contractor scouring and restoration of Cerne Abbas Giant carried out by E.W. Beard & Co., Swindon
- (1946: In June the Wells Natural History and Archaeological Society express disappointment the Giant is overgrown and only parts could be seen)
- 1945: Piggott restoration. Wartime camouflage removed
- (1934: Following threats in 1932 Giant remains unrestored)
- 1929: The fund established in 1924 to maintain the Giant cited in argument to preserve chalk regimental badges
- 1924: Giant 'cleaned' – The Secretary of the National Trust [1911–1934], Sam Hamer
- 1920: Giant 'cleaned' – National Trust acquisition from Pitt Rivers Estate
- 1905: (October) – Giant 'cleaned'. Following the visit of Sir Frederick Treves in early August, this cleaning appears to be an initiative instigated perhaps without permission by visitors with gardening tools. The cleaning as reported was executed in a single day … it has been suggested that it was during this period the Giant lost his navel, which was then incorporated into and thus extending the penis (Castleden 1999, fig. 23)
- 1908: (January, reported February) – Several inhabitants giving the Giant its periodical 'cleaning' (*Dorset County Heard*, February 1908)

1887: Scourings (Keithley *et al.* 1999, 18). Colley March reported in 1899, from 'Dorset Folklore Collected in 1897', that the Giant was 'last renovated in 1887, under the direction of General Pitt Rivers, and in part for Queen Victoria's jubilee

1868: Scourings (Keithley *et al.* 1999, 18). Restoration (descriptions before the restoration 'a shabby appearance on account of the trenches being choked up with weeds and rubbish')

1694: Scourings (Keithley *et al.* 1999, 18)

This list revises and amplifies information published by Keithley (1999) and elsewhere.

The Giant in history

There are two aspects briefly considered here. First the history of the development and activity at Cerne and Cerne Abbey (see also Chapter 8), and secondly the history of the first documentation and depiction of the Giant himself. There are a number of books and records of the history of Cerne and the abbey and in particular readers are directed to the slim *Cerne Abbey millennium* lectures edited by Katherine Barker and published by the Cerne Historical Society, and the summary of the history of the town on their website (https://cerneabbashistory.org/village-history/). The history of the Giant, or Giant in history, and antiquarian references and letters have been published in numerous places, not least *The Cerne Giant: an antiquity on trial* (Darvill *et al.* 1999, see especially Bettey, Morgan Evans and Vale, but also Bettey 1981; Castleden 1996). Whilst readers are directed toward those, and references cited therein, a brief résumé of the key points, and a list of key dates summarises some of these events.

History of Cerne

The interest in the history of Cerne, as far as we are concerned is principally that of the abbey (see Chapter 8), and is summarised above and in the key dates listed below.

Records of the Giant

The earliest clear record of the Giant's existence is the churchwardens' accounts of 4 November 1694 when a sum of 3s 0d was paid for repairing the Giant. There are other fleeting references in the 1730s to 1750s (see Darvill *et al.* 1999 esp. Bettey, Morgan Evans, and Vale). Between 1733 and 1737 Rev John Hutchins wrote to the Bishop of Carlisle 'with regard to the Giant at Cerne Abbas' (Bettey 1981, 121). We know about this because about 30 years later the Bishop presented this information to the Society of Antiquaries on 9 February 1764 (Morgan Evans 1999, 110). Meanwhile in 1751 Hutchins wrote with some information to Dr Charles Lyttelton, Dean of Exeter (later to become Bishop of Carlisle, and president of the Society of Antiquaries 1765–1768) who in turn asked Rev Francis Wise FSA in 1756 for his view on the Giant. He'd asked Wise, also a Fellow of the Society of Antiquaries, as in 1742 Rev Wise had published an account in a pamphlet on *Further Observations upon the White Horse* in which he records 'a Giant cut on a sidelong hill, but to be seen at any great distance' (Wise 1742, 48 cited by Bettey 1981, 121; Castleden 1996, 18). Meanwhile in 1754 Richard Pococke FSA had also toured Dorset noting the Giant. Exchanges followed between William Stukeley and Hutchins, and then to Lyttelton (then Bishop of Carlisle). However, probably in part inspired by Rev John Hutchins, William Stukeley FSA, Richard Pocock FSA, and the interest of others, there was a flurry of publications in the mid-eighteenth century, and the earliest known surviving crude sketch of the Giant is in 1763, which was between two pages of the minute book of the Society of Antiquaries (November 1763), see Morgan Evans (1999, 113; Fig 1.12). This interest was promoted by, or promoted discussions amongst; antiquarians, men of the cloth, and scholars, and also fieldwork was undertaken recording the size and nature of the Giant more accurately.

The first published *account*, however, of the Giant seems to be an anonymous letter in the *Royal Magazine* from September 1763, followed a month later by a similar, again anonymous, letter in *St James's Chronical* in October 1763, both of which were accompanied by the same set of 29 measurements. Only in July 1764 when the same letter of October 1763 was published again was it accompanied by a measured plan of the Giant (Morgan Evans 1999; Fig. 1.13). As we have seen, the Giant was discussed at the Society of Antiquaries, London and recorded in their minute book on 9 February 1764. So during the eighteenth century the Giant had certainly awaked the interest of antiquarians and gentlemen.

Whilst all this prominent action of survey, letters and accounts had circulated between gentleman, clergy and antiquarians, two great antiquarians preceding them astoundingly seemed to know nothing of the Giant. John Leland (1502–1552) stayed 7 miles from Cerne in 1542 but makes no mention of the Giant (Chandler 1993; Bettey 1999, 77), and just as surprising is John Aubrey FRS (1626–1697) author of *Monumenta Britannica* (1665–93) and the county histories of both Wiltshire and Surrey, who'd been schooled in Dorset and spent much of his life in south Wiltshire not far from Cerne Abbas does not recall the Giant (Bettey 1981, 120). Perhaps this suggests that he was not (readily) visible, physically nor consciously, at this time. A point we return to later in this book, and see *A seventeenth-century Giant*, below.

Lampooning

Rev John Hutchins claimed that the Giant was cut by Lord Denzil Holles (1599–1680), when he was Lord of the Manor and owned and managed the land from 1642 until 1666. But Hutchins may have confused the scouring events by a team of locals, with the construction of the figure itself (Marples 1981, 165).

FIGURE 1.13. The Cerne Giant's debut as a published drawing; the earliest published measured plan of the Giant (*Gentleman's Magazine* August 1764)

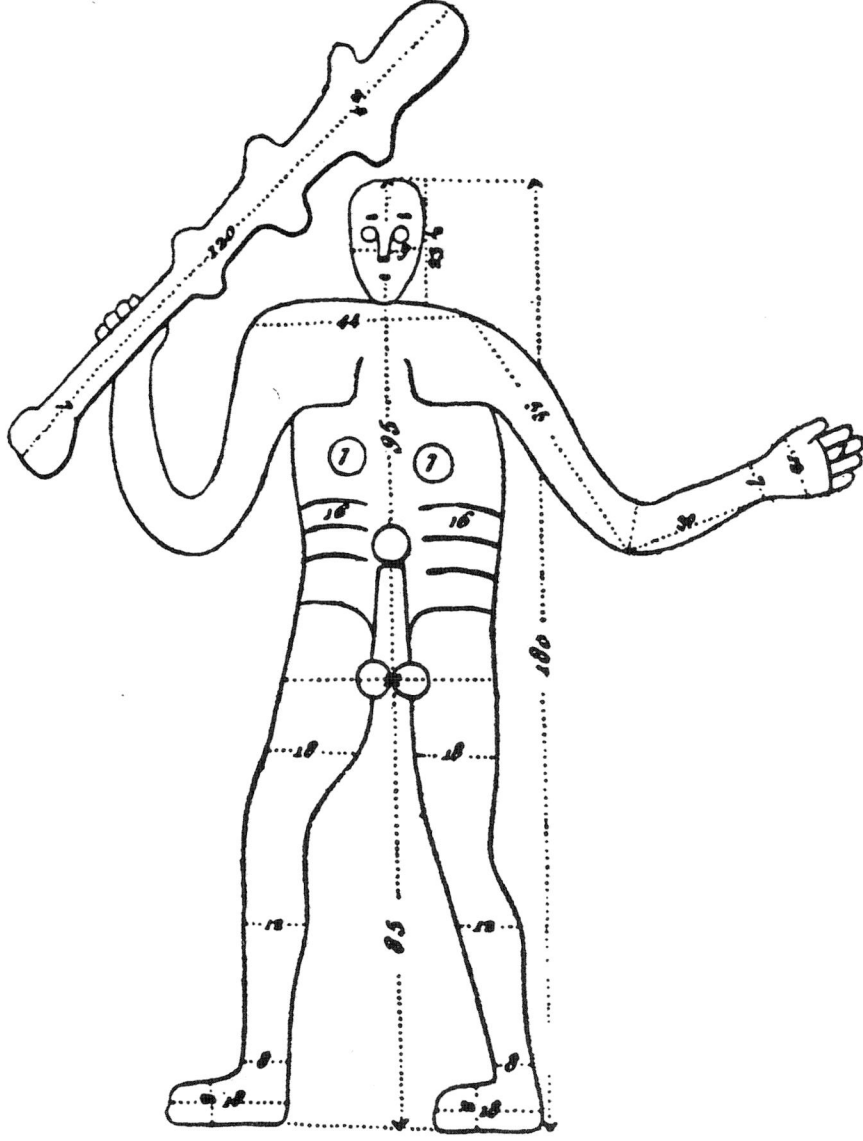

However, because of Holles' known disgust of Cromwell, Bettey (1999, 85) has suggested he could have been cut by him as parody of Cromwell (1599–1658). Others have suggested that the villagers cut him at this time, ridiculing Holles himself (Castleden 1996, 46). The main depictions of the Giant are: the sketch in 1763, publication in the *Gentleman's Magazine* 1764, Hutchins' emasculated drawing of 1774 (Fig. 1.14), Sydenham's similar illustration 1842 (see Fig. 1.15d, below) also lacking genitalia, William Plenderleath's similar genitalia-less illustration of 1885 (and in the second edition 1892, p. 39), and Petrie's resurvey (1926, pl. III; see Fig. 1.5).

32   *Michael J. Allen*

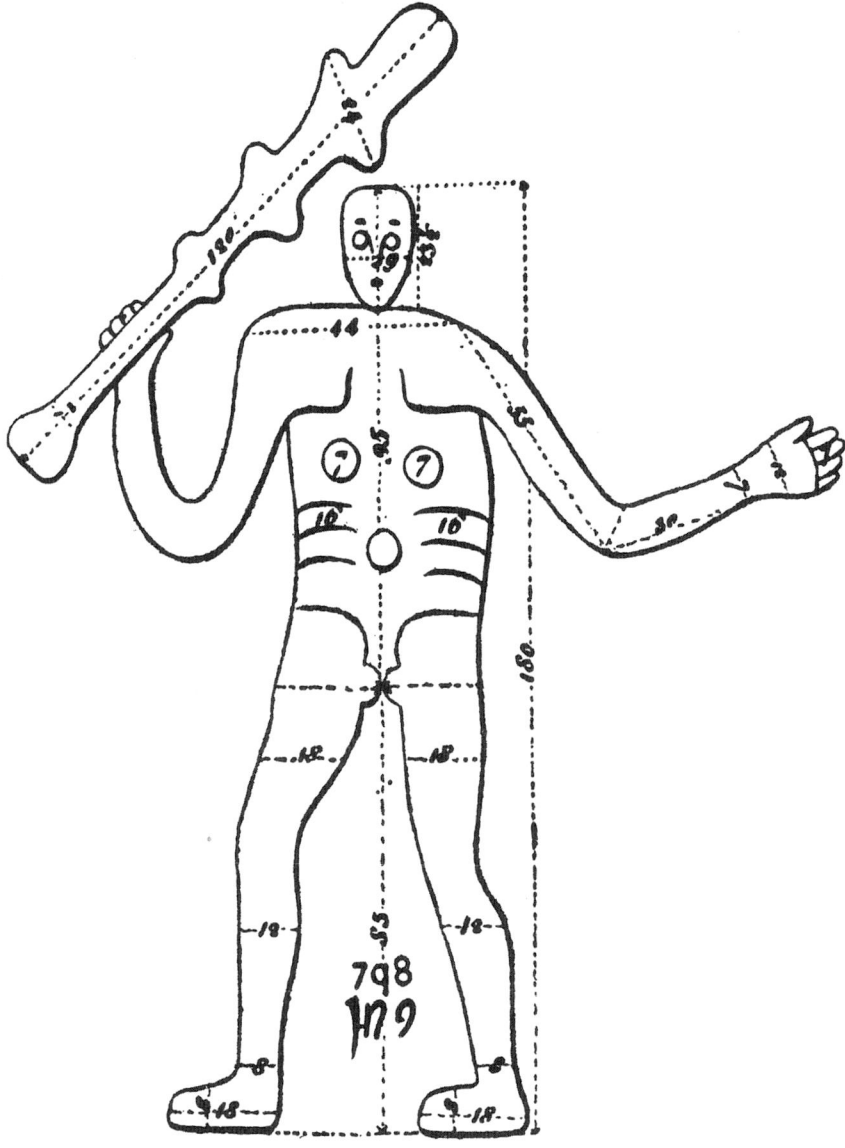

FIGURE 1.14. Hutchins' 1774 emasculated version of the Giant with symbols between his legs (see Fig. 1.9)

A brief history of Cerne

c. 870–885	John Leland (1503–52) noted that the *vetus codex* had stated there was a coenobium (monastery) for three monks at the site of the well at Cerne (Leland 1770)
888	Founding of Shaftesbury abbey; Ælfgiva (Æthelgeofu), daughter of King Alfred appointed the first Abbess (Kelly 1996)
c. 970–975	Cerne considered as a site for a Benedictine monastery in the reign of King Edgar (Squibb 1984)
987	Charter – Cerne Abbey established as a Benedictine monastery (Squibb 1988) by Æthelweard ealdorman of the western provinces, or his son Æthelmær the Stout (Licence 2006)

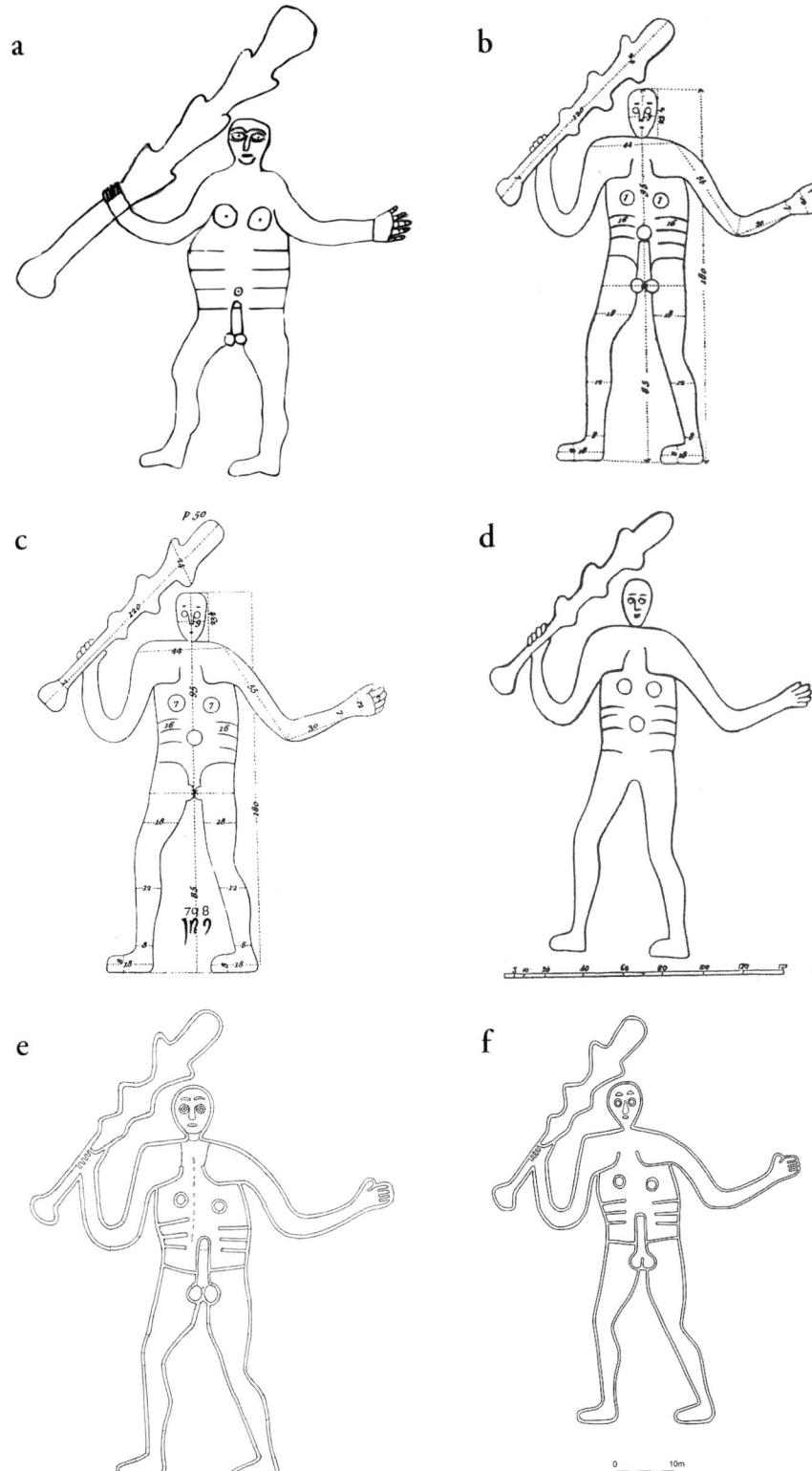

FIGURE 1.15. The main illustrations of the Giant from the seventeenth to twentieth century; a) Society of Antiquaries minute book 1763; b) *Gentleman's Magazine* 1694; c) Hutchins 1774; d) Sydenham 1842; e) Petrie 1926; and f) National Trust 2020

987–1002	Ælfric of Eynsham (955–1010) serves as abbot at Cerne Abbey. Benedictine reformer, one of the earliest translators of the Bible into English, committed to making complex religious writings accessible to ordinary people (Barker 1988a; 1988b), dedicated to accessibly formulating an English Christian identity (Stafford 1978), but noted little of his surroundings (Yorke 1988; Barker 1999, esp. 91)
1017–1080	Cerne Abbey acquires relics of St Eadwold (Licence 2006; Faulkner 2008). Translation of St Eadwold in recognition of miracles *c.* 1020
1090s	Goscelin of Canterbury's *Historia translationis St. Augustini et aliorum sanctorum*. Story of Augustine founding a spring/well, also of a worn out parish priest at Cerne Abbey who was restored following a vision of Augustine (Faulkner 2008)
c. 1125	William of Malmesbury (*c.* 1095–1143), *De Gestis Pontificorum Anglorum* – History of the English Bishops (Hamilton 1870; Preest 2002; Winterbottom 2007). An account of Eadwold. According to this source, Augustine was dismissively received in Dorset, having fish (or cow's) tails to his clothes he retreats 3 miles, then cried out '*Cerno deum!*' ('I see God!'), in response to which the locals begged him for forgiveness and he named the place 'Cernel' by joining the Latin 'cerno' ('I see') and Hebrew 'El' ('God'). Note: this story of Augustine naming Cerne after encountering antagonism is reminiscent of and therefore potentially borrowed from Bede's account of Gregory encountering barbarians in Rome prior to becoming Pope (Wood 1994; Latham 2015)
c. 1137	Geoffrey of Monmouth's *Historia Regum Britanniae* projects foundation myth surrounding giants as indigenous occupants of Albion defeated by the Trojan coloniser Brutus, a descendant of Aeneas after whom Britain would be named, this fable being further enlivened by an arranged fight between Brutus's champion Corineus, after whom Cornwall was named, and Gogmagog, leader and sole survivor of the giants wiped out by Brutus. Despite receiving several broken ribs in the struggle, Corineus manages to hurl Gogmagog from coastal cliffs.
1145	A Cerne Abbey rebellion (Foliot *et al.* 1948)
1214	Cerne granted the right to hold a market by King John (Gibbons 1962, 92)
1280	Cerne referred to as Moneke(s)cerne, Monecherne – 'Monk's Cerne' (Assize Rolls)
1288	Cerne first referred to as Cerne Abbas (Assize Rolls)
c. 1293	In *Memoriale*, apparently based on William of Malmesbury's earlier account, Walter of Coventry (Stubbs 1872–73) baselessly suggests the population at Cerne worshiped a god named Helith (Hutton 1999a)
1460	Cerne Abbas granted fairs annually (Gibbons 1962, 92)

Records of the Giant

1542   John Leland makes notes about Cerne Abbey when staying nearby – there is no mention of a hill figure (Smith 1907–10)

1617   Detailed survey of Cerne manor by John Norden and his son refers to 'Trendle Hill 130 acres' but no mention of the Giant (Bettey 1981, 119; PRO E 36/157 ff 112–123)

1924   Giant scheduled on 15 October (Scheduled Monument 1003202)

Information supplied by Brian Edwards.

Folklore and the Giant

A figure so rampant as the Giant cannot escape entering local myth, legend and folklore. Of the two most commonly recited, the first, understandably, is about fertility and pregnancy. That is if a woman wants to become pregnant she should sit (or sleep) on the Giant's phallus and Newman reports of tales that 'young brides, on the night before their wedding, climbing the slopes Giant Hill, throwing off their clothes, and prostrating themselves between the massive chalk thighs' (1987, 99). In the same vein it is believed that if a couple want to conceive they should have sex on the Giant's phallus and was recorded as such by Udal in 1922 (although this may have been in verbal circulation long before that). Similarly, a woman should walk around the Giant to prevent losing her boyfriend or husband, and women about to marry often make a pilgrimage to the Giant for this reason.

The second main tradition is not strictly of, on, or with the Giant, but is the erection of a maypole and maypole dancing on Giant Hill above the Giant. It is reported that a maypole or fir-bole, was erected in the Trendle at night on May Eve and decorated to dance around the following morning (March 1901), however, whether a maypole was ever erected on the Trendle is a matter of debate. The village itself had its own maypole on its green beside the Town Pond at the north end of Abbey Street (Legg 1986, 10) until it was taken down in 1635 (Castleden 1996, 102–3). Nevertheless, today the Wessex Morris Men go up to the Trendle at dawn on every May Day morning with the Dorset Ooser (an ox-like god … a figure of a large horned Giant; Fig. 1.16) and dance to welcome in the May. After the dancing on the hill, a procession forms beside the Village Hall, then the dancers make their way in processional dance along the village to perform outside the Royal Oak, Cerne Abbas, before breakfast.

There are vaguer, largely lost, folklore and legends about the Giant himself (see Doel & Doel 2007). These include stories that the rude man was an outline that was cut around the body of a Giant who had laid down on the hill after feasting on grazing sheep at Blackmore, and was killed by the angered peasants (Hutchins 1774, 229). And that the Giant would leave the hillside each night to drink in the River Cerne was reportedly recorded by Darton (1922, 119) whilst Morris Marples also says the Giant would devour maidens but does not elaborate, nor attribute this story (1981, 165).

FIGURE 1.16. The Dorset Ooser with the Wessex Morris Men make their way in processional dance along the village to perform outside the Royal Oak, Cerne Abbas, on May Day before breakfast. Image © Jane Tearl

Needless to say that both traditions are linked with fertility and procreation – the maypole itself often being seen as a fertility symbol, and dancing representing, or leading to, promiscuous behaviour. More detail of folklore associated with the Giant can be found elsewhere (Dewar 1968; Newman 1987, 97–100; Castleden 1996, 16–7, 102–2, etc.).

Considering the date (and identity) of the Giant: a summary

We have outlined some of the key presumed or considered dates for the Giant above, and of course the trial held in 1996 deals with a number in detail. Here we briefly review both archaeological and historical considerations of the date of the Giant, and possible incentives, before providing the results of the 2020 research. For more information the reader is directed to the discussion in Chapter 7, Brian Edwards' more detailed and considered discussion (Edwards 2020; Chapter 6) and former publications such as Darvill *et al.* (1999) and Castleden (1996) in particular.

The prehistoric origin for the Giant has been summarised above, needless to say that apart from Sir Flinders Petrie who, in 1926, considered him to be Bronze Age, most archaeologists proffering a prehistoric origin have generally considered him to be Iron Age.

For the historical records we must avoid confusing the dates of publication of early reports on the Giant (i.e., *Gentleman's Magazine* 1764) with the date of his origin as some scholars have done; these dates indicate a time at which he existed, but not was created. There is a flurry of historical documentation and publication in the eighteenth century, some of which we have alluded to earlier. During this period the date of the Giant is strongly linked to who the image was thought to depict, or under whose auspice he was made.

Camden/Gough	1687/1789	Saxon (Heil/Helith)
RM/SJC/GM	1763/4	Prehistoric (Ancient Britons)
William Stukeley	1764	pagan-Roman (Hercules)
Steward of the manor	1772	Modern (1642–1666 Lord Denzil Holles)
Rev John Hutchins	1774	Saxon (may be a work of Saxon Age)
John Sydenham	1842	Iron Age (Celtic; Baal/Bel/Belinus)
Dr Wake Smart	1872	medieval ('monkish work')
Sir Flinders Petrie	1926	Bronze Age
O.G.S. Crawford	1929	Prehistoric
Stuart Piggott	1932; 1938	Romano-British (Heth, Helis; Hercules)
F.J.H. Darton	1935	old
Morris Marples	1949; 1991	–
Bettey	1981	post-medieval
Paul Newman	1987	Celtic or Romano-British
Paul Newman	1996 (pub 1999)	prehistoric (Bronze Age) to Romano-British
Tim Darvill	1996 (pub 1999)	Later prehistoric (LBA–IA)
Rodney Castleden	1995; 1996	Iron Age (Celtic)
Bill Putnam	1996 (pub 1999)	Iron Age (Durotriges)
Ronald Hutton	1996 (pub 1999)	17th century
Vivian Vale	1996 (pub 1999)	17th century
Katherine Barker	1996 (pub 1999)	Post-medieval/17th century
Dai Morgan Evans	1996 (pub 1999)	17th century
Joseph Bettey	1996 (pub 1999)	17th century

RM; Royal Magazine / STC St James's Chronical / GM; *Gentleman's Magazine*

A seventeenth-century Giant

One of the most common alternatives to a prehistoric Giant, is a seventeenth-century one (Barker 1999; Hutton 1999b); the strongest plank in the argument for a seventeenth-century Giant is the total lack of records prior to the church-wardens' report of 1694, and other records not until 1763, 1764 and 1774 as we have seen. Ronald Hutton, Joseph Bettey and the late Dai Morgan Evans all use this in their arguments (in Darvill *et al.* 1999). Key to the concept of absence, is the fact that in puritan Dorset, especially around Dorchester, in the sixteenth to eighteenth centuries were a number of highly notable Antiquarians, none of whom mentions the Giant in their writings. Most notable amongst these, as Hutton points out (BBC TV programme 1996), is that father and son John and John Leland do not mention the Giant in their survey of Cerne in 1617. It is remarkably detailed, describing everything around where the Giant is but not the Giant himself. The Nordens are known for mentioning prehistoric monuments in the surveys; they mention the well (St Augustine's), the Trendle on top of the hill, but not the figure right in front of their eyes, which later gave its name to the whole hill. John Aubrey (1627–97), one of the most celebrated antiquarians, went to grammar school locally at Blandford before he went to Trinity College Oxford in 1642, yet in none of his writing does he mention the Giant. Neither does 'Coker's' systematic description of the history, topography of the County: *Survey of Dorsetshire* (1732), actually written by Thomas Gerrard and misattributed (Gerrard 1980; Legg 1980) or published under the name John Coker in 1732.

An Identity? Hercules, Helith and Cromwell (and Holles)

Stukeley was one of the first to suggest an identity (and thus date) of the Giant as Hercules, or Heracles in 1764 and also suggested the affinity with Helith (Piggott 1932, 215; Darvill 1999a, 11), and Tim Darvill intimates oblique references to local pagan beliefs. However, Richard Gough (editor of Camden's 1637 *Britannia*) linked the Giant with the heathen Saxon idol Heil (Helith) (Camden 1789a), taking his inspiration from William Camden's own assertion that the abbey was founded by St Augustine to commemorate Heil (Camden 1637, 2221, quoted by Darvill 1999a, 12). Over 50 years later Sydenham suggests Baal (Bel/Belinus) as the identity of the Giant, a Celtic deity representing the sun god (1842), once again placing the Giant in the Iron Age but with an association that does not seem to have found favour with subsequent scholars.

There was a rumour that Lord Holles may have created the Giant in the mid-seventeenth century, and there was speculation that this was even a caricature of Cromwell. Unfortunately, this uncorroborated speculation has entered the literature all too frequently, and it is likely that recounting of the cleaning and refurbishment of the Giant by the employees of Holles was mistaken by the Rev John Hutchins as his creation (Hutchins 1774; Marples 1981, 165) as discussed above (*The Giant in history*). The new date (Chapter 3) will allow us to reconsider his identity and discount many.

Previous archaeological work

The previous archaeological work has primarily been survey, either plotting the Giant himself, or a geophysical record. The first measured survey, and real fieldwork on the Giant, was the detailed plan published in the *Gentleman's Magazine* in 1764. It was republished by Hutchins in 1774 with the addition of figures and symbols beneath his legs, but this latter report does not indicate any additional fieldwork or survey, just the addition of the elements between the Giant's legs. The surveys included both the illustration (Figs 1.13 & 1.14), and a list of measurements (summarised in Table 1.1).

The only other full and accurate measured record was the taped survey (by triangulation) by Sir Flinders Petrie with his wife and son in the early 1920s and published in 1926 (Figs 1.5 & 1.15e). No other more modern full record of the Giant was undertaken or existed, other than tracings from aerial photographs, until the detailed aerial photogrammetry commissioned by the National Trust in 2020 for this project (see Chapter 3, below). However, detailed contour survey of the knoll below the left hand in 1995 by Martin Papworth, became especially informative when processed by Rodney Castleden (1996, 162–5) where he felt that it was a severed head dangling from the Giant's hand and that he could even discern a mop of hair, two eyes and mouth.

No formal archaeological excavation had been undertaken prior to 2020; a trial trench had been conducted by Stuart Piggott in 1945 through the 'severed head' (see below). Auger soundings had been undertaken by Alister Bartlett for Tony Clark in 1979 to ground truth geophysical results (Clark 1983; Chapter 3). The infill of deep scouring trenches had also long been known, observed but not formally recorded; nevertheless, Rodney Castleden made some very astute observations of the erosion and sedimentation within these trenches (Castleden 1996, 31, and figure).

Plans

Survey (full)	Anon	1764	–
Republish	Hutchins	1774	–
Survey (full)	Petrie	1920s	1926
Survey (below left arm)	Papworth & Gale	1995	Castleden 1996, 162–3
Survey (full)	Downland Partnership	2020	Chapter 3, see Figs 3.1, 3.2 & 3.3

Geophysical Record

Resistivity	Clark, Bartlett & David	1979	Clark 1983; Chapter 3, Fig. 3.6
Resistivity	Castleden Phase 1	July 1989 July 1990	Castleden 1993; 1996, 156–60 & 173–6; 2000, 66–7
Resistivity	Castleden Phase 2	April 1995 July 1995	($9m^2$ between lower legs – 'symbols') Castleden 1995; 1996, 156–60; & 173–6; 2000, 66–7
Res & Gradiometer	Gale (with Papworth)	1996	Gale 1999
Dowsing	Emery	2021	Emery 2021

Auger Reconnaissance / Survey

Systematic survey	Clark/Bartlett *et al.*	1979	Bartlett & Allen (Chapter 3, below)
Probabilistic augering	Allen	2002	Allen (Chapter 1, below)

Intervention

Small trench – 'cloak'	Piggott	1945	Unpubl. (see Chapter 1, below)

Measured survey

Several surveys of the dimensions of the Giant have been undertaken and include those printed with the plan illustration of the Giant in the *Gentleman's Magazine* in 1764 (but written in 1763; Anon 1763b) and reproduced by Hutchins in the *History of Dorset* in 1774 (though he excluded the phallus both illustratively and in measurements). Although the first illustration of the Giant and associated letter was published in the *Gentleman's Magazine* in August 1764, the same account (but no illustration) had appeared a year earlier in the *Royal Magazine* in September 1763 (Anon 1763a). Measurements had been made of the Giant by John Hutchins that year as he recorded in a letter to William Stukeley on 22 October 1763, to which Hutchins later wrote a rejoinder revising some of his measurements and apologising for 'an inaccurate description' (letter 29 November 1763). The fact that Stukeley submitted a note to the Society of Antiquaries (recorded in their Minute book for 15 March 1764) suggests that the anonymous writers of the same two articles may have been John Hutchins or William Stukeley. Although we have no firm evidence or proof, it seems likely, therefore, that the first main measured survey and plan seems to be that of, or arranged by, William Stukeley or John Hutchins leading to the plan of 1764, accompanying a republished anonymous letter. Many of Hutchins' measurements are listed in Newman (1987, 77). Some of the key measurements taken from these sources (along with authors own records) are given in Table 1.1.

No further detailed or recorded survey was undertaken until that by Sir Flinders Petrie in 1919, just after the National Trust had obtained ownership of the Giant, and published in 1926 (Petrie 1926, pl. III; Fig. 1.5). Petrie surveyed the Giant with his wife and son, laying out a grid he undertook a full measured survey and using triangulation. Although Newman says he recorded about 220 points, allowing him to plot them on graph paper (Newman 1987, 77), there are about 326 points, marked in the field and on his plan. These survey points can clearly be seen in his plan (Fig. 1.5), and this provides an outstanding plan; this is the image, or the basis of the image, that is published and republished until now.

No comparable measured survey providing a plan of the Giant has, astoundingly, been undertaken since Petrie's 100 year old survey, and it is this plan on which all modern depictions of the Giant have been based. Subsequent surveys have only been of specific parts of the Giant – typically the area below the left arm, the phallus and navel and the numbers and symbols between the Giant legs. The latter have largely been the preserve of Rodney Castleden's topographic ground observations. A topographic survey was conducted by Martin Papworth and John Gale of the mound beneath the outstretched left arm in 1995. This was conducted using an 'electronic distance measuring device' and levels were taken across the mound at 1 m intervals (Papworth pers. comm.). This provided topographic records (not reported), which enabled Rodney Castleden's 3-D surface model (1996, 165), from which he suggested a tilted head and face with hair, mouth, nose and right eye (1996, 164–5) as discussed above. Further

TABLE 1.1. Dimensions of the Giant after *Gentleman's Magazine* in 1764, and Newman (1987, 77) and author's own records

Element of the Giant	Metres	Feet
Height (head to soles of his feet)	55	180
Shoulder width	13.4	>44
Waist and torso	c. 13.4	c. 44
Club	36.5	120
Club width	7.3	24
Head	7.2	23½
Eyes	0.75–0.9	2½–3
Phallus	4.8	15.5
Navel	2.5	23.5
Phallus and navel (post-c. 1920)	7.2	23½
Phallus, testes and navel (post-1926)	9.1	30
Total width of figure	*c.* 60	197
Total height of figure	*c.* 100	328

small-scale micro-topographic survey, or observations, were made by Castleden in 1995 of the symbols between the legs with unqualified success (1996, 173–6). Other than this no full survey of the Giant was undertaken until 2020 when the drone photogrammetry survey was completed for this project (Chapter 3).

Geophysical survey campaigns

Four main geophysical survey campaigns have been conducted over the Giant, and were principally resistivity surveys. The first formal geophysical survey by Tony Clark and colleagues in 1979 was a resistivity survey of the whole figure (Chapter 3, Fig. 3.6) and was one of the first pieces of evidence to suggest the presence of a skin or cloak beneath the Giant's left arm. It has been written up for publication here for the first time by Andrew David *et al.* on behalf of his former colleagues (Chapter 3). Rodney Castleden conducted two campaigns of resistivity survey; the first in July 1989 and July 1990 where he examined the left elbow and a swathe from his left side to the knoll under his left hand, down to an area level with half way down the thigh to include the cloak and head, and second in April and July 1995 concentrating on the symbols between the legs, which was largely a micro-topographical record, but also included resistivity looking at the area above the head for horns/horned helmet and phallus/navel where readings were 10 cm apart. The most recent work, until our GPR survey (Cheetham, Chapter 3), was by John Gale in 1996 where he covered most of the Giant with both resistivity and magnetometry, and a 2021 dowsing survey by P. Emery (Papworth pers. comm.).

The surveys by Rodney Castleden (Fig. 1.17), tended to be of specific areas to amplify and confirm ambiguities in the outline, such as the cloak beneath the outstretched left arm, the numbers and symbols between the legs, horns on the head/helmet and the phallus and navel (Castleden 1993; 1995a; 1996, 156–60 & 173–6; 2000, 66–7). The results of these are summarised above. The magnetometer and resistivity survey of the whole figure by John Gale in 1996

 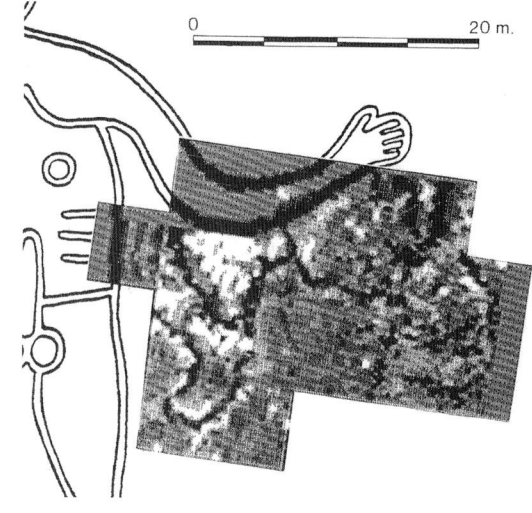

FIGURE 1.17. (left) Rodney Castleden undertaking resistivity survey of the Giant; (right) The plot from under the left arm. Images © Rodney Castleden 1996/2000

was actually not very successful nor useful (Gale 1999). Ironically the most informative survey was actually the first geophysical survey. This was undertaken in 1979 by Tony Clark assisted by Alister Bartlett and Andrew David for Yorkshire TV, but only summarily published in the résumés of the 1983 summer meeting of the Royal Archaeological Institute at Weymouth (Clark 1983); and the survey is only one paragraph (p. 30). That survey revealed the whole Giant, as proudly displayed on the AML geophysics section wall at 23 Savile Row, London, in 1983–4 (Fig. 1.18; note a younger Mike Allen working on other geophysical results in the foreground and Alister Bartlett next to the window). That astonishing survey is published here for the first time in Chapter 3.

List of previous geophysical survey campaigns

Campaign	Date	Survey	Equipment	Area
Clark *et al.*	1979	Resistivity	Clark prototype (twin probe-array)	Whole site
Castleden P1	1989–90	Resistivity	2-probe soil resistivity meter*	4 targeted areas
Castleden P2	1995	Resistivity	2-probe soil resistivity meter*	5 targeted areas
Gale	1996	Resistivity	Geoscan RM15 meter twin probe	Whole site
		Magnetometer	Geoscan FM36 fluxgate gradiometer	Whole site

*see Castleden, Chapter 17

Castleden's resistivity survey was conducted using a 'simple two-probe soil resistivity meter, designed with the help of an ingenious Hungarian physicist from Sussex University, Dr Micha Ertl' (1996, 156–7). However, unfortunately, more recently geophysicists Paul Cheetham (Bournemouth University) and Paul Linford (Historic England) indicate that making the potential measurements with just the two current injection electrodes means the main measurement is the contact resistance between the electrodes and the soil. This changes for every

FIGURE 1.18. Alister Bartlett and Mike Allen in the Ancient Monuments Laboratory (English Heritage) geophysics section in 1983 or 1984, with the survey of the Giant on the wall behind. Photo © A.E.U. David

insertion so there will be a lot of measurement noise. For this arrangement, the depth sensitivity falls off exponentially and, coupled with the short electrode separation, means the vast majority of the current travels through only the top *c.* 2 cm (perhaps 4 cm) of soil directly between. All that is being measured, therefore, is the resistivity of the ground surface electrodes. That may be suitable for a building surveyor's damp meter but not for archaeological surveys where the requirement is to detect what's beneath and below the topsoil not just the topsoil (Linford pers. comm. 13/12/23).

Consequently Castleden's 1989/90 surveys of the area beneath the left arm have only recorded resistance variations in the surface soil, and the results in the greyscale plot might bear this out; they've strongly detected the bare chalk surface of the Giant's outline but the rest looks like possible 'noise' and near-surface animal runs. It cannot indicate the presence or absence of buried archaeology. Nevertheless, the fact that map of the 'noise' indicates an area of clear *surface* disturbance that is different from the adjacent soil is significant, and reflects the unprocessed geophysical results of Tony Clark's 1979 survey (Fig. 1.18; David *et al.*, Chapter 3). These near-surface variations are also seen in the results from the 1979 auger survey (which indicate thicker soil and the presence of chalk stones absent in the adjacent soils; Bartlett & Allen, Chapter 3), and the results of the 2023 radar and resistivity surveys (Cheetham, Chapter 3). Ironically, therefore, most of the 'feature' under the left arm, as augered 1979, and geophysically surveyed in 2023 (Chapter 3), is essentially surface soil variations, so in this instance Castleden's survey is a real picture of that feature.

Auger survey 1979

As a part of the resistivity survey in 1979 (Clark 1983; Chapter 3), a hand auger survey was conducted only over an area 20 m × 20 m over and below the left arm. A series of 125 hand auger holes were recorded in six main transects (and auxiliary holes) and soil records and total soil depth recorded in an attempt to 'calibrate' the resistivity results. These are discussed in relation to the resistivity survey (David *et al.*, Chapter 3), and fully reported for the first time in Chapter 3 (Bartlett & Allen).

Probablistic augering 2002

Twenty years prior to this project I was invited to discuss the possibility of dating the Giant with a television research team (BBC 2 and Open University) who did not wish to touch the figure itself as it was a Scheduled Monument and they were concerned about the length of time and effort that it might take to gain permission from the Secretary of State. I suggested that one way to date him might be by proxy; by examining colluvium at the foot of the hill below the Giant that might have its origin in the preparation of the land and hillside (clearance or woodland and scrub) and his initial cutting and creation. My aim was to locate hillwash at the foot of the hill, below the Giant's feet, by hand augering. Once located, the best deposits (i.e., deepest and best stratified), could be examined by targeted research excavation of the colluvium, recovering and recording the distribution of artefacts (cf. Bell 1983; Allen 1988; 2005; 2007), in combination with OSL. This could date the colluvium and thus, by proxy, the creation of the Giant on the Cerne Abbas hillside. The concept and prepared research design was accepted by the TV research team and the National Trust, and latterly I visited the site with Prof Aubrey Manning and a film crew in 2002.

The first phase of filming was the hand augering with Prof Manning on site. A morning was spent discussing, on camera, the Giant, the concept of finding the colluvium and of what it meant, and how the deposits relate to human activity and the Giant. This included them discussing with Hazel Riley the then recent RCHME survey of the earthworks to the east of the abbey site (Riley & Wilson-North 1999, fig. 28). A number of key locations of potential colluvial accumulation were hand augered with 5 cm diameter dutch combination auger. The locations augered included: at the foot of the hill, upslope of the footslope woodland, within the woodland, below the footslope woodland, and in the field beyond at the foot of the down (Fig 1.3). Despite the steep slope and obvious expectation of footslope colluvium, and the recorded and mapped colluvium on the valley floor (Geological Survey 1958; Bird 1995), below the Giant, the woodland and lower hillside, the lower footslope below the woodland, and the field below the woodland was test augered. Astonishingly (and embarrassingly), no hillwash was found. None of the auger holes revealed anything more than enhanced rendzinas, or possibly a shallow colluvial brown earth (in the field below the Giant). Only shallow rendzina soils from 20 cm to shallow colluvial brown earths in the woodland (35–40 cm deep) were present. The programme of work at the Cerne Giant was cancelled.

It was suggested to attempt the same concept for the Long Man of Wilmington, in my home county of East Sussex. Previous augering there in relation to my PhD research, and a trip specifically to undertake judicious augering in advance of any formal work, again confirmed the presence of shallow calcareous footslope colluvium over 0.8 m thick (not bottomed) below the Long Man. Unfortunately my then employer (Wessex Archaeology) did not give permission for me to continue with this non-funded exercise in work time; I had already spent over two days in the company's time, and a couple of days of my leave on the project to date. I suggested that the research team contact Prof Martin Bell at Reading University, who'd undertaken research hillwash in the county since 1975 (Bell 1977, 251–8; 1983) and who'd pioneered the method I was proposing for the Cerne Giant (cf. Bell 1983) and had even worked on the Long Man of Wilmington in 1969 (see Holden 1971). Martin took on the baton, and was successful in locating and dating the hillwash at the Long Man, and this instead formed a very successful and informative episode of BBC 2's *Landscape Mysteries* fronted by Aubrey Manning and aired in October 2003 (Bell & Butler 2014, 20; Chapter 15).

Fieldwork 1945

During a major scouring and rechalking campaign supervised by Stuart Piggott in 1945, he digressed momentarily from the key works, to place 'a trial cutting into "lion's skin" lump near his left hand' (20 August 1945) aka the 'severed head' (Appendix 2). No details were published about this exploration (i.e., size of the cutting, nature of the deposit), except that in his letters he records that nothing was found and the tests were negative, see below. No archive has been located to date relating to this small trial trench. Piggott was distracted soon after the fieldwork by his appointment as Abercromby Professor of Archaeology at the University of Edinburgh in 1946.

The National Trust's centenary of ownership

As the fieldwork we discuss in the next few chapters was prompted by the National Trust's 100 years of ownership, it seems appropriate to finish this relatively comprehensive introduction and review of the Giant by reflecting on some notable archaeologists and people-of-the-day, and archaeological events associated with the Giant during that century.

1920s; acquisition Pitt Rivers, Hoare, Petrie and Hardy

General Augustus Pitt Rivers on inheritance of the Rivers estate and Cranborne Chase also became the owner of the Giant. The Giant was neglected during the 1914–18 war, and in 1919 although Cerne Abbas and surrounding farms were sold by public auction, the Giant was excluded from the sale with right of public access (Legg 1986, 13; Castleden 1996, 27). The following year, however, the Giant was given to the National Trust on 20 July 1920 by the General's eldest sons Alexander Edward Lane Fox-Pitt-Rivers and St George Lane Fox-Pitt (Nature 1937a; 1937b). Four years later Sir Henry Hoare of Stourhead, great, great nephew of Sir Richard Colt Hoare (1758–1838) the great archaeologist, provided an endowment (of £150) for the upkeep and maintenance of the Giant. There is no doubt that the Pitt Rivers of Rushmore, Dorset and Hoares of Stourhead, Wiltshire (only 18 miles down the old coaching road) were strongly acquainted. In 1919 Sir Flinders Petrie at the age of 65 and in the year he was knighted came onto the scene undertaking an accurate survey of the Giant, publishing this in 1926. It cannot be a coincidence that only three years after the National Trust's acquisition this excellent accurate survey was undertaken, but whether this was as result of a direct approach from the Trust, or via a connection with the archaeological Pitt Rivers' and Hoare's dynasties we do not know. It would be likely that they moved in the same circles and luncheons, or to have met in London, or even at the Society of Antiquaries. The Giant was scheduled on 15 October 1924 (Scheduled Monument 1003202) with the coffin-shaped enclosure defining the area (as it still does even after the new fenced enclosure in 1977).

Two years after Petrie's survey, in 1925, and only a year after Henry Hoare bequeathed an endowment to the National Trust for its upkeep, there was consternation over the condition of the Giant. Sir Henry Hoare himself had complained to the National Trust that the Giant was overgrown, which their secretary, S.H. Harmer, refuted saying that it had been 'recently cleaned' (15 September 1925); it had been cleaned 1924 (see scouring record above). To his credit Harmer did contact H. Dominy of Cerne Abbas who claimed that the Giant was always as he had been (16 September 1925).

At the same time Alda Hoare (Sir Henry's wife) was also similarly concerned over the condition of their gift to the National Trust as too had an anonymous writer to *Country Life* (Anon 1924; Grinsell 1980b). Mrs Hoare had become acquainted with Thomas Hardy in Dorchester and previously corresponded with his first wife Emma, and in 1925 wrote to his second wife Florence to solicit his support. Florence Hardy replied the same day (21 September 1925) saying that she would get her husband to write 'a few lines'. He did so, but probably to the Hoare's dismay, as he did not concur that there was a problem.

This provides two important insights. Firstly that there was some confusion as to the complaint that the Giant was 'grown over'. The National Trust, their 'agent on the ground' (Dominy), and Thomas Hardy all confirmed that he was in good order, and Hardy specifically stated that he was never 'all over white' like other hill figures, and that he is defined by fairly deep trenches. Concluding that all that can be done is to 'keep the trenches cleaned out and spread white chalk over the bottom of them'. We suspect that the complaint was not that the Giant wasn't white all over, but that the Giant's outline became overgrown and difficult to distinguish. Secondly it suggests that the *trenches* outlining the Giant may have become overgrown.

1930s and 1940s: Bloomsbury School and Piggott – restoration and excavation

Eric Ravilious, the Bloomsbury artist living near Eastbourne at the time, visited the hill figures of the Long Man of Wilmington (Fig. 7.4) just down the road to him, and both Uffington White Horse and the Cerne Giant (Fig. 7.4) in 1939, producing paintings of each that year, for a proposed Puffin children's book. These too are informative, showing the Cerne Giant as a dark outline in clear contrast to the two other white hill figures (see Figs 15.1 & 16.1).

It is interesting to note that the National Trust's agent in the area and for the Giant was another Bloomsbury artist, Eardley Knollys, living at Long Crichel from 1945. The two, Ravilious and Knollys, may even have met. Knollys, however, had a key role in the Trust's 'cleaning out' of the Giant in 1945. The Giant had once again fallen into disrepair and was indistinguishable, possibly exacerbated by the Home Guard's covering of brushwood during the war (to conceal him from German bombers) but which promoted vegetation growth in the trenches. The Giant was in a sorry state; there was also damage caused by cattle and horses and parts of the outline had silted up (Knollys, 30 July 1945). This called for a major restoration job.

The clerk of the committee Cerne Abbas parish council (L.H. Shutter) complained to the National Trust (15 June 1945), who replied saying that this was in hand, and that they were discussing the matter with their archaeological advisor Sir Cyril Fox, then director of the National Museum of Wales, Cardiff. It was he who suggested Stuart Piggott supervise the works: 'he is the best man in England for the job' wrote Fox (3 August 1945), probably because Piggott had already published two papers in *Antiquity* on speculations of the origin of the Giant (Piggott 1932; 1938). Within a month of the parish councils' complaint, Stuart Piggott, then studying the work of William Stukeley at Oxford, but based at home in Rockbourne about 40 miles away on the Hampshire/Wiltshire border, was indeed 'on the case', arranging to meet up with Eardley Knollys (17 July 1945), and by August had visited the Giant and work was 'going well' (16 August 1945).

Restoration or excavation? The lion skin

Although Piggott was tasked with cleaning and restoring the Giant, his archaeological interest and desire led him to some more clearly archaeological investigations. While pondering the lion skin, which he was becoming less convinced about, he proposed to dig it towards the end of the works (16 August 1945).

A trial cutting into the lion skin lump near his left hand (i.e., the 'severed head') was proposed (20 August 1945), and after 'carefully examining the lion's skin' they found 'no evidence that it had ever been artificially formed or outlined by trench' (25 August 1945) and tests were 'wholly negative' (4 September 1945). He reported that there were other (natural) 'bumps' on the hillslope beyond the Giant, the inference being that this was just another one of those. The work produced no 'archaeological results' (Piggott, 16 August 1945), but this whole encounter did convince Piggott of its Romano-British origin. Work was concluded, and the fence reinforced with a third strand of wire to keep the cattle out, while admitting sheep.

Although Cyril Fox was sure that Piggott would publish a 'scientific account' of the works (3 August 1945), and Piggott even suggested that 'a report on the work, plus this detailed analysis might well be given to the Society of Antiquaries this winter' (16 August 1945) the report never materialised. His appointment as Professor of Archaeology in Edinburgh in 1946 may have played some part in that.

Fending off complaints

The National Trust has received complaints about the condition or visibility of the Giant ever since they've owned him (see Alda Hoare etc. 1925, above). Even recently (October 2023, four years after refurbishment and rechalking in summer 2019) a disappointed visitor complained via a letter in the *Dorset Echo* (6 October 2023; Lumb 2023) and picked up by the *Daily Telegraph* (7 October 2023; the combined online comments totalled more than 800 in four days). After the major restoration in 1945 supervised by Stuart Piggott the National Trust tried to entice him to return to the Giant to supervise restoration in 1953. This may have been prompted not only because the Giant was again looking tired, but that the Uffington White Horse was being spruced up that year. Piggott declined, in part as he was now ensconced in Edinburgh (rather than more locally in Rockbourne, Hampshire, or Oxford University as he was in 1945), and he'd seen the Ministry of Works arranged restoration at Uffington, which he saw as a new standard in conservation. Beard & Co., who'd done the work at Uffington were contracted to do the same for the Giant, despite the much steeper and severe slope on which he was sited. The Beard's work in 1956 and 1979 also played a key role changing of the Giant's image, as we shall see in Chapter 7.

1990s: research design, geophysics and trials – Castleden and Barker

All seemed to go quiet on the archaeological front save major industrial-scale scourings by E.W. Beard in 1956 and 1979, and the recognition of the loss of his navel by Pitman in 1978 also reported by Grinsell in 1980. A resistivity survey by Tony Clark and colleagues (A. Bartlett and A. David) undertaken on weekends for a television programme in 1979, was the only archaeological research, but not fully reported until now (see David *et al.*, Chapter 3).

Ten years later the National Trust were keen to encourage and assist Rodney Castleden's approaches to undertake geophysical survey of four small targeted areas in 1989 and 1990 (Castleden 1993), then five further targeted areas in 1995 along with selected survey of the Trendle. Martin Papworth contributed by undertaking a close-contour survey of the severed head for him (Castleden 1995a; 1996, 163).

Elsewhere excavation and OSL dating of the Uffington White Horse (1993–4) was being undertaken. This may, along with the inspiration of meeting the participants of a conference on Cerne Abbey in 1987, have been the impetus for the instigation of the Trial (a one day conference debating the origin of Giant in 1996) organised by Katherine Barker (Chapter 10; Darvill *et al.* 1999). A flurry of activity temporarily ensued. Geophysical survey was undertaken by John Gale specifically for this, and it is no coincidence that later that year (1996) a research design for archaeological investigations was commissioned by the Trust and delivered by the Oxford Archaeological Unit in 1998. Unfortunately funds could not be raised to realise these proposals. Nevertheless, interest as a result of the trial led to the Giantess being born, and standing fleetingly, next to the Giant for a few days in 1997 (see above, Barker 1997; Chapter 10).

2000s: attempts and failures

Leading on from OAU research design in 1998 several attempts were made to start field projects including a suggestion to excavate through the sediments against the left foot (Allen 21 June 2002, National Trust Archaeology, Cerne Abbas correspondence, National Trust archive), which were realised in 2020. Limited auger surveys had failed to find deposits of interest outside the scheduled area in 2002, and the archaeological interest in the Giant waned, and instead archaeological interest was focused on the hill figure of the Long Man of Wilmington (2002 and 2004).

2020

The Trial of 1996 and fieldwork outlined in the Oxford Archaeology Units 1998 research design were the genesis of Martin Papworth's and the National Trust's long desire to date the Giant. Following field reconnaissance in 2019, this was realised in 2020, which this book discusses (Chapters 2–3, 7 & 14). This was the first archaeological excavation of the Giant, other than Piggott's unpublished trench through the severed head in 1945.

Note

1 William is Keighley, not Keithley as published in 'Owning and managing a Giant', pp. 18–26, in *Cerne Giant: an antiquity on trial* (Darvill *et al.* 1999), however references are cited as published. He is referenced here as Keighley as in that publication and subsequent citations.

CHAPTER TWO

The new research: dating the Giant; reconnaissance, aims, and methods

Michael J. Allen

There was a clear archaeological desire to date the Cerne Abbas Giant (cf. Castleden 1996; Darvill 1999a), especially after OSL dating of both the Uffington White Horse (Miles *et al.* 2003), and the Long Man of Wilmington (Bell & Butler 2014). In both cases comprehension of deposit taphonomy (i.e., formation and relation to the event of creating the Giant) was crucial and geoarchaeological investigation and land snail analysis were key elements of both projects. The presence, however, of suitable deposits was not guaranteed. Nor could it be guaranteed that if any deposits survived that OSL dating would be successful. The fieldwork conducted was, therefore, targeted, clinical and minimally invasive, and a very high proportion of the monument remained untouched for future generations to explore.

Aims and objects (raison d'etre)

The excavation was led by Martin Papworth of the National Trust and conducted for five days from 16–20 March 2020. Four small clinical incisions were hand excavated into the Giant (Fig. 2.4; Trenches A–D) with the clear primary aim:

- To enable sampling to provide a *terminus ante quem* date for the construction of the Cerne Abbas Giant, via optically stimulated luminescence (OSL) dating.

And with subsidiary aims defined as:

- To examine the geoarchaeology of the chalk figure (and its associated deposits);
- To attempt to examine the palaeo-environment contemporary with the construction of the Giant (and its hillside land-use history) (Allen 2020a, 3).

2. The new research: dating the Giant; reconnaissance, aims, and methods

The dating aims were to indicate if the Giant was prehistoric or historic (cf. Darvill *et al.* 1999) i.e., broadly one of:

- Prehistoric (typically Iron Age, cf. Darvill 1999a)
- Roman
- Anglo-Saxon
- Medieval, or
- Post-medieval i.e., a seventeenth-century figure (Hutton 1999c)

At the end of the 1996 'Trial' 50% of the audience believed the Giant was prehistoric or Roman, 35% seventeenth century or later, and 15% were 'date-neutral' (Barker & Darvill 1999, 162).

One other major funded element was an invaluable aerial high-resolution photogrammetric survey, which would, ultimately, provide the first accurate true plan of the figure (Fig. 3.5) since Sir Flinders Petrie's published nearly 100 years ago in 1926.

Reconnaissance and implications: walkover survey

The Giant was visited on by Martin Papworth and Mike Allen 18 September 2019 to examine the potential for the presence of suitable deposits, or sediment accumulations, directly related to the Giant, which could be sampled for OSL dating. The freshly completed rechalked outline (August to September 2019), and the geomorphology of the figure was carefully studied during a detailed walkover examination. A number of key observations were made relating to sedimentation associated with the chalk outline (Allen 2020a). They were:

- There was clear build-up of soil/sediment at the soles of both feet and the left elbow;
- There was small build-up and accumulation of soil and sediment on the upslope side of both feet;
- There were clear near level areas (soil accumulation) in several places around the cut figure, and;
- That a low mound was present below the left hand

Even after only a few weeks the recently emplaced, packed and rammed chalk of the figure outline was itself already suffering erosion. There were rills in the new (two weeks old) packed chalk surface (Fig. 2.1). These were created by rainwash, and small chalk stones had been washed downslope to the figure's extremities, accumulating in particular at the elbows and feet, but also at the base of the club. In other words, at the lowest end of any long downslope outline. Coarser chalk gravel (to about 2 mm) was observed at the ends of the rills (e.g., ankles), whilst a finer chalk gravel (*c.* 1 mm) formed small fans at the

FIGURE 2.1. The left leg showing rainwater scouring in September 2019 (as recorded previously by Keithley *et al.* 1999, 19). Image © M.J. Allen 2019

2. The new research: dating the Giant; reconnaissance, aims, and methods 53

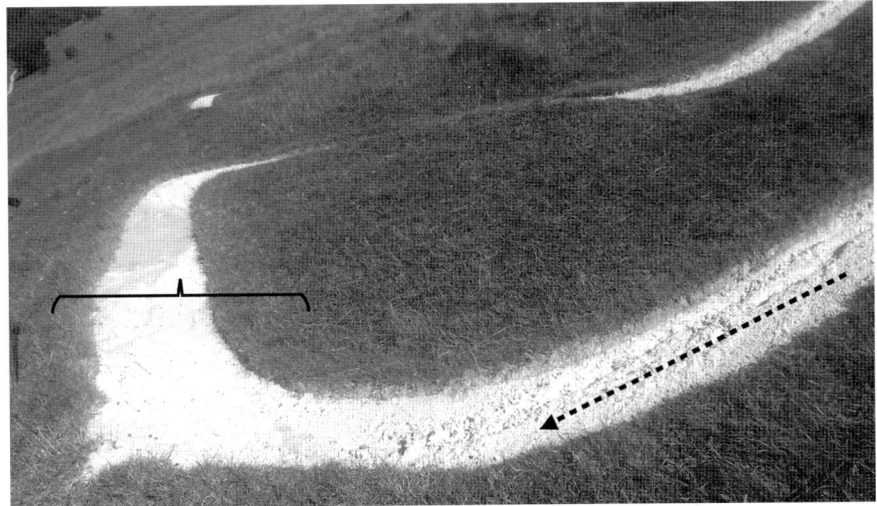

FIGURE 2.2. The Giant's left foot with rills (arrow) and chalk accumulation, September 2019. Image © M.J. Allen 2019

ends of the rills, and fine chalk mud accumulated against the lowest edge of the chalk lines against the turf, and spilled over the grass especially at the soles and elbows. This showed typical lateral sorting of fine chalk pieces being carried downslope by water: the coarse chalk pieces dropped first and the finer chalk gravel and mud carried further. Some fine chalk mud and small chalk pieces had been washed out of the figure and into the grass to become incorporated into the present-day downland rendzina soil adjacent to the figure.

There was also a clear build-up of soil and sediment against the soles of both feet and the left elbow in particular (Fig. 2.2). In all cases there was an accumulation of sediment on the upslope side of the chalked line on top of the feet, but especially at the soles, and crooks of the elbows. The combination of the chalk wash and chalk spill of the erosion of the figure outline, and colluviated soil building up against the upslope side of the chalked lines has resulted in some portions of the figure becoming elevated, and the associated 45° slope tending more towards the horizontal (Fig. 2.3).

Finally, the low (0.4 m) and broad (*c.* 8 m) mound below the left hand, which had been reported previously (e.g., Castelden 1996) and surveyed (by Papworth in 1995), was a clear feature.

The fact that the chalk-packed limbs of the Giant suffer erosion is far from a new observation. Castelden has pointed this out (1996, 30–1), and Keithley *et al.* (1999, 19) described the rainwater scouring of the packed chalk we also observed (see Fig. 2.1). Separate to this was a clear build-up of deposits *against* the chalked lines, and the creation of a changed micro-topography around the figure (Figs 2.2 & 2.3). From these field observations alone we could conclude there was a clear potential for the presence of sediment receptor points directly associated with the Giant, and thus the possibility of obtaining samples for OSL dating seemed real. However, this had to be tempered by the possibility of some deliberate or inadvertent sculpting of the land surface associated with

54 *Michael J. Allen*

FIGURE 2.3. The Giant's left foot showing 1) A clear flat-level zone upslope of the foot may indicate deliberate sculpting in the creation, or re-creation, of the figure, or a product of rechalking of the original line (Darvill 1999a, 10), and 2) An enhanced and raised area downslope of the foot which may be a) sculpted man-made/ deliberate and thus seal 'ancient'/near original deposits possibly suitable for OSL dating OR natural sedimentation and accumulation (hillwash) potential buried earlier deposits, and the micro-hillwash or rill-wash itself is potentially suitable for OSL dating. Image © M.J. Allen 2019

the outline, perhaps during scouring and rechalking. This seemed a possibility, especially in those flat almost level areas we had identified around some of the outlines (i.e., Fig. 2.3).

Some of the features we observed had been described by Castleden as 'hard shoulders' (1996, 31), which Darvill considered the result of 'successive re-cutting that have not exactly followed the lines of earlier cuts' (1999a, 10). If correct this would suggest that the figure we see today is not just a plain cut chalk outline, but is in part sculpted, and may include deliberately dumped deposits built up adjacent to the figure. Whilst the colluvial soils may be appropriate for OSL dating (as had already been demonstrated at both Uffington and the Long Man of Wilmington), anthropogenically dumped deposits had a significantly different taphonomy and their mode of deposition and burial would require careful and precise consideration.

The observation of potential sediment receptor zones (i.e., localised colluvial build-up against the chalked lines), however, is key. This potentially provided the opportunity to examine both natural modifications and sediment accumulation (for suitable OSL) and the possibility of anthropogenic sculpting of the slope above and below cut lines to emphasise the figure or to enable chalk packing infill. These observations provide the opportunity to examine both natural modifications and sediment accumulation (for suitable OSL) and the possibility of anthropogenic sculpting of the slope above and below cut lines to emphasise the figure or enable chalk packing infill. The three main conclusions of the walk-over survey were:

1. Possibility of potential sediment receptor points (i.e., suitable for OSL sampling)
2. Active erosion and redeposition of the chalk outline just 2 weeks after completion, and
3. Possible indications of sculpturing and accentuation of the figure

2. The new research: dating the Giant; reconnaissance, aims, and methods

Concerns

The fact that the Giant has been scoured (and possibly remodelled) on a number of occasions, and inevitable changes had occurred to varying extents on each and every rechalking, is well known (e.g., Castleden 1996; Darvill 1999a, 10). We therefore would need to be exceptionally careful that any accumulations exposed in excavation and sampled, related to the Giant in antiquity (whatever that is) rather than any subsequent rechalking and remodelling events. Any date obtained, will *always* have to be considered a *terminus ante quem*, and potential of actually dating the Giant from 'original' cutting or deposits (with our current understanding of the figure) is negligible. Clear attempts would be needed to mitigate against sampling obviously later material, and this was undertaken via historical research, and careful geoarchaeological observation and examination.

Fieldwork approach and methods

Field visits identified four potential locations on the Giant that might contain deposits suitable for OSL dating the construction of the figure. The precise location of those investigations was determined by hand augering using a 4 cm diameter dutch auger and a 3.5 cm diameter gouge auger to locate more precisely suitable deposits. Up to eight auger holes were undertaken in the target locations through the chalk fill, and the adjacent turfed areas. The auger profiles were fully described and recorded and any soil/sediment returned to the hole and heeled in. These assisted in the precise location of the four trenches.

Surveys

The specific methods of the photogrammetric and geophysical surveys (1979 earth resistivity, and 2023 ground penetrating radar and resistivity) are provided with the results in Chapter 3.

Hand-excavated trenches

Four small trenches were hand excavated (Fig. 2.4). Each was (only) 2 m long and very narrow at merely *c.* 0.6 m wide, only being increased in width with a slot to enable horizontal augering to facilitate gamma spectrometer insertion during onsite measurement associated with OSL sampling (Trench B). Full archaeological (Papworth) and geoarchaeological (Allen) recording was made of the profile in every trench, with soil and sediment descriptions following standard notation (e.g., Hodgson 1997), to assist in the interpretation of taphonomy and formation of the deposits. The trenches labelled clockwise from the Giant's left foot (A) to left elbow (D; Fig. 2.4).

Sampling

The principal sampling was that for optically stimulated luminescence (OSL) dating. The profiles and deposits in each of the four trenches was reviewed with the OSL team (Prof Phil Toms/Dr Jamie Wood, University of Gloucestershire),

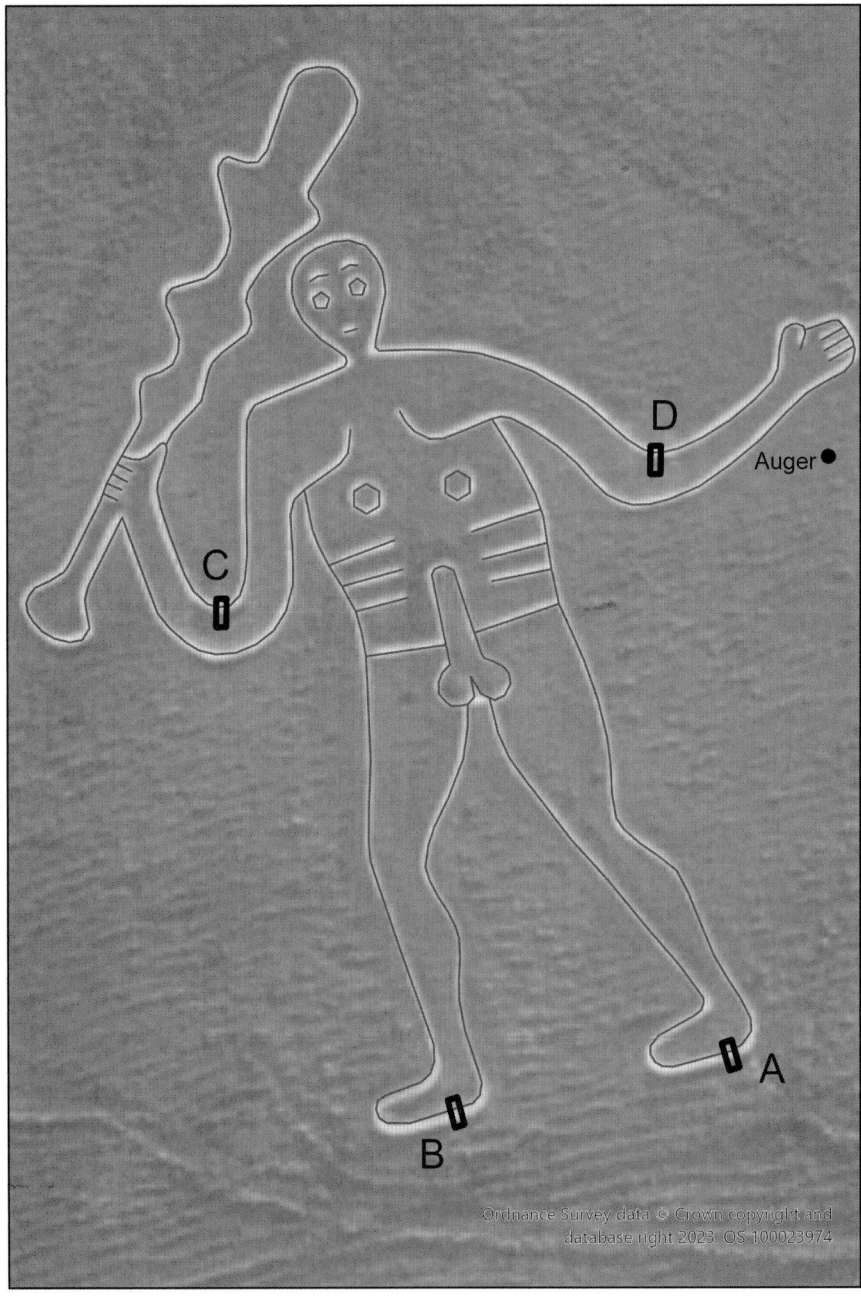

FIGURE 2.4. Plan of the Giant showing the accurate location of the four 2020 trenches and the auger hole. Image: National Trust

and together the most appropriate profiles were selected for OSL dating. These were Trenches B, C and D. Optically stimulated luminescence sampling by Prof Phil Toms and Dr Jamie Wood included on site gamma spectrometer readings (Fig. 3.20). Detailed methods of measurement are given below. Samples from Trenches B and C were advanced and dated (Chapter 3).

2. The new research: dating the Giant; reconnaissance, aims, and methods

The dating was augmented by sampling for land snails and soil micromorphology, and sampling was of the principal profiles selected for OSL dating (especially Trench B, left foot). Soil/sediment micromorphology (and soil chemistry) block and small samples were taken of the key deposits with the potential that these may improve the understanding of the deposit formation and chalk preparation for the figure. Sampling was undertaken in conjunction with geo-archaeological description and sampling for land snails. Methods follow French (2015). The deposits sampled for soil micromorphology were evaluated by the field team and Prof C. French, and any appropriate samples advanced for thin section manufacture and analysis. Sampling for land snails had the potential to provide some information about the near contemporaneous environment and land-use – though this is potentially fraught with problems (cf. Bell & Butler 2014). Sampling methods followed Evans (1972) and Allen (2017a; 2017b). All samples were processed (Allen 2017a), assessed and selected samples analysed.

It was felt at the outset that the deposits would, in general, not be suitable for the recovery of charred plants and charcoal, due to the lack of human occupation, settlement, fire setting and crop processing associated with the Giant on the steep hillside. Nevertheless provision was made for sampling and processing.

Programme

The main phase of fieldwork (augering, four trenches and sampling) was conducted over just five days from 16–20 March 2020.

During the post-excavation phase, augering was conducted of the natural soil depths immediately outside the scheduled area and the possibility of determining the presence of the belt (and thus if the phallus and testes were part of the original design or a later addition) was undertaken in 2023 and 2024. Paul Cheetham undertook localised ground penetrating radar (GPR) survey (the latter already in our original already drafted further research work agenda) and resistivity surveys in December 2023, and scheduled monument consent was obtained for a single day's augering along with a section 42 agreement for the survey conducted in January 2024 (Allen 2024).

Optically stimulated luminescence sample preparation and measurement

Phillip Toms and Jamie Wood

Mechanisms and principles
Upon exposure to ionising radiation, electrons within the crystal lattice of insulating minerals are displaced from their atomic orbits. Whilst this dislocation is momentary for most electrons, a portion of charge is redistributed to meta-stable sites (traps) within the crystal lattice. In the absence of significant optical and thermal stimuli, this charge can be stored for extensive periods. The quantity of charge relocation and storage relates to the magnitude and period of irradiation. When the lattice is optically or thermally stimulated, charge is evicted from traps and may return to a vacant orbit position (hole).

Upon recombination with a hole, an electron's energy can be dissipated in the form of light generating crystal luminescence that in turn provides a measure of dose absorption.

For this study, quartz was segregated for sediment samples for dating. The utility of this minerogenic dosimeter lies in the stability of its datable signal over the mid to late Quaternary period, predicted through isothermal decay studies (e.g., Smith *et al.* 1990; retention lifetime 630 Ma at 20°C) and evidenced by optical age estimates concordant with independent chronological controls (e.g., Murray & Olley 2002). This stability is in contrast to the anomalous fading of comparable signals commonly observed for other ubiquitous sedimentary minerals such as feldspar and zircon (Wintle 1973; Templer 1985; Spooner 1993).

Optical age estimates of sedimentation (Huntley *et al.* 1985) are premised upon reduction of the minerogenic time dependent signal (optically stimulated luminescence, OSL) to zero through exposure to sunlight and, once buried, signal reformulation by absorption of litho- and cosmogenic radiation. The signal accumulated during the burial period acts as a dosimeter recording total equivalent dose absorption, converting to a chronometer by estimating the rate of dose absorption from radiation emanating from within the surrounding lithology and streaming from the cosmos:

$$\text{OSL Age} = \frac{\text{Mean Equivalent Dose (De, Gy)}}{\text{Mean Dose Rate (Dr, Gy.ka}^{-1}\text{)}}$$

Aitken (1998) and Bøtter-Jensen *et al.* (2003) offer a detailed review of optical dating.

Sample preparation
Conventional OSL sediment samples, those located within matrix-supported units composed predominantly of sand and silt, were taken during the excavations and processed for dating. To preclude optical erosion of the datable signal prior to measurement, all samples were prepared under controlled laboratory illumination provided by Encapsulite RB-10 (red) filters. Samples were subjected to acid and alkaline digestion (10% HCl, 15% H_2O_2) to attain removal of carbonate and organic components respectively.

Fine silt sized quartz, along with other mineral grains of varying density and size, was extracted by sample sedimentation in acetone (<15 μm in 2 min 20 s, >5 μm in 21 mins at 20°C) from each sample. Feldspars and amorphous silica were then removed from this fraction through acid digestion (35% H_2SiF_6) for two weeks (Jackson *et al.* 1976; Berger *et al.* 1980). Following addition of 10% HCl to remove acid soluble fluorides, grains degraded to <5 μm as a result of acid treatment were removed by acetone sedimentation. Limited quantities of fine silt quartz were recovered, with none found within sample OSL 8. Four multi-grain aliquots (*c.* 1.5 mg, 10 mm Ø) were then mounted on aluminium discs. These were used to assess the mean D_e value within the fine silt quartz component of each sample. Though small in number, these aliquots should

2. The new research: dating the Giant; reconnaissance, aims, and methods

provide a good estimate of the mean D_e value given the large number of fine silt quartz grains per aliquot.

For fine sand fractions a further acid digestion in HF (40%, 60 mins) was used to etch the outer 10–15 μm layer affected by α radiation and degrade each samples' feldspar content. During HF treatment, continuous magnetic stirring was used to effect isotropic etching of grains. 10% HCl was then added to remove acid soluble fluorides. Each sample was dried and resieved. The quartz component was isolated from the remaining heavy mineral fraction using a sodium polytungstate density separation at 2.68g.cm^{-3}. No quartz was found in this fraction for sample OSL 8. Where the quantity of datable material was sufficient, 2 mm multi-grain aliquots were constructed to undertake diagnostic tests to establish optimum, average measurement conditions for single grain aliquots. Up to 1800 sand grains of each sample were located individually in 300 μm (diameter and depth) holes drilled as a 10×10 grid into anodised aluminium discs (Duller *et al.* 1999). These were then used to evaluate the inter-grain D_e distribution.

Acquisition of the D_e value
All minerals naturally exhibit marked inter-sample variability in luminescence per unit dose (sensitivity). Therefore, the estimation of D_e acquired since burial requires calibration of the natural signal using known amounts of laboratory dose. D_e values were quantified using a single-aliquot regenerative-dose (SAR) protocol (Murray & Wintle 2000; 2003) facilitated by a Risø TL-DA-15 irradiation-stimulation-detection system (Markey *et al.* 1997; Bøtter-Jensen *et al.* 1999; Duller *et al.* 1999). Within this apparatus, optical stimulation of single grain aliquots emanates from a focused solid state 532 nm (green), 10 mW stabilised laser (Laser 2000 LCL-LCM-T-11ccs) scanned across grains by means of mirrors mounted on and moved by motorised linear stages. Optical stimulation of fine silt multi-grain aliquots was provided by an assembly of blue diodes (5 packs of 6 Nichia NSPB500S), filtered to 470±80 nm conveying 15 mW.cm^{-2} using a 3 mm Schott GG420 positioned in front of each diode pack. Infrared (IR) stimulation, provided by 6 IR diodes (Telefunken TSHA 6203) stimulating at 875±80 nm delivering ~40 mW.cm^{-2}, was used to indicate the presence of contaminant feldspars (Duller 2003). Stimulated photon emissions from quartz aliquots are in the ultraviolet (UV) range and were filtered from stimulating photons by 7.5 mm HOYA U-340 glass and detected by an EMI 9235QA photomultiplier fitted with a blue-green sensitive bialkali photocathode. Aliquot irradiation was conducted using a $^{90}Sr/^{90}Y$ β source calibrated for 5–15 μm quartz multi-grain aliquots and single 180–250 μm quartz grain aliquots against the 'Hotspot 800' ^{60}Co γ source located at the National Physical Laboratory (NPL), UK. In calibrating single sand grain aliquots, no significant spatial variation in dose rate from the source was found.

SAR by definition evaluates D_e through measuring the natural signal of a single aliquot and then regenerating that aliquot's signal by using known laboratory doses to enable calibration. For each aliquot, up to six different regenerative-doses were administered so as to image dose response. D_e values for each aliquot were then interpolated, and associated counting and fitting errors calculated, by way of exponential plus linear regression. The accuracy with which D_e equates to total absorbed dose and that dose absorbed since burial was assessed. The former can be considered a function of laboratory factors, the latter, one of environmental issues. Diagnostics were deployed to estimate the influence of these factors and criteria instituted to optimise the accuracy of D_e values.

Acquisition of the D_r value
Lithogenic D_r values are defined through measurement of U, Th and K radionuclide concentration and conversion of these quantities into α, β and γ D_r values. Radionuclide contributions and level of U disequilibrium were evaluated from sub-samples by laboratory-based γ spectrometry using an Ortec GEM-S high purity Ge coaxial detector system, calibrated using certified reference materials supplied by CANMET. Measured radionuclide concentrations were converted into D_r values (Adamiec & Aitken 1998), accounting for D_r modulation forced by grain size (Mejdahl 1979), present moisture content (Zimmerman 1971) and, where D_e values were generated from 5–15 μm quartz, reduced signal sensitivity to α radiation (a-value 0.050 ± 0.002). Internal dose rate arising from U and Th within quartz was assumed to be 0.010 ± 0.002 Gy.ka^{-1} (Vandenberghe *et al.* 2008). Cosmogenic D_r values were calculated on the basis of sample depth, geomagnetic latitude and matrix density (Prescott & Hutton 1994).

Estimation of Age
OSL ages provide an estimate of sediment burial period based on mean D_r values and D_e values derived from Minimum, Finite Mixture (FMM_{Min} and FMM_{Major}) and Central Age Models, along with associated, 1σ analytical uncertainties (Galbraith & Laslett 1993; Galbraith & Green 1990; Galbraith *et al.* 1999). For multi-grain aliquot analyses, only the Central Age Model can be applied.

CHAPTER THREE

Research results: fieldwork, dating and analysis

Michael J. Allen

'Scope for excavation'; Working hand-in-hand with soil experts and archaeologists to open up strips around the hillside sectioning the Giant itself (Hoade 1978, 29)

As we have seen from the pre-excavation reconnaissance walkover survey and observations (Chapter 2), there are potential sediment accumulations like those at Uffington, but the Long Man of Wilmington, for instance is significantly different. A programme of minimally intrusive fieldwork was designed to expose suitable deposits to provide a date for the Giant and to maximise information recovery about the environment in which he was built and the history of the hillside, together with the history of the Giant's rechalking and development. In addition a high resolution aerial photogrammetry survey enabled modelling of the surveyed topography and provision of rectified images to give a new plan of the Giant's outline (Fig. 3.5). A targeted auger survey was conducted to ensure the correction location of the four trenches, and to test the stratigraphy of the low mound below the left hand (the putative severed head). Four small clinical incisions were then excavated through the soles of both feet and crooks of both elbows aimed to obtain samples for dating and palaeo-environmental research as well as recording the history of rechalking. During the post-excavation and reporting stage two further auger surveys were conducted; one to examine the natural soil depth on the slope adjacent to the scheduled area and a second, in 2023, to test the publicised, but unsubstantiated, suggestion of the presence of a former belt under the phallus, and its implications. The possibility of examining and dating the chalked infill was plausible, but in view of the history of scouring, and the industrial scale of that exercise from at least 1887 (Chapter 1), whether any of the 'original' chalking survived was unknown. Indeed whether any chalking could be identified that was original was in serious doubt. Hence the importance of the recognition of deposits accumulated against the Giant (see Chapter 2, and below).

High resolution photogrammetric survey (3-D ground surface model)

Michael J. Allen

The 2020 survey of the Giant was not done by traditional survey methods; gridding, triangulation, use of total station or GPS equipment. The slope was steep and the grass with natural terracettes, enhanced by animal tracks, rabbit holes and tussocky ant mounds. None of these is conducive to an accurate survey and creation of a new plan of the Giant; the first for essentially 100 years. Photogrammetric survey via drones was far more accurate, and could compensate for the steep slope. The survey data would allow digital manipulation of the images to enhance the surface topography, increasing our ability to make out subtle hidden changes on the hillside which might represent changes in the form and nature of the figure. A vertical ground image (Fig. 3.1) was rendered (Figs 3.2 & 3.3) to enable archaeological observation and interpretation. The survey could also create an accurate plan of the Giant perpendicular to the slope, rather than a vertical overhead (slightly foreshortened) one (Fig. 3.4). In 2023 the Downland Partnership remodelled the vertical data to create an image perpendicular to the slope (rather than direct vertical) thus creating a true plan image of the Giant from which a new plan of the Giant was then made (Fig. 3.5), and provides the first accurate plan of the Giant since Petrie's survey published in 1926.

FIGURE 3.1. (opposite) Colour image of the drone survey in 2020, modelled to give a vertical view; the recent (2019) outline shows up clearly. Downland Surveys 2020

Survey methods

Tony Davies

An accurate ground surface model enabling a virtual reality photo-rendered image of the Giant was undertaken by Downland Survey & Measurement. This was conducted using a UAV (drone) in conjunction with a high definition camera. Although laser scanning was considered, because of the nature of the terrain and the size of the site, this was neither cost effective nor would provide as good a result as high definition rendered photographs. The photographic survey was undertaken with multiple overlapping images to give a strong photogrammetric solution (Fig. 3.1). A series of pre-marked control points were surveyed using a Leica total station to provide accurate scaling and orientation to the 3D model. This process is sometimes referred to as SfM (Structure from Motion). The models (Figs 3.2 & 3.3) were generated using Bentley ContexCapture software. The software is capable of data outputs in many different formats although it can be most efficiently used in its own .3mx configuration.

Archaeological interpretation: survey results and archaeological comments

Martin Papworth and Michael J. Allen

The photogrammetrically produced digital terrain model of the survey area (Fig. 3.1) produced by Downlands Survey was further processed by Keith Challis (National Trust digital data/remote sensing coordinator) using ESRI's ArcGIS Pro and the Relief Visualisation Toolbox extension (https://github.com/EarthObservation/RVT_py). This allowed the generation of a number of different renderings of the terrain data, each accentuating different aspects of relief and microtopography as an aid to interpretation (Figs 3.2 & 3.3).

64   *Michael J. Allen*

The rendered images produced by the National Trust were examined from an archaeological perspective. These images allowed a better appreciation of the earthworks. The best of the processed images has accentuated the contour differences and shows the Giant in darker (blue) and lighter (yellow) colours. The outline of the double bank and ditch of the Trendle is clear and the coffin shape fence line that once surrounded the National Trust ownership boundary of the Giant is visible. In earlier years this existed as iron railings installed in 1886 (Chapter 1, *contra* Darvill 1999a, 9; Vale & Vale 2000, 15), and, together with soil development under the fence and pathways along its boundary, it shows up quite clearly (Figs 3.2 & 3.3).

Figures 3.2 and 3.3 showing the Giant and Trendle are rendered and modelled from a vertical image, while Figure 3.5, has been drawn from the remodelled Downland Partnership perpendicular view, thus providing the first accurate surveyed outline since Petrie's 1926 outline (Fig. 1.5) was published.

The Giant's outline
The Giant's outline is clearly visible in the survey images. The pronounced earthworks from soil settling on his horizontal lines can be seen in the rendered images (Figs 3.2 & 3.3) on his elbows and feet as lines of light grey (yellow). His nose, recreated in 1993 and reinforced with chicken wire mesh, is pale grey (glows bright yellow). Raised areas of soil can be seen in the crook of both arms, soles of both feet, top of mid- and fore-foot, and the pommel of the club. These confirm that choice of location of the four trenches seems to be well founded. There seems to be no indication that the head was a complete circle. The slight groove for the chin recorded by Petrie (1926, 10) that we thought we could observe on the ground in 2020, does not seem to be depicted in the survey images. The navel at the top of the phallus is however detectable (Fig. 3.3).

Below his left hand
Another interesting revelation is the blobs of material surrounding the Giant, the largest of which is below his outstretched hand. This has been interpreted as the earthwork of a severed head, once held either in a bag, or dangling by his hair, from the outstretched hand (Castleden 1996, 181). All these blobs look more like the leftover remnants of chalk brought to the Giant over the centuries to rechalk him, and the 'head' is just a large leftover spoil heap, as also suggested from the three auger profiles (see below, auger surveys). There is also no obvious nor clear indication of a cloak or lion skin below the left arm. It is possible to convince oneself that there are some vague lines or indications here, but the variations here are no different, nor stronger, than anywhere else around the outline.

Symbols
There are possible traces of letters or numbers between his legs but nothing is legible nor obvious. Rev John Hutchins could apparently see them more clearly in 1774 than the twenty-first-century drone.

A pre-existing belt?
Looking at the centre of the Giant below his ribs, there are two chalked lines at the top of his legs, separated by the phallus. A possible very faint yellow line across the phallus linking these two probably represents a shallow groove (or terracette) but the dark line may be where sediment has settled against the horizontal groove suggesting a former cutting. The yellow line is faint but continuous, and could, argues Martin Papworth, represent the belt (and even trousers). If this is correct, is it possible that the phallus was not there when he was first created. He has perhaps had a makeover in later centuries, and a typical 'cock and balls' added. However, although this faint line is probably a product of the natural variability in the slope (there other similar patches elsewhere), and here this represents a small terracette (or even a small former animal trackway) that can be seen running between the waist lines and continues beyond the Giant in both directions, this suggestion was examined as its implication were significant. There is no clear record of this 'pre-existing belt' on aerial photographs earlier than the mid-1970s, and the Aerofilms transparency used for the cover image of the 1978 special issue of *Dorset Life*'s 'Dorset & Cerne's god of the Celts' (Legg 1979) shows no such line, but it is potentially visible as a possible line in some aerial photographs from the 1990s. The aerial photogrammetric terrain model (Fig. 3.1), and more significantly the digitally modelled image of the Giant (Fig. 3.3) was specifically examined by a number of archaeologists, including Bob Bewley, experienced aerial archaeologist, former head of survey and aerial survey at English Heritage (and before that the Royal Commission on the Historical Monuments of England) who is also familiar with the site having undertaken several drone flights for this project. He reported that 'try as I might ... and I have been looking on and off for a few hours, I can't see anything that could be confirmed as a belt' and concluded that there was 'nothing to see' (pers. comm. 19/10/23).

The terracette/animal track could have been exacerbated by a temporary fence line erected across the lower part of the Giant by the National Trust in 1989 over the 'the lower half of the naked Cerne Giant ... because visitors are wearing it out', as reported, for instance in the *Daily Mirror* (Tuesday 29 August 1989, 13) in an article entitled 'Private Parts'. After investigation of dozens of aerial photographs from the 1920s, as well as the photogrammetry survey, and further field visits, the suggestion that the top of the legs originally formed a single complete line representing a pre-existing belt was tested by geophysical survey (radar and resistivity) and augering, see below, and shown to be untenable.

The Trendle
The processed image shows the Trendle earthwork on the hillside above the Giant. A rectilinear structure, probably building footings, can be seen in the centre of the earthwork, above which (outside the fenced area) is what look like prehistoric rectilinear field boundaries approaching the enclosure.

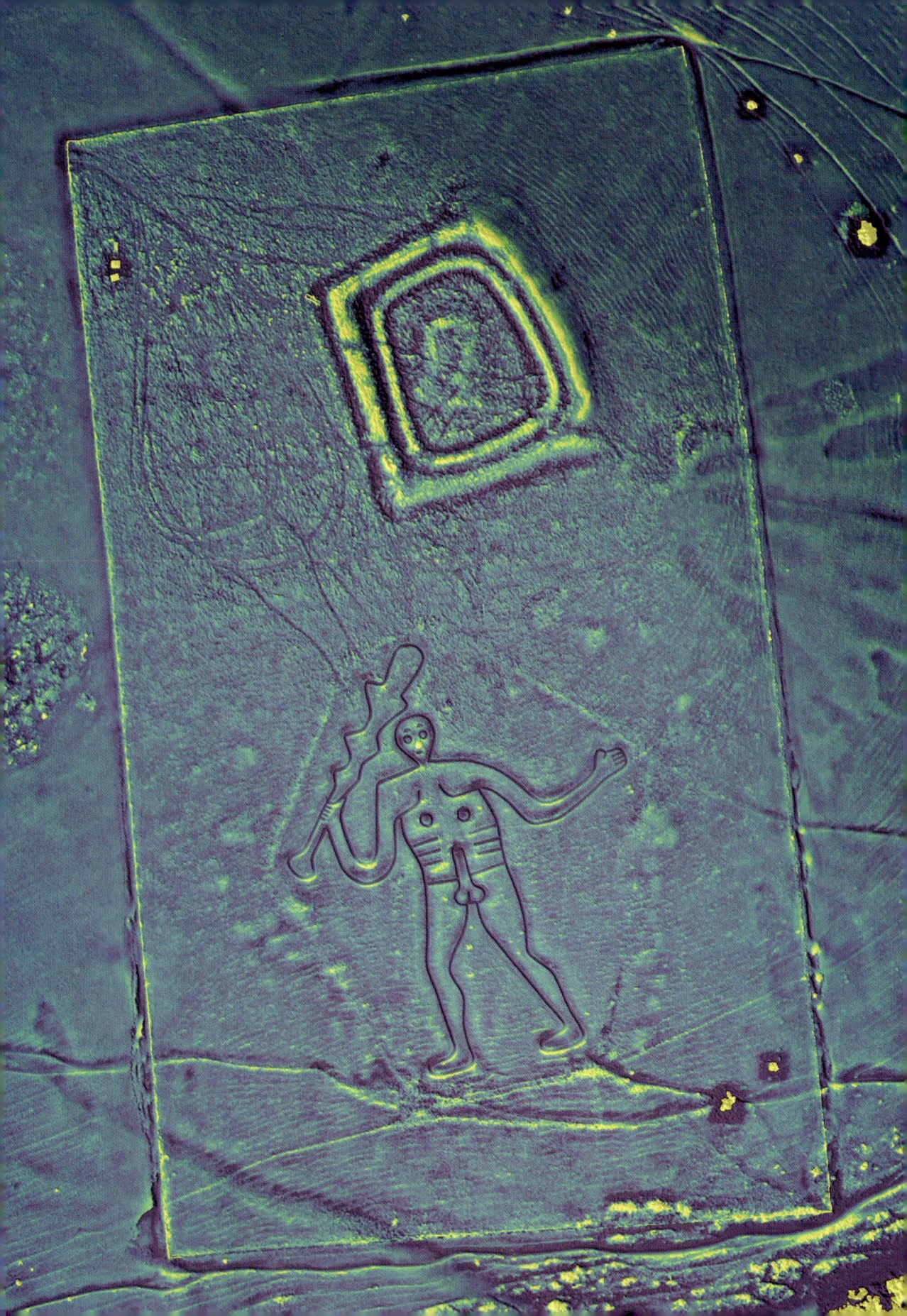

3. *Research results: fieldwork, dating and analysis* 67

FIGURE 3.2. (opposite) The processed and rendered image derived from the digital data. Image: National Trust 2020

True Plan Image

A rectified image perpendicular to the slope allows the production of a new scale plan of the chalked outline of the Giant (Fig. 3.5); the first to be generated since the 1920s. It shows remarkable similarity to Petrie's. Obvious, but subtle, changes are the testes, which are no longer complete circles. More indistinct points include the chin (which did not show up on the rendered images; Figs 3.2 & 3.3) but may be present in the survey image according to Martin Papworth (Fig. 3.1), and faint lines representing the edge of the pectorals, and the bottom of the former navel. No obvious medial line of the sternum, as dotted by Petrie,

FIGURE 3.3. Detail of the Giant in the rendered image. Image: National Trust 2020

can be seen. Overall, the image scoured and rechalked in 1924, just after the Trust's acquisition of the Giant, and the survey by Petrie are essentially what we see on Giant Hill 100 years later. Surprisingly, and reassuringly, there are no major differences between the two indicating that, overall, under the National Trust's ownership care and management, his outline has essentially been retained. Changes in the figure were more obvious from the excavated results (see below).

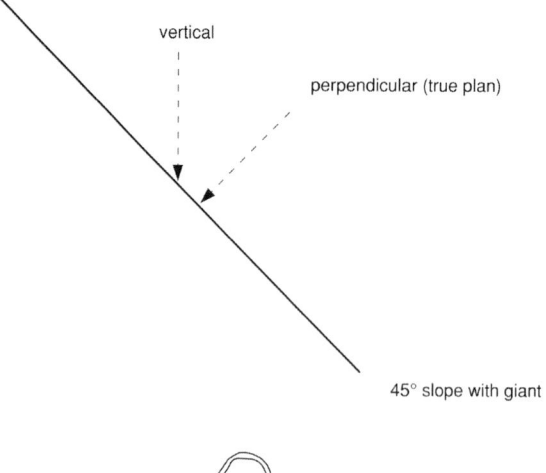

FIGURE 3.4. Vertical vs perpendicular view to give a true plan image of the Giant. Image: Justin Russell 2024

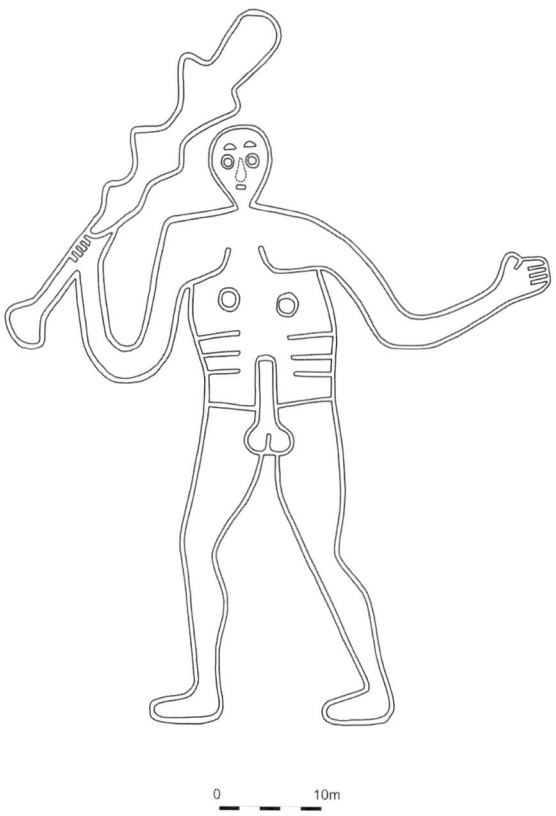

FIGURE 3.5. New outline of the Giant drawn from rectified photogrammetry survey. Image: Justin Russell 2024

Geophysical surveys

1979–80: whole figure earth resistance survey

Andrew David, Tony Clark†, Alister Bartlett†, Paul Linford and Megan Clements

Forty-five years ago, in 1979, Yorkshire Television approached the late Tony Clark, then Head of the Geophysics Section of the Ancient Monuments Laboratory (AML) of the Department of the Environment, to enquire whether he might contribute his expertise in geophysical survey towards a speculative investigation of chalk hill figures, concentrating primarily on the Cerne Giant of Dorset. This would be part of a short series of programmes investigating unusual and unexplained phenomena, to be entitled *Arthur C. Clarke's Mysterious Worlds*. Intrigued, not by the mysterious worlds, but by the rather unusual archaeological interest and associated challenges, Tony agreed to

3. Research results: fieldwork, dating and analysis

be involved. However, as there was no pressing conservation imperative, and mindful of possibly adverse sensationalism, this needed to be in a private capacity and undertaken in his own time, assisted likewise by colleagues and family.

The brief was to use earth resistance survey in an attempt to define any lost features of the Giant, especially any that might indicate what may once have lain beneath his outstretched left arm – that may, or may not, support an iconography related to the legend of Hercules (Piggott 1938). Earth resistance measurements provide an indication of the relative moisture content of the subsurface and, if recorded over an area in sufficient density, can very successfully map the extent and pattern of high-contrast features such as walls – perfect, for instance, for mapping Roman villas or medieval monastic sites, when conditions are suitable. The presumption in the case of chalk hill figures is that the original trenches cut to expose the chalk, subsequently infilled, will also present a detectable resistivity contrast. This presents something of a challenge, though, as such contrasts were expected to be very slight, shallow, and complicated by recutting and other interferences as well as by the steep topography and the effects of associated downward soil movement.

Trying to be optimistic in the face of these challenges, not to mention a film crew sometimes in attendance, the survey was conducted initially on 4 August 1979 by the late Tony Clark and the late Alister Bartlett, with follow-up visits over two weekends in November by Alister Bartlett and Andrew David. Return visits for the purpose of corroborative augering took place the following year. A magnetometer survey was also undertaken over the entire figure but was uninformative, showing probably modern ferrous material only, and will not be considered further here. Until now the only formal reference to this geophysical endeavour was a brief note consequent on a visit to Cerne by members of the Royal Archaeological Institute in 1983 (Clark 1983). In this, Clark struck a cautiously optimistic note, suggesting that the survey data did indeed provide 'tentative' evidence for a feature such as a lion skin in the expected area. This was supported by reference to initial and unpublished plots of the survey data (Fig. 1.18).

Once baselines were established and measured into local features earth resistance readings were taken using the Twin Electrode configuration with a mobile probe spacing of 0.5 m, at intervals of 0.5 m over an irregularly shaped grid covering much of the figure and the open area under the left arm (Fig. 3.6). The resistivity meter was a prototype of Clark's own construction and – ever resourceful as he was – the two mobile probes were fixed to a frame attached to a pair of lawnmower handles. The meter readings were written down by hand as the survey progressed and were transferred to computer at a later date. Data processing was done initially by Alister Bartlett at Surrey University using his own bespoke software, and subsequently by SIA Services, London.

Unfortunately the original processing notes and intermediate versions of various plots are no longer available. One version, perhaps a composite, that

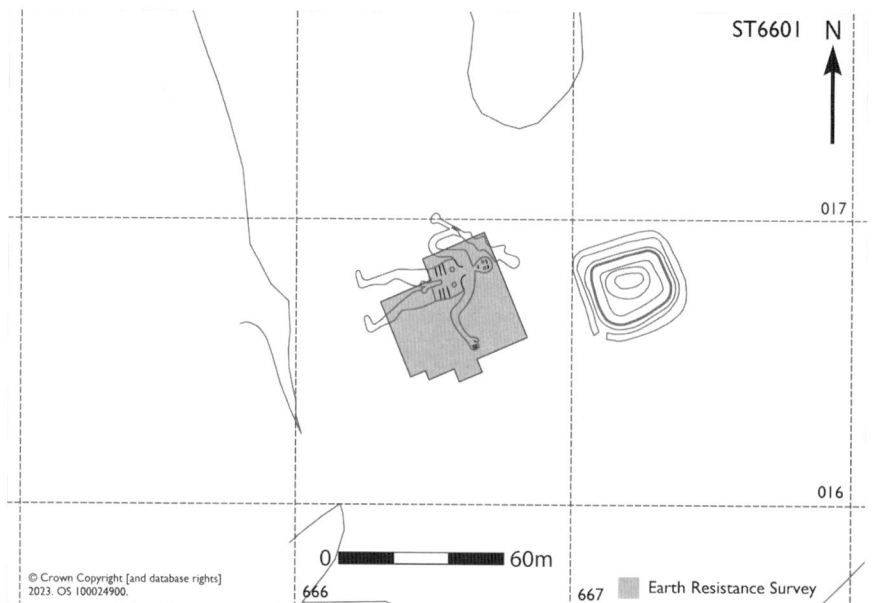

FIGURE 3.6. Location of the 1979 earth resistance survey (grey polygon) superimposed on Ordnance Survey base mapping. Image: A. David, P. Linford & M. Clements

may be responsible for Tony Clark's initial optimism is shown in Figure 3.7c (and 1.18). However, reprocessing of the raw data by one of us (PKL) in a similar manner has not been able to reproduce this image, resulting instead in the data plots depicted in Figure 3.7a and b. To produce these, values from the block of data surveyed in August when the ground was dry were multiplied by a factor of 0.49 to match its absolute deviation with that of the other survey areas measured in November when higher soil moisture content had reduced the resistivity contrast. The edges of blocks surveyed with different remote electrode positions were then matched using the method of Haigh (1991). For the greyscale image (Fig. 3.7b) the dynamic range of the dataset was equalised across different regions of the survey using the contrast enhancement algorithm of Wallis (1976) with a window radius of 15 m and a detail to background ratio of 0.95.

The results of the survey discussed here are those shown in Figure 3.7a and b. Figure 3.7d shows the current outline of the Giant superimposed on the recently computed greyscale (Fig. 3.7b). It is immediately clear that the method has at least clearly identified the general outline of the main linear elements of the extant figure, as far as these were covered by the survey. The Giant's outline had then very recently (May–July 1979) been scoured and restored with the result that the compacted and damp chalk in the Giant's features is clearly apparent in the data as linear low resistance anomalies. The trackway that serviced the 1979 restoration works is also clearly visible extending upslope from the Giant's left shoulder; and the position of the former protective perimeter fence has also been detected.

The delineation of details such as the features of the left hand and the face are fuzzy although the eyes and mouth can be made out. Also somewhat fuzzily defined, but in accordance with the depiction of 1764 in the *Gentleman's Magazine*, the outline of the head seems likely once to have been a closed oval, suggesting that the open neck is a later modification. The nipple and ribs are defined clearly on his right side, but less so on the other. The phallus and testes are more-or-less clearly defined; the resistivity response over the end of the phallus is equivalent to that over the nipples and hence supports the earlier depiction of a separate bellybutton that has since been incorporated into the Giant's already lively manhood (Pitman 1978; Grinsell 1980a). Elsewhere, similarly confined areas of slightly low resistance that might be worth drawing attention to are those that coincide with the ends of the upcurving lines that define his pectoral area, although what significance they may have – if any – is unclear; a similarly unexplained circular area of distinct low resistance occurs midway between the Giant's head and the recent workers' trackway. Anomalies such as these (and there are others) may be little more than irrelevant declivities with marginally deeper topsoil, but perhaps worth bearing in mind in any future investigations.

There is some evidence that the current outline of the Giant is slightly displaced from an earlier outline. This is most apparent when viewing the imagery of the shoulders and his upper left arm where there is a 'ghost' of less intense lower resistance parallel to and upslope of the more intense low values of the current demarcation. This suggests that these parts of the figure, at least, have migrated downslope, westwards, by as much as 2 m in places and is evidence either for remodelling and/or the effects of progressive recutting over time. This phenomenon is also apparent on aerial photographs as previously interpreted and figured by Rodney Legg (1978; reproduced in Darvill *et al.* 1999, fig. 44), although the 1979 resistance data is unclear concerning the suggested shift northwards along the contour of the Giant's left side and leg (*ibid.*). An exception to the downslope movement of features may be the Giant's left nipple where a poorly defined circular low resistance anomaly is apparent below the present-day circular depiction. If not the response to a downslope accumulation of soil, this might be evidence for a minor anatomical 'correction' to level up the two nipples? A further detail to note is that the putative 'belt' feature has not been detected underneath the phallus.

As for the all-important area beneath the outstretched arm, the anomaly suggested by Alister Bartlett's initial processing extending some *c.* 12 m beneath the left elbow (Fig. 3.7c), and then speculated to represent a possible lion skin, could not be reproduced. Inspection of the raw field measurements over 40 years later suggests this may have been an artefact of the more limited computer processing and display capabilities available in the 1980s. Instead, there is an irregular area of higher resistivity *c.* 8 m across immediately beneath the left elbow but no coherent pattern is discernible within this. A discontinuous and narrow sinuous low resistance anomaly visible in places may indicate some sort of outline to the higher resistance values but this is only very weakly defined and of uncertain significance. To the south there is further variation in resistance values downslope and away from the hand but, again, without distinct linear

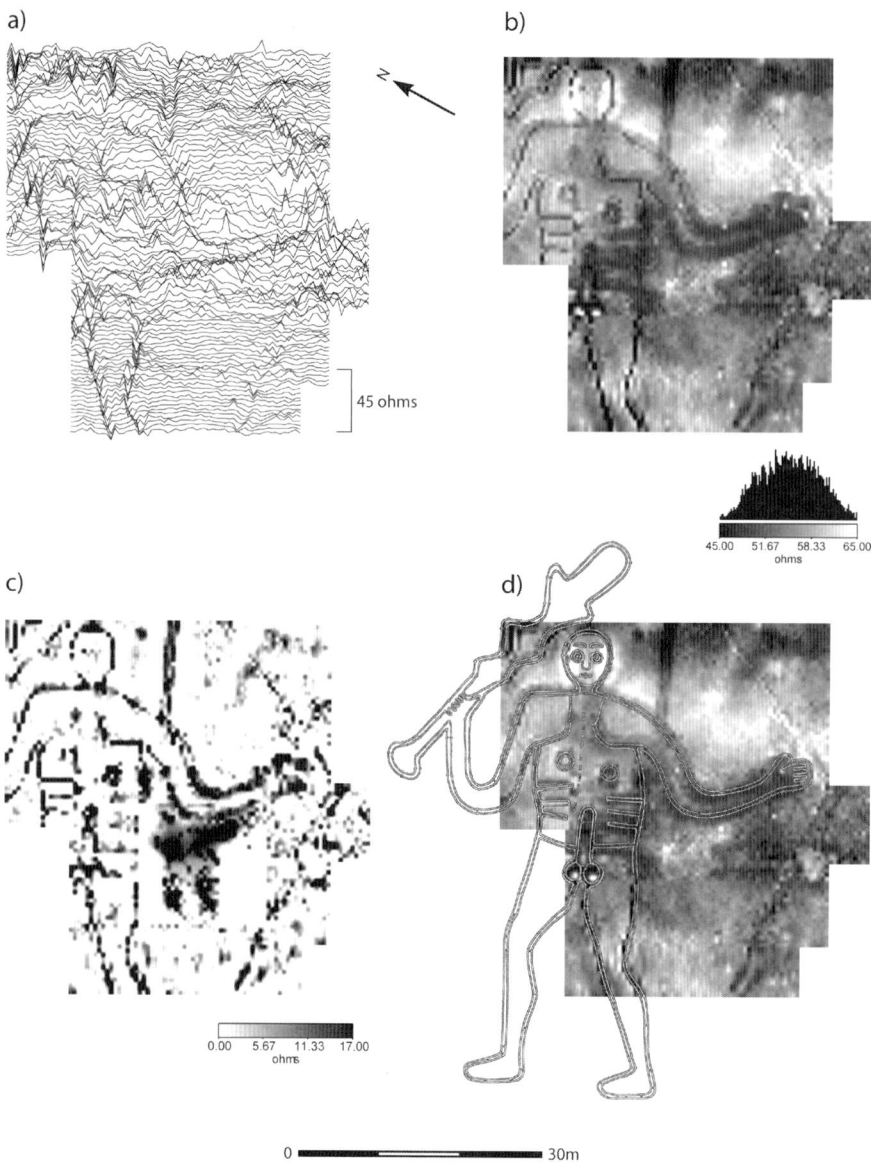

FIGURE 3.7. a) Trace plot of the 1979 earth resistance survey after rescaling of the measurements collected in August and edge-matching of the survey blocks; b) Linear greyscale plot of the same dataset with additional processing using Wallis contrast enhancement; c) representation of Alister Bartlett's original plot of the data (the processing used to create this is unclear); and d) greyscale plot with the Giant outline superimposed. Image: A. David, P. Linford & M. Clements; 3.7c, A. Bartlett

elements. An area of higher resistance *c.* 5 m in diameter, and *c.* 10 m below and to the south-west of the hand, lying mostly outside the remains of the recent perimeter fence, coincides with the 'knoll' described by Castleden (see below).

Several transects of auger holes were made parallel to the contour, and also at right angles to it, both over the outstretched arm and below it over the postulated possible skin/cloak. These show significant soil depths (45–75 cm) over the arm that may result from accumulation up against the outlines that here run approximately parallel to the contour, hence possibly caused by natural processes as well as those associated with reworking the features and/or their deliberate sculpting (see Bartlett & Allen, and Cheetham, below). Further downslope, over the position of the supposed cloak/skin there are isolated instances of significantly deeper soil (45–60 cm) but these only partly coincide with the putatively significant linear low resistance anomalies seen in the initial data processing.

While undertaken in less than ideal circumstances and before the technical improvements that have since so much advanced the efficiency and resolution of earth resistance survey, these results are informative yet frustrating: informative in that they tend to confirm details of modifications to the recorded outline of the Giant and hint at anomalies on which to focus future attention; frustrating, in that absolute clarity is still elusive, including in the area below the arm, and explanations for the geophysical response are left highly ambivalent. This remains the case even after the subsequent surveys by Rodney Castleden in 1989–95 (Castleden 1993; 1995a; 1996). These latter confirm the evidence for the modification to the penis and the migration of the outline; they also confirm that suggestive anomalies are appended below the outstretched arm although their exact position and definition varies somewhat from those apparently detected in 1979. The Castleden surveys extended further southwards beneath the extended left hand to investigate a raised area (or knoll) that was speculated to represent a possible severed head or other feature such as a purse suspended from the hand – but neither topographic mapping here, nor the 1979 survey, yielded conclusive results. Magnetometer and earth resistance surveys by John Gale in 1996 also had limited success (Gale 1999, 62).

These geophysical enterprises, and – more recently – high resolution photogrammetric survey (3-D ground surface model, see above) cannot be said to have resolved the issue of what may or may not have once been depicted alongside the Cerne Giant. On balance, the earth resistance data, as replotted here, does not provide reliable evidence for hidden outlines suggestive of a skin or cloak; while resistivity contrasts are apparent these are generalised and irregular in distribution and lack convincing linear elements. Any hope of improvement on this situation, short of the ideal, which would be further excavation, should include a more up-to-date survey, preferably utilising multiple methodologies extending widely over the Giant and his immediate surroundings at a time of year of optimum soil moisture contrasts. Earth resistance survey is still likely to be the best approach and could now take advantage of methodologies that allow high resolution coverage and computation of variation with depth.

2023: targeted ground penetrating radar and resistivity surveys

Paul Cheetham

The tops of the legs are clearly marked in the Giant's outline, interrupted by the erect phallus. However, the presence of the continuation of these lines, as a belt, across the phallus has a major significance as it could indicate that the phallus and testes were a later addition to the Giant. Noted by Martin Papworth, his reporting on the National Trust blog/website led to its significance being highlighted and questioned across social media, and latterly by archaeologists and academics. The possible belt in the photogrammetry survey (Fig. 3.1) is a shallow groove or line across the phallus between the chalked lines marking the top of

74   *Michael J. Allen*

the legs. Looking at the National Trust rendered image of the survey (Fig. 3.3), in the centre of the Giant the lines of the top of his legs (or belt) are clearly visible. This was examined as area A (Fig. 3.8).

In addition, the opportunity was taken to examine the area of the putative lion skin/belt (Area B), and the figures and symbols between the Giant's shins (Area C; Fig. 3.8).

In December 2023 several additional geophysical surveys were conducted to help confirm or clarify previous survey results and test recent theories concerning 'lost' elements of the Giant (Fig. 3.8; Allen 2023b). The work included the first use on the Giant of ground penetrating radar (GPR) and electrical resistance imaging (ERI). In addition, the work included repeating some previously surveyed areas with electrical resistance (ER) survey using the multiple-potential electrode twin array (MPET) (Cheetham 2001; 2003), which provides higher resolution of near-surface anomalies than the more conventional twin-probe array previously used on the site. The surveys were conducted after an extended period of heavy rain. This can be advantageous for the discrimination of topsoil

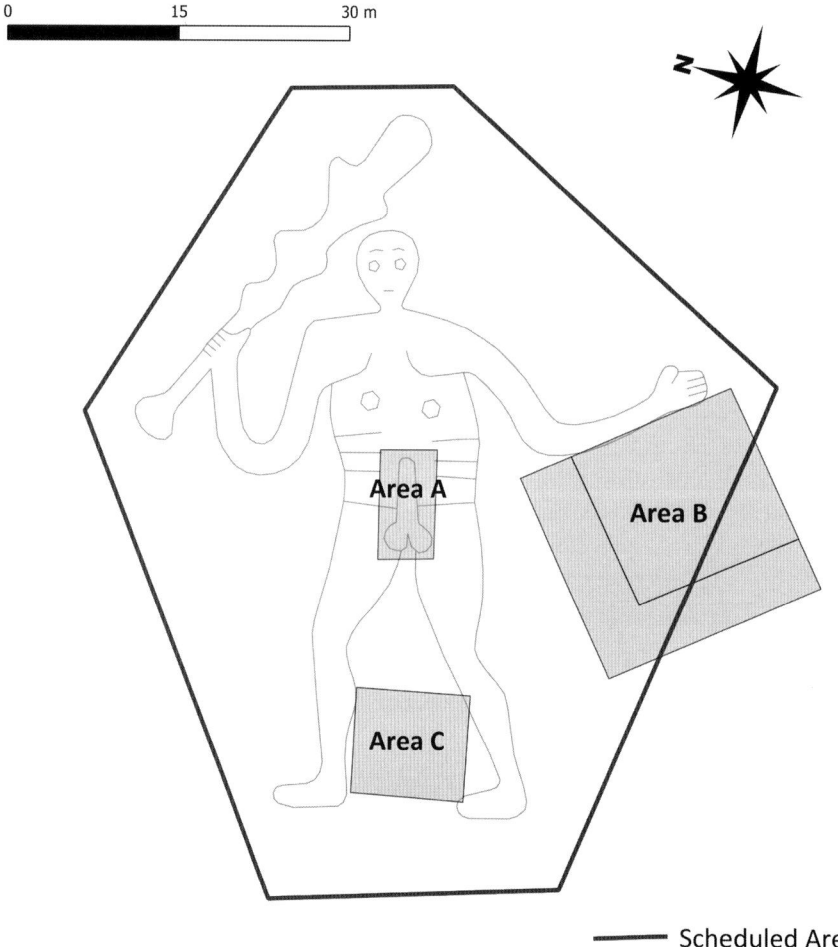

FIGURE 3.8. Location of the three 2023 geophysical survey areas. The smaller square within Area B is the 15 × 15 m GPR survey area. Image: P. Cheetham

and shallow features by electrical methods (Cheetham 2014). Three areas were chosen for survey and are shown in Figure 3.8. This commentary only summarises and discusses the results as a full technical survey report is available (Cheetham 2024).

The survey included the first ground penetrating radar (GPR) imaging of the Giant combined with earth resistance and electrical resistivity imaging to try to improve upon previous surveys (see Castleden 1996; Gale 1999). Except Tony Clark's 1979 survey (Fig. 1.18), previous geophysical surveys have had limited success. Whereas Areas B and C were surveyed with both GPR and ER, Area A was additionally surveyed using ERI. Areas to each side and over the phallus (Area A) were recorded at 20 cm traverses by ground penetrating radar (ground coupled, impulse) using a Mala RAMAC X3M, and resistance (resistivity) with an Allied Associates Tigre Resistance meter using 10 cm separation pole-pole and Wenner arrays. Subsequently four 4 m long pole-pole profiles were manually recorded at 10 cm intervals to the north and south of the phallus and two within the phallus, using a Geoscan Research RM15 resistance meter.

Area A (overall extent 10 × 5 m), covering the phallus, was to test the suggestion that the 'belt' trench may have originally been continuous, with the phallus added later, and so the remnants of the refilled belt trench should be detectable beneath the phallus turf. High frequency (800 MHz) high-resolution GPR, and high-resolution MPET ER and ERI surveys were conducted over this area. While all the techniques clearly detected the belt trench to either side, none convincingly detected any remnant trench under the phallus (Fig. 3.9). That a refilled trench may not be detectable by these techniques must be considered,

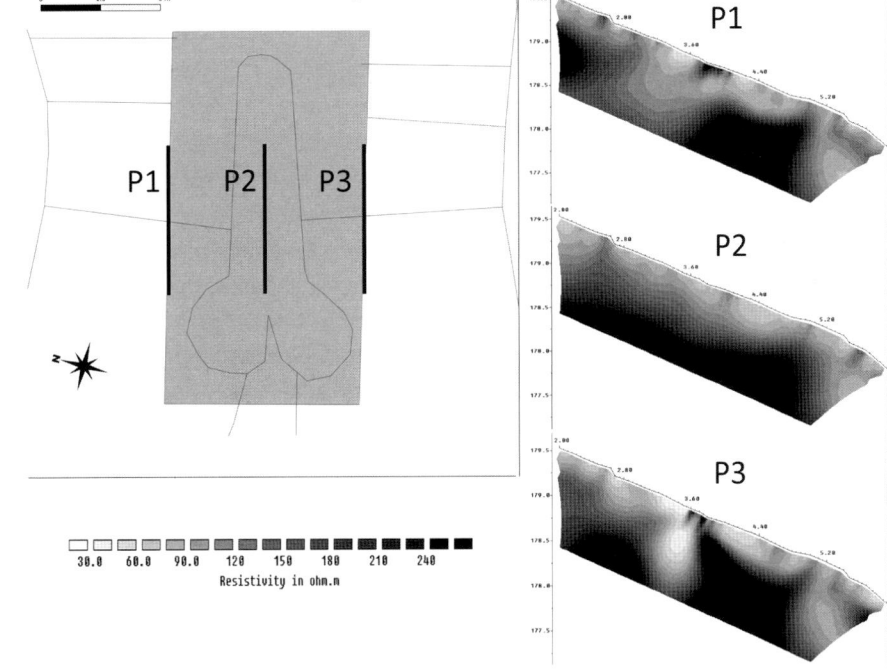

FIGURE 3.9. Electrical imaging profiles across the top of the legs/belt (P1 & P3) and down the phallus (P2). P1 and P3 show anomalies at the existing trench positions whereas in comparison P2 is more uniform across the centre of the profile indicating that any infilled trench is not present or at least not detectable. The slight near-surface anomaly at around 4.0 m in P2 corresponds to a sheep terracette that crosses the phallus at this point. Image: P. Cheetham

76 *Michael J. Allen*

however, auger evidence (see Auger Surveys Allen below) supports these results confirming that the belt trench did not continue under the phallus and so the phallus was part of the original form of the figure.

Area B covered the area under the figure's outstretched left arm to test the nature and form of the topographical feature previously suggested to be a severed lion's head and the area of the draped cloak or lion skin. Both GPR and MPET ER were conducted over a 15 × 15 m area overlapping with the ER covering a larger 20 × 20 m area. The results failed to detect the cloak described as being apparent in the 1979 survey. The topographical 'head' feature was demonstrated to be a heterogeneous mix of low and higher resistance deposits with high reflectivity GPR hyperbolic responses suggesting chalk or flint rubble patches. In the experience of the author, this heterogeneity characterises the mix of topsoil, subsoil, bedrock of a typical excavation spoil heap, which, in this case, may have been in use during the original digging, the 1946 restoration, or any rescouring of the figure. The outline of the disturbance, however, largely follows the areas previously plotted by Tony Clark and Alister Bartlett in 1979 (Fig 1.18) and by Rodney Castleden (Fig. 1.17b).

Area C, situated between the legs, was to investigate the existence/survival of the symbols evident on Hutchins' 1774 drawing of the Giant. Both GPR and MPET ER showed clear detail in the near-surface deposits between the legs including detecting the remnants of present and erased sheep tracks/terracettes (Fig. 3.10), but the GPR failed to detect any evidence of the figures or symbols. However, the high-pass filtered MPET ER survey did reveal indistinct anomalies between the knees that, under parching conditions, could have resembled the features Hutchins records. That said, these results suggest that these anomalies do not result from distinct cut features, therefore, any symbols were surface only features that have now been largely eroded away.

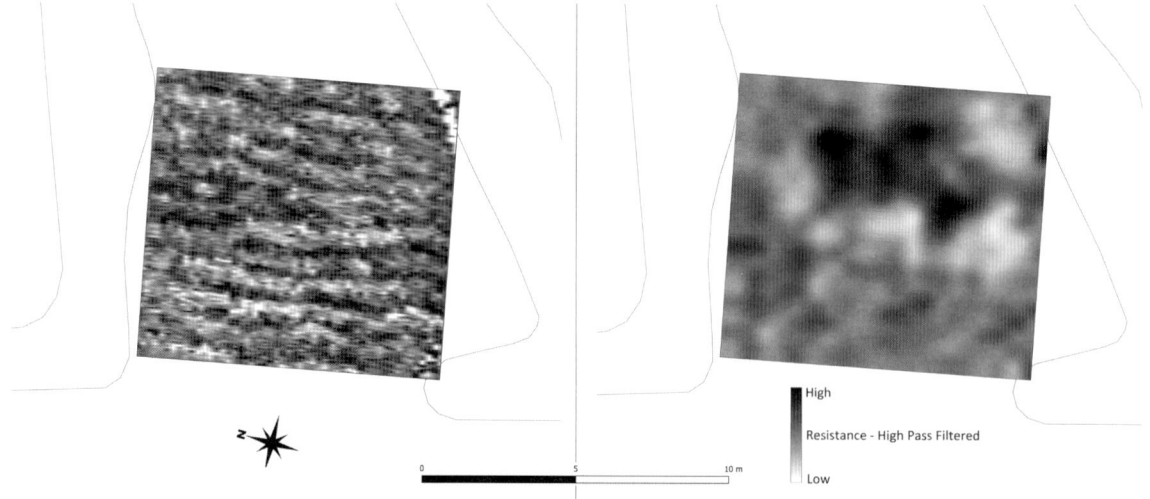

FIGURE 3.10. Area C: GPR (left) and MPET earth resistance (right). The GPR image is a raw data slice at sample 81 having estimated depth of 0.12 m, which shows clearly the extant and former terracettes but no other clear anomalies at this depth. In contrast, the resistance survey clearly shows some high resistance anomalies in the position the symbols recorded by Hutchins (1774) would be expected – see the text for the discussion regarding these results. Image: P. Cheetham

In summary, this recent work suggests that, in terms of additional 'lost' features considered here, within the limitations of the methods employed, the figure's cut features and outline are today largely as they were in the past.

Excavation results

Michael J. Allen and Martin Papworth

A pre-excavation auger survey of the proposed trench locations confirmed the presence of variably 40 cm to 70 cm of colluvium, confirming Castleden's hypothesis and indicating that these 'sediment traps may come to play a key a role in the future, as they possibly hold the silts that in their turn hold the secret of the date of the Giant's creation' (Castleden 1996, 29). It was clear from the outset that the construction of the Giant, and its survival on the steep hillside, was probably more akin to the Uffington White Horse where local colluvial accumulations were noted (Miles *et al.* 2003) than the Long Man of Wilmington where OSL dates were obtained from associated footslope colluvium (Bell & Butler 2014). No previous excavation had been conducted of the Giant (other than minor exploration of the severed head by Stuart Piggott in 1945, see Chapter 1, *The National Trust's centenary of ownership*). An early call for excavation was, however, made by Bill Hoade in 1978. Whilst principally looking at methods to confirm former outlines he, nevertheless, suggested that investigations 'would have to work hand-in-hand with soil experts and archaeologists' and open up strips around the hillside sectioning the Giant itself (Hoade 1978, 29); and over 40 years later that is what Martin Papworth (with Mike Allen) intended to do. Four locations were specifically selected where the maximum potential depth of deposits were noted in the walkover survey (Chapter 2) as *in-situ* chalk deposits, and sediment accumulations (colluvium) directly associated with the figure itself. They were examined by minimally invasive and targeted 'surgery' with four small hand excavated test pits or trenches each only 0.6 m and between 1.9 and 2.25 m long to minimise impact, while being long enough to ensure they included the full stratigraphy upslope and downslope of the chalking.

Four trenches were excavated across the chalk outline (16–20 March 2020; Fig. 3.11) at the points where the maximum sediment build-up had been noticed; i.e., the feet and elbows (Fig. 3.12), and confirmed by pre-excavation augering. Hand excavated trenches were ascribed letters A to D clockwise starting with his left foot. Hand excavation and recording followed standard methods (see Chapter 2) with trenches through the soles of his left and right feet (Trenches A & B) and crooks of right and left elbow (Trenches C & D). Note right and left refer to the Giant's right and left, not to their location on the hillside.

Blocks of context numbers were allocated to each trench; Trench A 1–99, Trench B 101–199, Trench C 201–299, and Trench D 301–399, but importantly parity was maintained where contexts were perceived in the field to represent the same event or deposit; i.e., 27 = 127 = 227 = 327, see Table 3.1.

78   *Michael J. Allen*

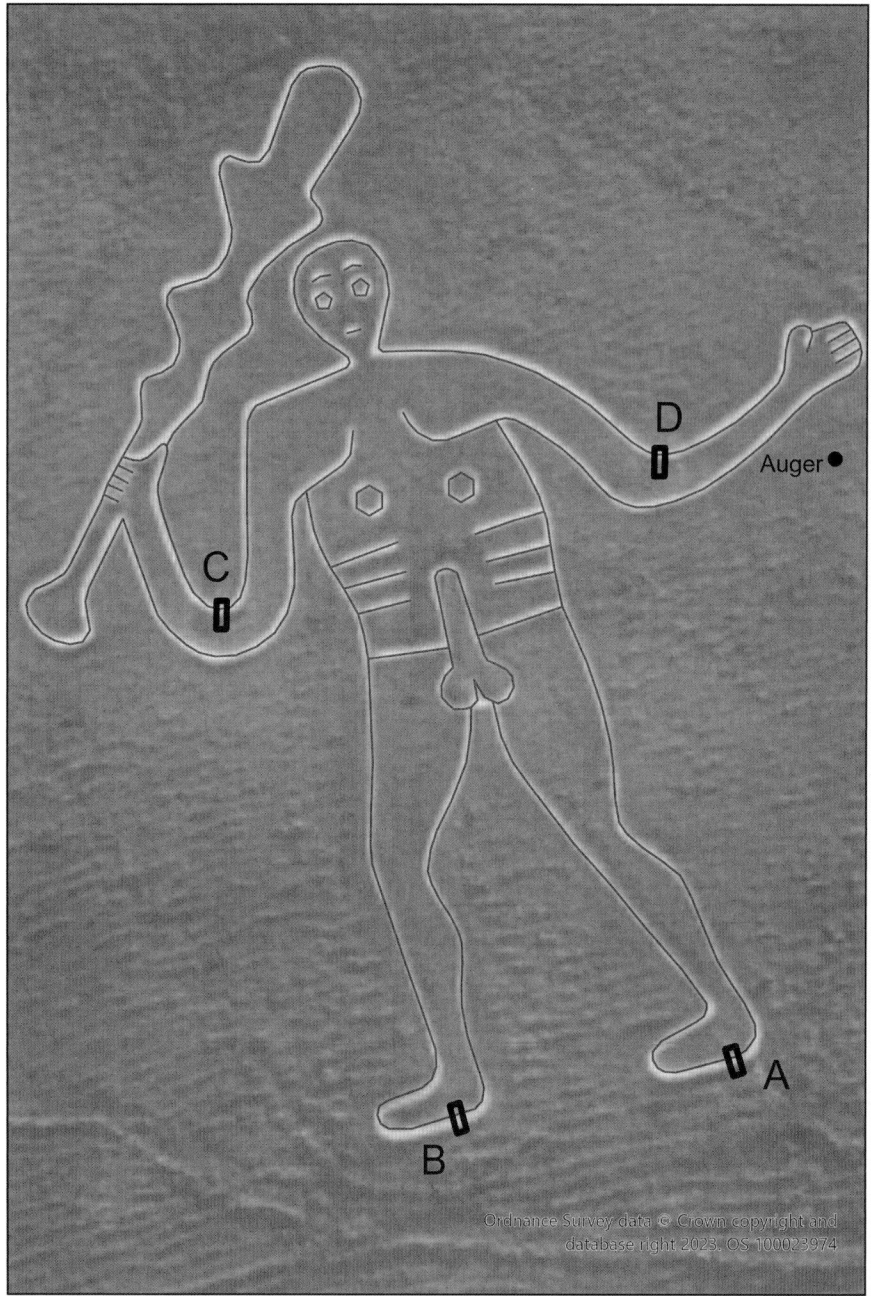

FIGURE 3.11. Plan of the Giant showing the accurate location of the four 2020 trenches and the auger hole. Image Gareth Davies, National Trust

Although widely separated (Fig. 3.13), the stratigraphy of each of the four trenches was closely comparable. They revealed remarkably similar sections including those from the feet vs those from the elbows, reflecting the consistency of rechalking episodes across the figure during various phases of rewhitening/repair and, significantly, of the sediment accumulation both upslope and downslope of the chalked line. The chalk infill was so similar that the same

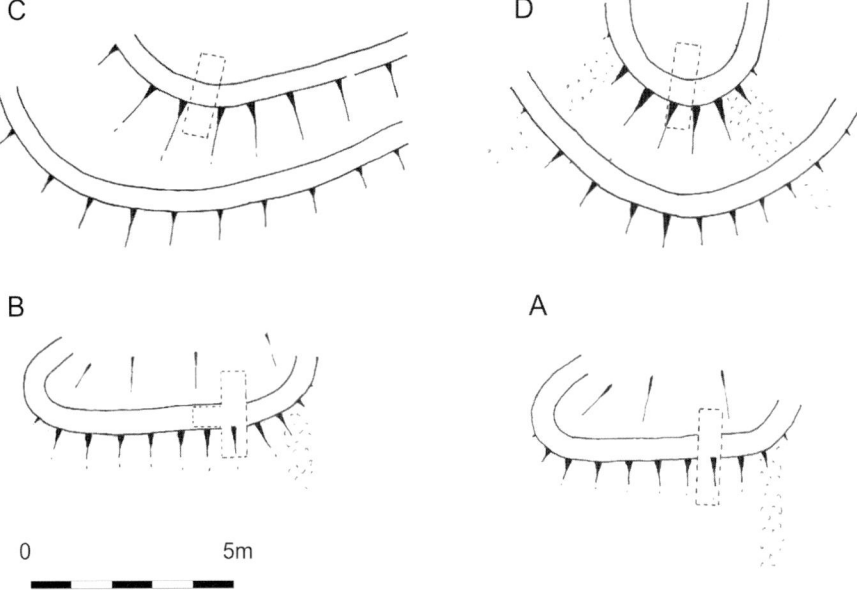

FIGURE 3.12. Plan of the locations of the four trenches: A: left foot, B: right foot, C: right elbow and D: left elbow. Image: Martin Papworth and Justin Russell

FIGURE 3.13. Excavation of the four small trenches on the Giant. Image © J. Chapman 2020, Cerne Abbas Historical Society

pattern of rechalking events could be identified in each of the four trenches. The comprehension of the deposits surrounding the chalking and their taphonomy is critical to understanding the dates (i.e., what OSL is dating); the history of the Giant in its present form; the change in the figure's outline; and the land-use history via the land snail evidence. These too were similar in all four trenches. Because of this the taphonomy of the upslope and downslope deposits (geoarchaeology) is carefully considered, following the results of the excavation of the chalk fills.

A series of nine OSL samples was taken by Prof Phil Toms in discussion with the field team and their understanding at the time of the formation and relation of the deposits (geoarchaeology) with the monument, see below. Four of those samples were selected for dating (see Table 3.3, below). In conjunction with this, a controlled programme of palaeo-environmental sampling was undertaken comprising land snail samples to provide a land-use history of the hillside, and four block soil (kubiena) samples to augment geoarchaeological interpretation if required.

Excavated sequence

Not only did the excavated sequences in the trenches at the two elbows mirror each other, and those from the feet mirror each other, but that the same set of events were represented in all trenches whether elbows or feet (Fig. 3.14). The sequence as a whole is, therefore, discussed regardless of their trench, with reference made to specific layers or deposits where better or different examples existed, or where significant absences were noted (Table 3.1). The four key elements were the two sets of primary cuts for the chalk fill, the chalk fill itself with the wooden stakes, and the upslope and downslope deposits; these are dealt with in those categories. The best representation of the deposits is seen in Trench D (left elbow). The context description summaries given are an amalgamation of the context records (archive, Papworth), and geoarchaeological records (Allen, see below).

Each of the four trenches contained three similar and distinct deposits (colluvium against the chalked line upslope, the chalk fill, and the downslope stratigraphy), and surprisingly were up to 0.85 m deep. Further, the sequence within each of the sediment profiles was generally consistent in all four locations. Clear accreted and successional rechalking phases were apparent and were surprisingly consistent between all four trenches. It was evident from each trench that the chalk figure comprised an accumulated succession of chalk deposits, typically (see Figs 3.14 & 3.16):

- a chalk rubble, successive chalk marls, kibbled chalk, and recent chalk marl layers (Fig. 3.16) initially cut into the hillside in a shallow cut and comprising up to 0.85 m of stratified chalk fills (some may have been revetted with wooden boards and stakes, Trench B, Fig. 3.17), and latterly forming a barrier to soil erosion;
- upslope stratigraphy comprising colluvial soils (with some spoil) accumulating against the successive chalked outlines, and with intermittent hints of a pre-colluvial soil beneath;
- downslope stratigraphy comprising a placed soil, colluvial soil, spoil and wash from the Giant developing below, but against, the chalking.

The three deposit types (fill, upslope and downslope) are not typical feature fills and cuts, and their accumulation is complex. Although the chalk fill is principally a result of scouring and deliberate infill, the deposits on either side are a combination of both *in-situ* deposits, which may have been cut through, but most are those accumulating and developing *during* the life of the Giant between episodes of scouring and rechalking. The boundaries between them define the edges of the chalk fill and are not, therefore, always simple 'cuts' in archaeological terms. In many cases they are purely the boundary of the eroding colluvial soil that has gently washed down slope and come to rest against the chalked outline, or represent spoil, spill, and soils developing against the chalking downslope. These processes occur during the 'life' of the Giant and over the decades *between* scourings. The scourings are clearly major punctuations in this process and tend to hide a more routine low-level maintenance. The fresh, even tamped down, chalk infill is also subject to rainwash (see Chapter 2, and Figs 2.1–2.3). The rainwashed chalk moves down the figure and can overflow the chalked lines and trickle over the grass (later to become incorporated within the soil profile, and bioturbated). In recent years it has been noted that this process of washing out generally takes place in the first three to four years after chalking until the chalk becomes frost-shattered, weathered and naturally bound and consolidated by weathering, wetting and drying. Some additional localised repair is required infrequently (Michael Clarke, National Trust Area Ranger, pers. comm.), as and when necessary between scourings. A trailer load of chalk is brought up every few years to rewhiten the chalk between the major chalking events (11 to 23 year intervals since the 1950s). Even these main scouring events since 1979 only remove 10 to 12 cm of chalk, and replace it with fresh quarried chalk. The old dirty chalk is full of seeds so is removed from site to prevent them rapidly recolonising, and the new surface watered and tamped down so that the chalk bonds, creating a layer more resistant to erosion (Clarke pers. comm.).

Despite Piggott's contention that it was 'impossible to get direct archaeological evidence by excavation [from the Cerne Giant]' (21 June 1946), the excavation results were clear, remarkable and informative. Two cuts into the natural chalk leaving distinct cut features (37 and 36). The 'cuts', and 're-cuts' (edges) cut through just the soil and hillwash accumulating against the outline are less easy to define as they are a combination of physical cuts through existing deposits, the cleaning of pre-existing cuts, and edges of deposits that had accumulated against the chalking since the last scouring. Unless clear, well-defined cut events (i.e., whole clear features) can be recognised these are hereafter called edges, rather than cuts.

Two Giants

Significantly, the excavations have identified that two giants can be recognised in the sections of both the elbows and the feet. The first is a wide shallow *cut* into the chalk, which removed the weathered chalk exposing bright, clean fresh chalk (chalk-cut Giant 1), and was probably defined with a shallow low upcast (unrecognised) bank. After soil had eroded into and against the feature a second figure was cut through the now deeper soil and through the original outline.

82   *Michael J. Allen*

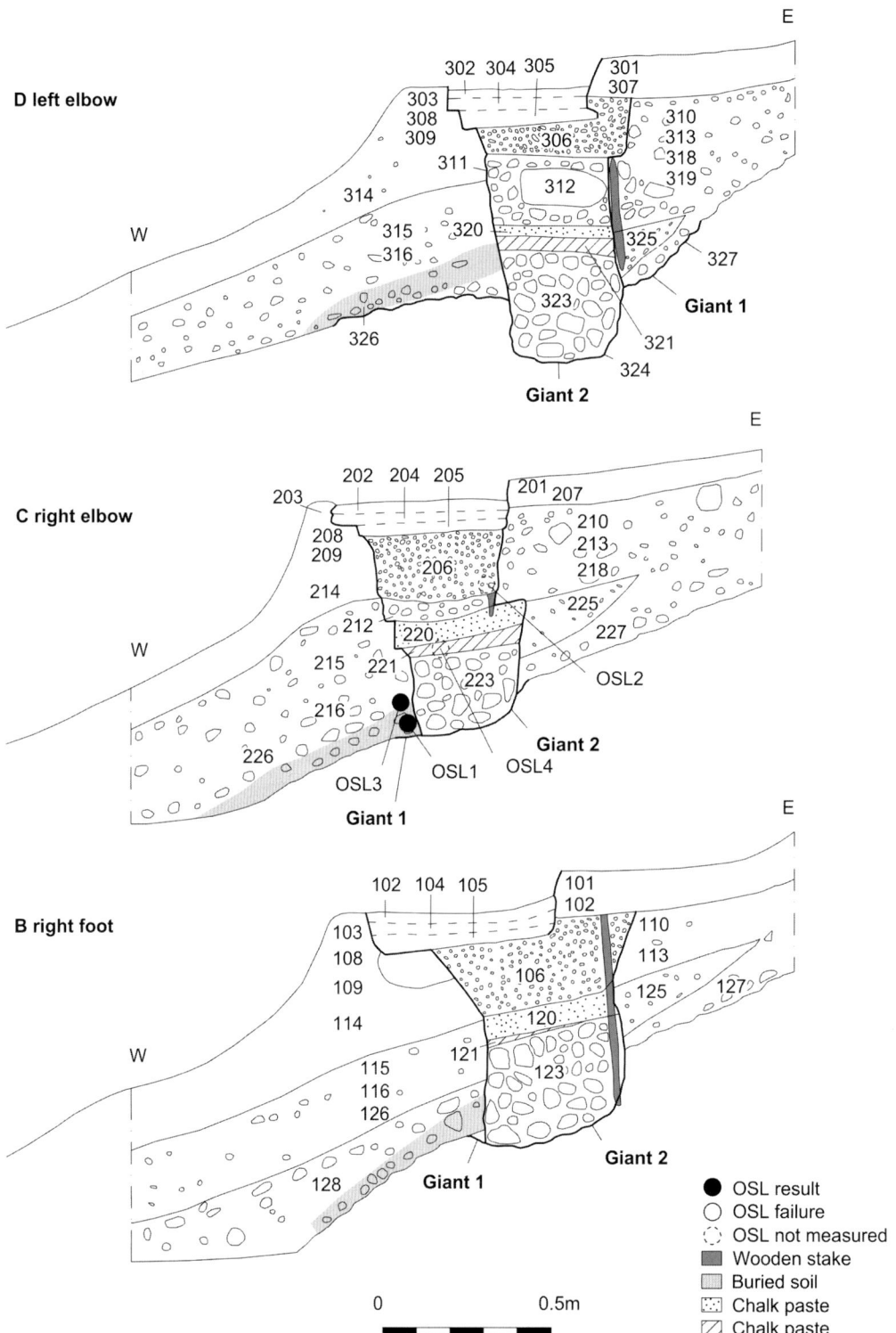

FIGURE 3.14. (left) Section drawings of Trenches B–D, north sections; (right) Section drawings of Trenches A, B and D, south sections, arrows mark profile descriptions. Images: Martin Papworth and Justin Russell

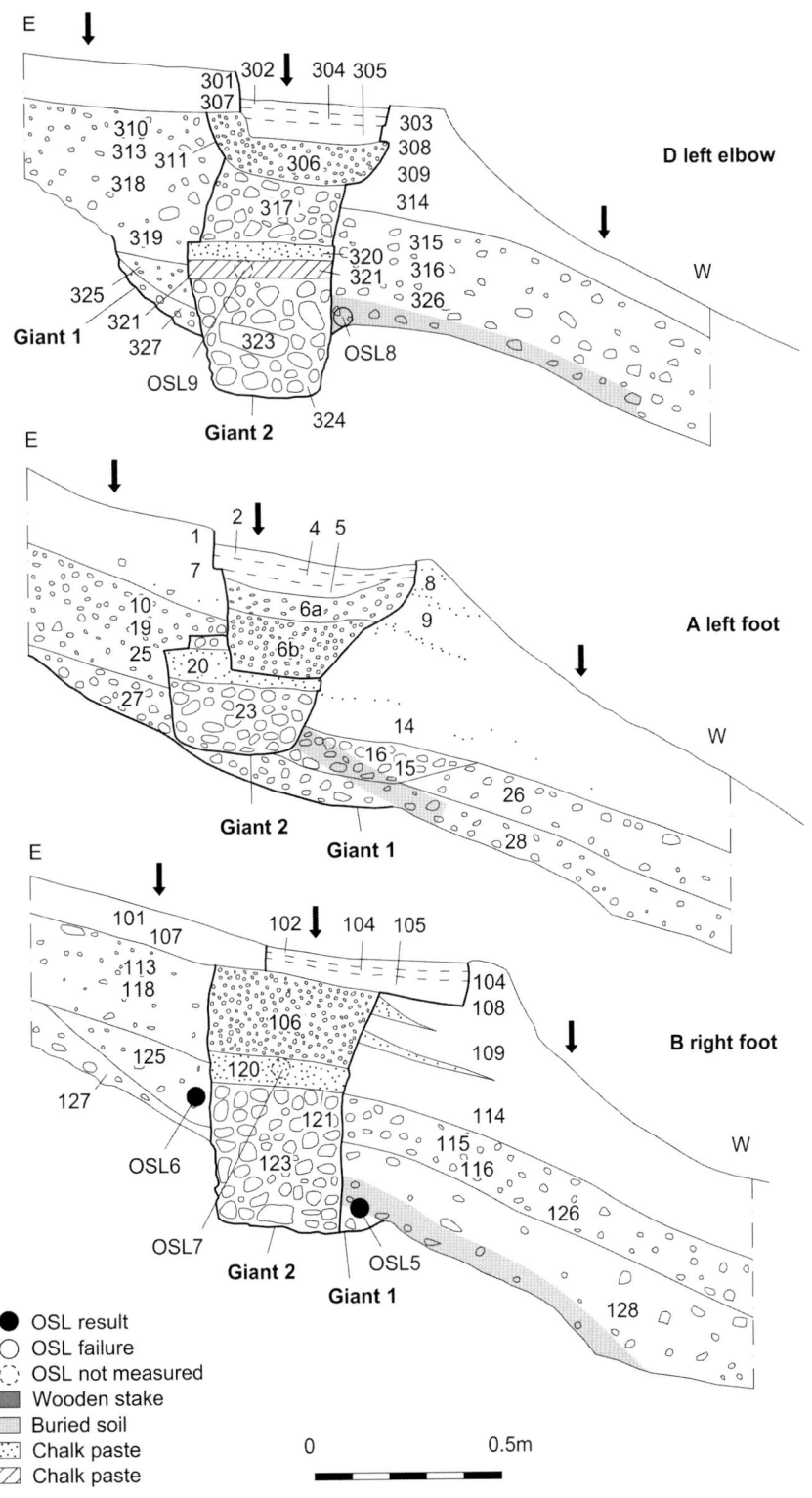

The resultant cut (Giant 2) was narrower and deeper, and the exposed chalk would not be readily seen from afar, so the cut was infilled with recently cut and imported chalk to create a chalk-*filled* Giant; the figure we essentially see today.

Giant 1: a chalk-cut figure
A shallow feature (cut 37) representing the first Giant and pre-dating the main column or stack of chalk infill can clearly be seen in the left elbow (Trench D) about 0.65–0.85 m wide and *c.* 20 cm deep. The edges of the cut only just remain, being less than 5 cm deep at the right elbow (C), and right foot (B); most seems to have been cut away by the stack of chalk fill of the later Giant 2. Only the left foot (A) does not clearly display this feature, although even here, there is a possible wide (almost 1 m) shallow (10 cm deep) weathered feature. These features are filled with a unique layer (27) and the upper part of the feature constitutes the lower part of the downslope colluvium 25 and 28. The basal fill (27) is greyish to greyish brown calcareous silt loam with varying small and medium chalk pieces (weathered soily primary fill); most showing a weak blocky structure suggesting colluvial soil infill and *in-situ* soil development. Some (i.e., 27) contained more frequent small and medium chalk pieces more reminiscent of a proto-primary ditch fill (*sensu* Evans 1972, 321–8; Limbrey 1975, 290–300; Allen 2017b, 38–41). Where clear deeper cuts survived such as Trench D (left elbow) two clear fills were present. The upper part of the features were filled with a colluvial tertiary fill/base of colluvium (context 25, base of 26), which was a greyish brown to yellowish brown calcareous silt, with fine chalk stones and again many displayed clear soil structure. The deposit adjacent to this (and to the second Giant cut), context 28, is a combination of the weathered chalk (A/C) horizon and yellowish brown silt with weak, but moderately developed, blocky to prismatic structure, representing the base of a former rendzina soil of the slope possibly contemporary with the first Giant.

All the fills were essentially soily, with no remnants in any of a packed chalk fill, indicating the first giant was just a chalk-cut figure, cleaned down to fresh, unweathered bedrock. This shallow (20 cm + soil depth 20 cm) broad (65–85 cm) outline would have been easily and rapidly infilled with soil and detritus washing down the steep hillside.

Giant 2: a chalk-filled figure
The chalk cut for Giant 2 (cut 36) is narrower than Giant 1, being initially about 30 to 35 cm wide, and up to about 20 cm deep into the chalk (Figs 3.14–3.16).

The stratified column of chalking events contained up to 11 recognisable, different contexts or infill events/deposits, eight of which were present in most of the trenches (Table 3.1). Where possible we have tried to link these with recorded recent and historical scouring and rechalking events. The chalk events below the upper chalk rubble (20 and 12) are difficult to ascribe dates to, or relate to specific historical scouring and rechalking events. Overall these deposits record the different scouring, cuttings and filling events conducted to create and maintain the outline of the Giant.

TABLE 3.1. Context categories for upslope, chalk fill, downslope and stakes for the four trenches

Upslope				Chalking				Downslope				Stakes			
L. foot (A)	R. foot (B)	R. elbow (C)	L. elbow (D)	L. foot (A)	R. foot (B)	R. elbow (C)	L. elbow (D)	L. foot (A)	R. foot (B)	R. elbow (C)	L. elbow (D)	L. foot (A)	R. foot (B)	R. elbow (C)	L. elbow (D)
				2	102	202	302								
								3	103	203	303				
				4	104	204	304								
				5	105	205	305								
				6	106	206	306								
7	107	207	307												
								8	108	208	308				
								9	109	209	309				
10	110	210	310												
				11	111	211	311								
				12	112	212	312								
13	113	213	313												
								14	114	214	314				
								15	115	215	315				
								16	116	216	316				
				17	117	217	317								
18	118	218	318												
19	119	219	319												
				20	120	220	320								
				21	121	221	321								
22	122	222	322												
				23	123	223	323								
				24	124	224	324								
25	125	225	325												
								26	126	226	326				
27	127	227	327												
								28	128	228	–				
												–	138	238	338
				140											
				141											

The depth of the chalking from the 2019 chalk surface was up to 0.80 m (Trench D) and 0.55 m (Trench A). The lowest basal fill, 324, was only recorded in D. It was less than 5 mm thick and was a dark yellowish brown loam – a thin trampled skim of soil originating from the former soil that fell into the outline trench, and possibly trampled by the original creator of the Giant. Occurring in all four trenches was a loosely packed lower chalk rubble, 23, with voids, comprising medium chalk rubble, some large flat chalk blocks (some to 210 mm) with rare nodular flints and black brecciated flints (all probably from the Lewes Chalk Formation). One chalk block was 21 × 14 × 11 cm (Trench D). This fill occupied the majority of the base of the chalk fill in all four trenches varying from 19 to about 45 cm deep. This is clearly locally sourced and loosely packed chalk rubble capped with a thin chalk marl paste.

Two thin layers (20 and 21) of chalk marl or a chalky paste both between 5 and 10 cm thick 'sealed' the chalk rubble. The lower, 21, was a pale yellow to light bluish grey soft paste with some fine laminations and sharp boundary with a yellowish brown (orange) skim on the surface, possibly suggesting either a former exposed or weathered surface, or subaerial *in-situ* weathering. This peeled off to the upper, 20, white to pale yellow, easy to trowel, marl with some fine pea-grit chalk pieces. These soft chalk mud 'plugs' were present in all trenches and may suggest heavy tamping of an exposed chalk surface, which we tentatively ascribe to the scourings of 1868. A third chalk paste (317) only 1 cm thick is only present in Trench D. The OSL dating of the chalk fill layers was not successful; a sample taken from the chalk paste, 121, in Trench B failed due to insufficient quartzite crystals.

The second major chalk rubble layer (12) marks a step and realignment of the outline seen in most of the sections (Figs 3.14 & 3.16). This layer of small and medium chalk rubble, up to 18 cm thick (and with chalk blocks up to 240 × 150 100 mm in Trench D), is present in all sections. In C and D (right elbow and left elbow) it marks a clear new-cut, misaligned over the figures' outline, and in Trench A (left foot) clearly cuts the chalk paste below. The source is thought to be New Pit Chalk (BGS Geology viewer Nov 2022; R. Mortimer pers. comm. to John Charman), which is flint-free.

A second realignment of the outline can be seen with the onset of the presence of kibbled (quarried chalk from the Shillingstone Lime Quarry, Dorset) relating to the E.W. Beard Ltd. (formerly Messer Beard & Co. now Beard Construction), major, almost industrial-scale, scouring events of 1956 and 1979. This is a very light brown, pale yellow or cream compacted chalk layer of small and medium rounded chalk pieces without flint, generally about 20–25 cm thick. It is clearly cut into the chalk rubble below, in some cases downslope of the previous line (e.g., A, left foot), and in others making a cutting wider than, and completely embracing, the previous chalk rubble line (e.g., D, right elbow). In some places the upper 10 cm of chalk fragments were concreted together – tamped down (Trench A, 6a), which contrasted with those in the lower 14 cm, which were loose (6b). These two events are also picked out by the lens of chalk wash *downslope* of the outline (in Trench B); one trailing from the top edges of the kibbled chalk 106 (i.e., 106a) and the other 10 cm below it (i.e., 106b). The kibbled chalk belongs the Beard Ltd. scouring and rechalking works, and it is tempting to ascribe these two layers of kibbled chalk to the 1979 and 1956 scourings respectively.

The final four layers (2, 4 and 5) are all brilliant white and very pale yellow to light grey heavily compacted (tamped) chalk; the lower two pale yellow layers lacked flint inclusions. The top layer, layer 2, was distinctive in that it included a number of small and a few medium black and dark grey sharp flint fragments. In layers 4 and 5 even occasional fragments of cement adhered to clusters of small chalk pieces were present. In all these layers the chalk had been heavily crushed so that individual chalk pieces could not easily be distinguished.

If these represent recent National Trust activity, then they can be dated to the rechalking events of 1995, 2008 and 2019; the most recent (layer 2, 2019) with chalk from the David Lush Quarry, West Grimstead (Michael Clarke, pers. comm.), and lower two (layers 4 and 5), with chalk imported from the Shillingstone Lime quarry, Dorset. Some of the National Trust excavators (staff and volunteers) had assisted with the 2019 chalking and infilling and tamping and then, less than a year later, were removing it under archaeological conditions with a hand trowel. These rechalking events all show a consistent drift of the outline downslope, and of a significant widening of the outline (at the feet and elbows at least). They were packed on the downslope side by a 5 cm deep brown to light greyish brown stone-free silt loam; soil and turf that had been infiltrated by fine lenses of rainwashed chalk that had overflowed and trickled over the grass (see Figs 2.2–2.3).

Wooden stakes (with descriptions by Nancy Grace and Martin Papworth and identification by Cathie Barnett)
Fragments of preserved wooden stakes were found in three of the four trenches (B, C & D); they were hard against the edge of the chalk fill (Figs 3.14 & 3.17). All were similar in size (2–3 cm wide and sharpened to a point). They had all been driven into the edge of the deposits forming the chalk outline. One, in Trench B, was present from the top of the kibbled chalk (106) to the base of the Giant 2 trench, running along the edge of lower chalk rubble (123) and through the 1956 and 1979 kibbled chalk (16; Fig 3.17). This stake (138) was 58 cm (*c.* 1 ft 11 in) long. The stakes from C and D survived below the kibbled chalk and against the upper chalk rubble (212 and 112), but may be contemporaneous; that in Trench D was identified by Cathie Barnett as larch (*Larix decidua*) or Norway spruce (*Picea abies*) and extended through the chalk marl and into the upper chalk rubble (323), and that in Trench C was identified as pine family. Initial observation suggests all three are cleft stakes.

The stake in Trench B passed though the 1956 and 1979 kibbled chalk layer (106); it could pre-date this, however, if it was still fast and the chalk was emptied around it and then refreshed with new chalk. The stakes in Trenches C and D were potentially driven into the chalk from below the level of the kibbled chalk (6) and therefore were considered to be pre-1956 (see Fig. 3.14). They could, however, have been broken off by the scouring and recutting in 1979. The stakes from C and D are thought potentially to derive from poorly documented works associated with scouring and rechalkings dating from the late nineteenth to late twentieth century. Although probably mid- to late twentieth century, this information is of some interest, but of no great significance in understanding the Giant himself.

The stakes were not considered key to the excavation research aims and detailed records were not made on site. Two were retained and the longest, 238 (Trench B; Fig. 3.17), was left *in situ*. Their dimensions were taken retrospectively from the section drawings, and subsequently (in 2023) measured and recorded (Table 3.2). The stakes in archive had degraded slightly, however, the shape when first seen from the trench edge reminded us of chestnut paling, which is triangular, but when taken out of the ground (238 and 338), they were more oblong and roughly hewn/split, not square or triangular; they all seemed to be cleft stakes.

Stake (138; Trench B) was up to 580 mm long and 30 mm × 20 mm with a decayed surface it continued from the upper surface of the kibbled chalk layer (106) to the base of the rubble chalk layer (123); i.e., below the topsoil (107). The stake continued to the upper edge of the kibbled chalk filling.

The point of a 60 mm long wooden stake (238; Trench C) was found below kibbled chalk layer (206) cutting through the upper rubble chalk layer (212) and penetrating the top of silt chalk layer (220). It measured 60 mm long and was up to 20 mm in diameter, and contained a knot expanding its shaft to 34 mm at that point.

Trench D (south-facing section) a stake (338) was found below kibbled chalk layer (306) and rammed chalk layer (311). It ran down the side of chalk rubble layer (312) and penetrated at an angle into upslope soil layers (318); (319) and (325), though or adjacent to chalk marls 320 and 321. It was up to 330 mm long and 30 mm × 20 mm in breadth and width, with its point resting on the base of the small upslope cut (327).

Flinders Petrie (1926) stated during his survey of 1919 that the trenches were 2 ft deep (61 cm), this depth could be a reason for remains of the three timber stakes. The stakes probably held revetment boarding during scouring events as recorded in the 1956 and 1978 rechalking specifications, and indicated in the schedule of works in the 1974 management plan (Appendix 3). This is supported by the fact that all the stakes were located hard against the edge of the figure.

The current National Trust area ranger reports that 'there as still a few places were wooden pegs from earlier rechalkings are to be found' (Michael Clarke pers. comm.), and he has used them since 2002 as a guide to the width of maintaining the chalk lines. The date of the stakes is uncertain; they (e.g., 123) could be a result of the 1979 rechalking but may have pre-existed this; they could be stakes assisting in the major 1945 restoration and refurbishment supervised by Piggott or, it has been suggested, possibly even from Petrie's 1919 survey. Although the latter is tempting as the occurrence of the stakes in Trenches B, C and D coincides with the position of Petrie's survey points as marked on his plan (1926, fig. III), these are probably too long (60 cm), and not square-enough, to be survey pegs.

TABLE 3.2. Wooden stakes, upper measurement is field record (measured from the section drawing by Papworth) and the lower measurement is that from the retained item (measured 2023). Measurements in mm

Stake/Tr	Description	Identification	Length	Point length	Width
138/B	split wood pointed stake	–	520, 540, 580		30 × 20
			–	–	–
238/C	split wood, tapered to point	Pinacaea	60	60	30 × 20
			180	–	20 × 16
338/D	split wood, tapered to point	*Larix decidua/Picea abies*	330	–	20 × 15
			280	70	30 × 20

Origin of the chalk (chalk fill)

Two samples of chalk were retained from Trench D. One from the upper (312) and one from the lower chalk rubble layer (323). Geologist John Charman identified that from the lower chalk rubble as from Lewes Chalk (232), which is a greyer flint rich geological deposit found in chalk outcropping from the top of Giant Hill. The upper white and flint-free deposit (312) is from the New Pit chalk found in quarries at the foot of Giant Hill. Small rounded chalk pieces, kibbled chalk, was a component of the fills and originated from Shillington Lime & Stone Company Ltd. (Keithley *et al.* 1999, 22).

Dating the chalking events

Martin Papworth and Michael J. Allen

Given that the four trenches are consistent in their stratigraphy and although rechalking, associated with Beard's scouring of 1956 and 1979, has removed an earlier episode in one or two trenches (e.g., upper chalk rubble (12) and a chalk paste layer (21) in Trench A, left foot), it seems unlikely that any significant rechalking events have been completely lost from the four trenches. We can see four main fill groups on the basis of the fill types and the morphology of the chalk fill 'stacks'. This is best seen in Trench C (right elbow) and A (left foot; Fig. 3.14). These groups are: i) the upper three crushed chalk layers (2, 4 and 5); ii) the kibbled chalk (6); iii) the upper chalk rubble (312); and iv) the chalk marl or paste lenses and lower chalk rubble (20, 21 and 23). The readjustment, or misalignment represented by each of these for 'stacks' may represent periods when the original outline may not have been not clearly visible (e.g., possibly the lower two stacks), or was undertaken on a large 'industrial' scale (e.g., Beard 1956 and 1979), or with large numbers of enthusiastic volunteers who were not all closely supervised (e.g., recent events of 1995, 2008 and 2019).

The consistent layers in the chalking stacks were then ascribed dates by Martin Papworth successionally relating to the history of the known scourings (see Chapter 1; maintenance and renewal); this simple but effective method does not take account of the possibility that later, larger scourings may have wholly removed some to many other earlier episodes of rechalking. The upper three layers (2, 4 and 5) were easily matched to known recent National Trust scouring events (2019, 2008 and 1995), but the chalking events below the upper chalk rubble (12) are more difficult to date. The kibbled chalk, by its nature (freshly quarried chalk graded by screening through various sized meshes), is easy to recognise (layer 6) and to equate it with the Beard & Co. chalking events of 1956 and 1979. Two sets of kibbled chalk comprise this layer; a concreted kibbled chalk (6a) and a loose chalk kibble (6b). Although the later chalking has, in most cases, removed all 6b as defined in the National Trust specification for work (specification 2 in the scheduled of works dated 18 May 1978 for scouring conducted in 1979; National Trust Tisbury Hub Archaeology Information file), it does survive in particular as a loose kibbled chalk (6b) in Trench A. This could, therefore, be attributed to the earlier event of 1956. The upper concreted chalk kibble (6a) seen in all trenches is a result of watering and tamping as specified that the 'new chalk which must be clean … will be spread in suitable areas in layers not less than 6 inches in thickness and well sprinkled with water … rammed with a heavy wooden punner until the whole layer is one homogeneous mass' (specification 5, *ibid.*), and dates to the 1979 Beard rechalking.

The two chalk rubble layers (12 and 23) both represent significant undertakings and the lower (23) represented a particular large rechalking project, the resetting of figure and the creation of Giant 2. The depth of this cut and of the lower chalk rubble fill is surprising, cutting in to the chalk by nearly 20 cm with fill nearly 45 cm thick. One of these might be attributed to General Pitt Rivers who carried out work on the Giant in 1886 (Gaze 1988, 78–9; Castleden 1996, 26–7; Darvill 1999a, 9). Given Pitt Rivers' reputation for major works on archaeological monuments, it might seem reasonable to suggest that the upper rubble chalk episode (12) might be attributed to him. If so we could tentatively ascribe the chalk marl below to a rechalking in 1868 (Keithley *et al.* 1999, 18), leaving the date of the lower chalk rubble, and cutting of Giant 2, unresolved, except that it post-dates the tenth century, and pre-dates an OSL date of *c.* AD 1250.

Before 1956, the outline of the Giant was generally not a chalk outline at near surface level, but trenches up to 0.6 m deep (see discussion in Chapter 7). The Giant we see today, although modelled on a tenth-century predecessor, is a twentieth-century reincarnation.

The events and rechalking were easily identifiable. The most recent chalking events seem to have been restricted to just the upper, almost surface, portions (i.e., just scraping away the top dirtied chalk to renew the chalk). The 1956 and 1979 chalkings were huge by comparison, with the cutting away of most of the old chalk infill, and then infilling with kibbled and quarried chalk. The continual scouring of the Giant in the past has created a trench, first described by Rev John Hutchins (1774, 292) as 'two feet broad and as many deep', and the deepening of the trench into the chalk (see Fig. 3.14 Trench D especially). Rechalking led to a progressive accumulation of chalk rubble, chalk marl, kibbled chalk and recent (twenty-first century) chalk marl, as thick as 0.6 to 0.85 m. This is comparable to the prehistoric Uffington White Horse with deposits of about 1 m thick (Miles *et al.* 2003, fig. 5.7; Fig. 3.12), but unlike the Long Man of Wilmington, which was not chalk-cut but comprised of painted bricks embedded only into the soil (Bell & Butler 2014, fig. 3.6; Bell pers. comm. 26/3/20). Some of the chalk marl stratified under the kibbled chalk at Cerne, for instance, may be a combination of rammed crushed chalk, and chalk mud washing from the figure (see Figs 2.2–2.3).

Geoarchaeology of the upslope and downslope deposits

Michael J. Allen

The geoarchaeology and land snail evidence are intrinsically part of the same study and together provide a history of the creation, development and modification of the Giant's environment. The geoarchaeology, however, is considered here, as the comprehension of the taphonomy of the deposits is key to understanding the significance of OSL results and what they have dated. The land snail evidence is presented following the description of the excavation and other fieldwork. The interpretation of the land snail assemblages is, however, inextricably a part of the interpretation of the sediment formation processes and taphonomy (cf. Evans 1972, 41–2; Allen 2017b, 31) so both technically are part of the same investigation despite being separated in this publication.

Twelve key profiles were described; three from each trench (upslope chalking, and downslope), and three auxiliary profiles (two in Trench B and one in Trench C). Descriptions of most of the chalk infills, and all of the auxiliary profiles, are in archive (see Allen 2020b). In each case representative portions of the trench sections were selected and carefully cleaned to expose any pedological structure (cf. French 2015; Allen 2017b) rather than trowelled to create smooth, polished and vertical face. Descriptions were made of the profiles in both fresh and weathered condition following standard pedological notation (Hodgson 1997), with Munsel colours recorded on moist sediment in daylight conditions. This was important as subtle, but highly significant, pedological structure was noted consistently in some deposits and not others. Key locations in Trenches B and C were selected in discussion with Prof Phil Toms, for the main OSL and palaeo-environmental enquiry (OSL samples were also taken from Trench D but not progressed; see Figures 3.14 and 3.20). Land snail samples were taken from same trenches; principally from Trench B (right foot), with spot samples from Trench C (right elbow; see Allen below). The full descriptions are provided in this report with auxiliary descriptions and those of the chalk stacks in archive (Allen 2020b), and the profiles shown in Figures 3.14–3.16. Those from Trench B and C, from which the snails and the land-use history is derived, are dealt with in more detail. In addition, four undisturbed soil block samples (kubiena) were taken for soil micromorphology to augment soil and sediments interpretation. After discussion with Prof C. French (University of Cambridge), he suggested that these were not progressed as they were deemed to add little to the geoarchaeological interpretation. They are in short-term archive store with the National Trust.

We can recognise a series of deposits covering the history of the Giants, which are discussed below. These can be seen as:

- The infill of initial broad cut (Giant 1)
- A relict intermittent survival of the base of a former soil
- Upslope colluvium accumulating against the chalking
- Downslope colluvial soil developing against the chalking
- Upper colluvium and spoil (both upslope and downslope), and finally
- Modern modification and rechalking

The basic stratigraphy of each sedimentary unit is outlined, and then the interpretation of their deposition and formation is discussed, and this provides a prelude to the land-use history derived from the land snail evidence. The geoarchaeology, land snail analysis and OSL dating concentrated upon the earliest deposits: the lower soils and earlier colluvium. Land snail analysis cannot assist in the interpretation, or the land-use at the time of the creation of chalk of the figure outline itself or Giant 2. The geoarchaeology predominantly concentrates on the formation and relative chronology of the two colluvial deposits and associated soils on either side of the chalk infill stack. The arguments presented here form the basis for the interpretation of the OSL dates and of the narrative of the sediment development of the Giant and, importantly, of the land-use history after his construction. Following careful consideration of the hillwash deposits, a narrative of their formation and chronology is put forward, which revises and modifies that considered on site, and provided in previous archive reports, blogs or lectures.

In each trench there was an upslope and downslope sequence of hillwash separated by the accumulated chalking of the Giant's outline. Within this, the same detailed sequence of contexts was generally recognisable in all four locations whether they were in the crook of the arms or soles of the feet. This consistency at least indicates that the deposits are likely to represent a history of the hillside as a whole rather than of each individual location. However, because of the intervening chalked lines there is no direct stratification between the upslope and downslope deposits, nor with the chalking events themselves. It was, however, clear that a shallow trench was cut (later to hold the chalking) and that the majority of deposits on either side had accumulated (colluvium) and developed (colluvial soil) against that chalk outline. Chalk was added to maintain chalk levels and provide a near-surface visible chalk outline (at most times).

The upslope colluvium (118, 125, 127) comprise dark greyish brown to grey calcareous silt generally nearly stone-free, with few chalk pieces, suggesting largely fine downslope soil wash, that might be expected under grassland, rather than arable conditions. Most of the stones present from the original soil profile (A/C) or upcast material during digging, chalking and rechalking events. Generally three or four thin colluvial horizons were present with the matrix becoming less calcareous up profile. Voids and macropores were coated with chalk mud or precipitated calcium carbonate possibly as result of chalk-charged water percolating through the deposits.

TRENCH B (right foot) north-facing: upslope

Depth (cm)	Context	Sample	Description
0–12	101+107		Dark greyish brown (10YR 4/2) to greyish brown (10YR 5/2) stone-free humic calcareous silt loam, common fine fleshy roots, massive dense, and seemingly apedal, clear to abrupt boundary 5. A (and redeposited Ah/A i.e., spoil): Recent remodelling
12–28	113		Dark greyish brown (10YR 4-5/2) silt loam with common small chalk pieces, rare medium subrounded chalk pieces, and very rare medium flints, clear to abrupt boundary Soil spoil or colluvium 4. Upslope colluvium and dumped soil

3. Research results: fieldwork, dating and analysis

Depth (cm)	Context	Sample	Description
28–31	118		Dark greyish brown (10YR 5/2) silt loam – with fewer chalk clasts, sharp to abrupt boundary Soil spoil or colluvium *4. Upslope colluvium and dumped soil*
31–51	125	10: 31–41 9: 41–51 OSL 6	Grey (10YR 5/1) to greyish brown (10YR 5/2) calcareous silt loam, weak medium structure, essentially stone-free (with sharp contact against chalk figure fill), some voids and manropes coated with chalk mud, clear to abrupt boundary Colluvium [OSL 6: AD 1240] *3.1. Upslope colluvium*
51–62	127	8: 51–61	Grey (10YR 5/1) calcareous silt loam, few to rare medium chalk pieces at base, some voids and macropores coated with chalk mud (with sharp contact against chalk figure fill) Colluvial soil (contemporary/pre-dating figure G2) *2. Upslope colluvium (base of)* *?1. Primary fill of cut of G1*
62+	Cw		Weathered Chalk Cw

The downslope colluvium (128, 226, 216) was a weakly banded yellowish brown to brown silt loam with common to abundant subrounded (weathered) chalk at its base. It was lighter in colour than the upslope colluvium presumably as a result of chalk wash, and it was characterised by weak structure throughout, in contrast to the upslope colluvium where structure was only present intermittently in the basal deposits.

TRENCH B (right foot) north-facing: downslope

Depth (cm)	Context	Sample	Description
0–32	109 + 114		Grey calcareous silt turf, stone-free massive *5. Topsoil and worm-worked soil, placed chalky silt revetment: Recent remodelling*
32–43	115 / 116 / 126	3: 33–43	Greyish brown (10YR 5/2) silt loam, with some medium and rare small chalk pieces, clear but weak to prismatic structure, abrupt boundary *4 and 3.2. Chalk rubbly soil / spoil*
53–70	128	4: 43–53 5: 53–63 6: 63–70 OSL 5	Yellowish brown (10YR 5/4) silt loam with common to abundant medium, chalk pieces, clear but weak blocky to prismatic structure, with moderate well-developed structure at base, abrupt boundary. At 38 cm *C. virgata* and *P. elegans* in the colluvium. The base of which (against the chalk stack, and not in the main description/sample column) was siltier with fine chalk flecks and is the fill of the Giant 1 cut [OSL 5: AD 910] Base of *in-situ* rendzina *2. Calcareous contemporary/prefigure stony rendzina/early colluvium* *1. Giant 1*

Depth (cm)	Context	Sample	Description
70–75		7: 70–75	Mixed layer of yellowish brown (10YR 5/4) silt loam with pale brown calcareous marl and common to abundant medium subrounded chalk pieces, with incipient weak blocky to prismatic structure A/C – weathered base and *in-situ* rendzina/colluvial soil
75+	Chalk		C chalk

At or towards the base, the downslope calcareous silt loam colluvium consistently displayed weak, intermittently present, medium blocky to prismatic structure suggesting the survival of the base of the former rendzina or thin brown earth soil (27; Bw or B/C). This was not recognised as a separate context in the sampled trench (Fig. 3.14), but this soil structure was also present at the base of 128.

The upper colluvium, both upslope and downslope was a dark greyish brown to brown highly calcareous silt loam with common small and rare medium generally subangular (unweathered) chalk pieces. The less weathered chalk pieces may be chalk excavated during rechalking events, or imported chalk for the later rechalkings, which has become incorporated in the contemporaneous soil, colluvium (upslope) and colluvial soil (downslope).

The upslope upper deposits and upper colluvium (118 and 113) included dumps of soil when rechalking and re-establishing, revitalising the Giant. Dumps of soil and chalky soil were also present on the downslope sides, but fine chalky washes (114, 109) were more common here.

The upper portion of the chalk fill is enveloped in colluvium, spoil, and what we think is muddy chalk soil and turf packed against the recent downslope edges (Figs 3.14 & 3.15). Against the downslope edges of what is now the horizontal chalking of the foot and elbow (Fig. 3.14) were highly calcareous grey and greyish brown silt loams and even chalk paste (5) generally with rare chalk pieces, some with pseudomycelium. In particular thin lenses of fine chalk pieces were noted between 109 and 114, and within 109, 209, and 214. These represent chalk spill or rill deposits of chalk mud washed off the chalk outline as noted in 2019 (Figs 2.2–2.3). Also clear evidence of placed turf (stone-free deposits with small to medium crumb structure), and occasional brown rotted grass, was present in the upper 'triangle' of deposits (Fig. 3.14) on the downslope side suggesting the emplacement of turves supporting or revetting the chalked outline.

When originally cut, the chalk outline would have followed the lie of the slope (45°; Chartrand 1999, 4) and the original cut for Giant 1 cut into chalk bedrock and was found to be 50–60 cm wide (at the sole of the feet at least). The lower rubble deposit of Giant 2 (23 & 123) was 35–40 cm at the base and 40 to nearly 60 cm on the surface. At present, however, the chalk outline at the soles of the feet and crooks of the elbows are horizontal sitting on accumulated rechalkings, 40 cm wide and are subtly, but significantly, displaced further downslope (Figs 3.14, 3.16 & 2.3). The increased width of the chalk outline, especially at the soles of the feet, in recent years may, in part, be to facilitate its observation from a distance, now that the figure's chalked lines are horizontal

FIGURE 3.15. Trench B; north-facing section showing the upslope, chalk fill, and downslope deposits. Image © M.J. Allen 2020

rather than in line with the very steep chalk scarp slope. Although the available correspondence, particularly since the National Trust ownership, concentrates on 'rewhitening', there seems to be no discussion of the widening or levelling of the lines (M. Papworth pers. comm.), but this has clearly occurred (Fig. 3.14). Again although there is no record of the placing of turf against the downside edges (M. Papworth pers. comm.), this too has clearly occurred (unless regular weeding of the chalk figure occasionally leaves clumps of grass downslope). Although some soil may be natural soil wash, there is defined structured soil (small turf like peds) in the downslope 'triangles of topsoil' (see Fig. 3.14; and descriptions), some of which undoubtedly represent turves, and some are relatively recent (<100 years) as brown dried partially rotted grass (Trench C, context 209).

Geoarchaeology of the two hillwash deposits

Clearly the comprehension of the history of the Giant was important for the OSL dates and the interpretation of the accumulating deposits on upslope and downslope side of the chalked lines (Fig 3.19, below). In each trench the upslope deposits seem to indicate colluvium accumulating against the upslope edge of the chalk figure with some local spoil from rechalking and restoration events. The downslope deposits are also colluvial but are lighter (chalkier) and consistently have some pedological structure, which was originally taken to suggest the presence of a former soil, but now is believed to indicate the formation of a colluvial soil on the downslope side and in contrast to episodes of more rapidly accumulating colluvium on the upslope side.

96 *Michael J. Allen*

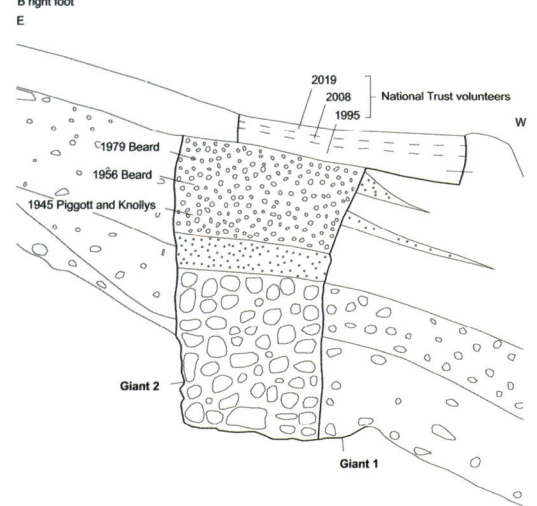

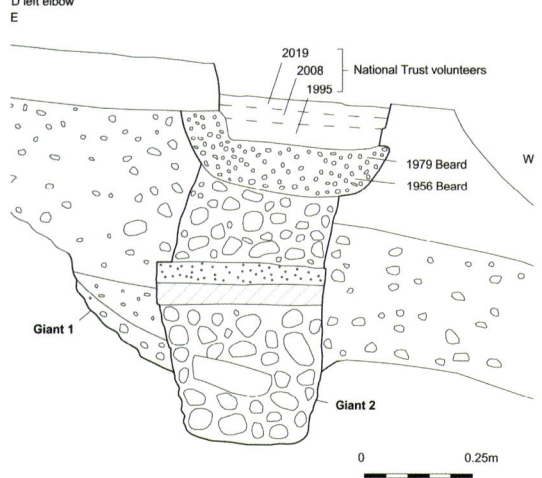

3. Research results: fieldwork, dating and analysis

FIGURE 3.16. (opposite) (top) a) Trench B; north-facing section showing the successive chalk fill deposits, and b) the dates of the chalk stack infill as proposed by Martin Papworth; (bottom) c) Trench D south-facing section showing the successive chalk fill deposits, and d) the dates of the chalk stack infill as proposed by Martin Papworth. Photographs © M.J. Allen 2020, graphic: Justin Russell, based on field drawing by M. Papworth

Taphonomy of the deposits: As discussed above, careful examination of the deposits, and deposit descriptions, resulted in revision of our original thoughts of their development and chronology as suggested in previous archive reports. The earliest deposits post Giant 1 seem to be the base of a former soil downslope (B/C horizon in Trench B, the chalky elements of 128) preserved by later soil accumulation and hillwash. The first phase of hillwash, however, accumulated against the upslope side (227, 127, 125). Soil development occurred on downslope side and included overspill soil wash, hence the colluvial soil on the downslope side consistently showed better, albeit weak pedological structure, than all the upslope lower colluvium. Subsequently the upslope upper colluvium (118 and 113) formed, which included dumps of soil when rechalking and re-establishing, and revitalising the Giant. Dumps of soil and chalky soil, and fine chalky wash (114, 109) were more evident on the downslope sides. It is on this basis that the chronology of the land-use history is based, and not the changing mollusc assemblages.

Discussion: twenty-first-century survival and the two Giants
Michael J. Allen

The fact that the chalk-packed limbs of the Giant suffer erosion is far from a new observation. Castleden has pointed this out (1996); Darvill explains some of the points observed in September 2019; and Keithley *et al.* (1999, 19) describe the rainwater scouring that we also observed (see Figures 2.1–2.3). Two important erosion and sediment accumulations are noted:

FIGURE 3.17. The wooden stake 138 in the south-facing section from Trench B. Image © M.J. Allen 2020

1. chalk in the figure is unstable, and rain-washed chalk is carried downhill along downslope limbs accumulating as sorted chalk at locations such as the crook of the elbows and soles of the feet. The redeposited eroded chalk was clearly sorted with small rounded pieces to 2 cm adjacent to the base of the rills (ankle), rounded 1 cm chalk stones (heel) and chalk mud (sole) (see Figures 2.1 and 2.3), and

2. the effect of the combination of the nature of the 'embanked' chalked outline and of the erosion is to increase the level of chalked outline at these locations, creating a barrier to soil wash. Fine rain splash colluviation and fine surface soil wash (cf. Kwaad 1977; Kwaad & Mücher 1977), lead to clear colluvial silts accumulating on the upslope edges of the chalked outline, and especially in the locations of chalk accumulation (Figs 2.1 & 2.3).

The consequence of these two observations is that colluvium in the form of soil wash, chalk mud and chalk silts accumulate against the chalk figure and ultimately subtly affects his shape and configuration at a local scale. The deposits comprise a combination of colluviated soils from upslope accumulating against the chalked lines, spoil from cutting and creating the bed of the chalked lines, small 'revetments' especially on the downslope edges to support the chalked lines, chalk rills of chalk spilling downslope from the chalked lines, and downslope colluvium some of which may pre-date the figure. No former soils were present, but some of the lower colluvium may be the basal horizon (A/C or B horizon) of a former colluvial soil.

Although the deposits upslope and downslope of the chalk outline superficially look very similar (as they would if they are erosion products from the same source), we can consider them potentially to have different accumulation histories. The upslope deposits have largely accumulated against the chalking. This upslope colluvium is eroded topsoil material with slightly more angular chalk fragments than that on the downslope. This is in keeping with the suggestion that on the whole these are a combination colluvium accumulating against the chalk outline and chalk spoil creation of the chalk outline. As such these are primarily contemporary with the existence of Giant 2 and post-date his initial creation, though hints of the base of a former soil is present (layer 28, Trench A and layer 128, Trench B). In contrast, the downslope deposits were lighter and in some instance a more yellowish brown colour, and the chalk fragments in the lower deposits tended to be more rounded than those on the upslope side. There was some soil structure noted in these which we take to be the development of an *in-situ* colluvial soil and subsequent soil deposits (colluvial and dumped) and chalk wash from rechalking events and immediate chalk wash off (see *Reconnaissance and implications: walkover survey*, Chapter 2).

One of the most significant finds from the excavation is the discovery of two Giants; the first being outlined as a shallow scoop cut into the soil, with a little of the weathered chalk removed to expose a bright white fresh chalk. The spoil was, presumably, largely spread either side of the fresh cut. These wide trenches created a chalk outline, which varied between 0.65 m and 0.85 m deep, up to 0.2 m deep and is best observed in Trench D (ironically the section paid the least attention in the field, and not sampled), see Figure 3.14. There is no evidence that these hollows were chalk-filled; in fact quite the contrary, and being both shallow and wide would have allowed the exposed chalk to be seen from a distance.

3. Research results: fieldwork, dating and analysis

This shallow hollow would, however, have rapidly become infilled with soil. Continued regular clearing of any soil would have been necessary to maintain visibility of the chalk trench. This first outline (Giant 1), or even marking-out trench, was replaced by a narrower *c.* 1 ft 5 in (*c.* 40–45 cm), clearly cut and well-defined cutting (Giant 2). Being too narrow and deep to enable the chalk base to be observed from a distance, it existed initially either as an unfilled open trench or the cutting itself was filled with chalk rubble creating a new Giant (Giant 2) that could readily be seen on Giant Hill from afar (see Chapter 7). Our limited excavations, the photogrammetry and geophysical surveys (above and Chapter 1, *Previous archaeological work*), all indicate that only a single Giant outline exists confirming the excavation results that he was recut throughout on the same line.

Key soil profile descriptions

TRENCH B (in text, above)

TRENCH C: right elbow (bottom)
TRENCH C: north-facing: upslope

Depth (cm)	Context	Sample	Description
0–9	203+207		Dark brown/brown (10YR 4/3) humic silt loam, stone-free, many fine fibrous roots, some chalk flecks in lower, no obvious structure, abrupt boundary A horizon (topsoil) *5. Recent remodelling*
9–21	210		Dark greyish brown (10YR 4/2) silt loam, common small subangular chalk pieces, several medium flints abrupt boundary A ?spoil and colluvial wash *4. Upslope colluvium and dumped soil*
21–29	213/218		Brown (10YR 4/3) silt loam, common to many chalk pieces, including some medium flints, vert stony, abrupt boundary ?stony spoil and rill deposits *4. Upslope colluvium and dumped soil* *3.1. Upslope colluvium*
29–42	225/227	1: 32–42	Yellowish brown (10YR 5/4) silt loam, fewer small and medium stones, but more (common), medium to large chalk pieces, with weak blocky structure in lower part; base of soil former profile/top of Rw and colluvium *2. base of colluvium and former (Giant 2) soil* *1. Base of former soil/pre monument colluvium* and
42+	Chalk		C chalk

TRENCH C: north-facing: downslope

Depth (cm)	Context	Sample	Description
0–8	?214		Dark brown/brown (10YR 4/3) humic silt loam, stone-free, many fine fibrous roots, some chalk flecks in lower, no obvious structure, abrupt boundary A horizon, topsoil downslope from soil revetment bank *5. Recent remodelling*

Depth (cm)	Context	Sample	Description
8–22	215		Brown (10YR 4/3) highly calcareous silt loam, psuedomycelium present even in short exposure, common fine fibrous roots, variable (heterogeneous) stone content with a predominantly small subangular (little weathered) chalk pieces, abrupt to sharp boundary A horizon material and spoil *3.2 Downslope colluvial soil*
22–29	216	OSL 3	Dark grey to dark greyish brown (10YR 4/1-2) – the colour reflecting less chalk in the matrix) - silt loam, common fibrous roots, abundant small and medium chalk pieces, weak small to medium blocky structure, abrupt boundary [OSL 3: AD 1250] *3.2. Downslope colluvial soil (and dumped soil)*
29–41	226	2: 31–41 OSL 1	Brown to yellowish brown (10YR 5/3-4) silt loam, variable stone content but generally common to many subrounded chalk stones, weak small to medium blocky structure in places Contemporaneous colluvial soil ?B (?some spoil) [OSL 1: AD 980] *3.2 Downslope colluvial soil*
41+	Chalk		C chalk

Kubiena (soil block) sample K1 30–44 cm (216/226)

TRENCH D: *left elbow (top)*
TRENCH D: north-facing: upslope

Depth (cm)	Context	Sample	Description
0–9	A 301/307		Dark greyish brown (10YR 4/2) stone-free humic silt loam, common fine fleshy roots, massive dense, and weak indistinct crumb structure, clear to abrupt boundary A horizon topsoil and soil packing *5. Recent remodelling*
9–27	310+313		Browner grey dark greyish brown (10YR 4/2) silt loam, weak small to medium blocky structure abrupt boundary, some to rare small and medium chalk pieces, common fine fibrous roots, abrupt boundary A (or redeposited Ah/A), spoil and colluvium *4. Upslope colluvium and dumped soil*
27–53	313+318+319		Brown (10YR 5/3) silt loam, some medium chalk pieces, fewer fine fibrous roots, fewer small stones Spoil and colluvium *3.1 Upslope colluvium*
53–68	325+327		Greyish brown silt loam (more calcareous than above) with blocky medium chalk lumps, an intermittent sharp contact may represent a recut between these two otherwise almost indistinguishable contexts Infill of Giant 1 cut *2. Pre-figure soil* and *1. Giant 1*
68+	Chalk		C Chalk

TRENCH D: north-facing: downslope

Depth (cm)	Context	Sample	Description
0–8	301+314		Dark brown/brown (10YR 4/3) humic silt loam, stone-free, many fine fibrous roots, some chalk flecks in lower, no obvious structure, abrupt boundary A horizon topsoil *5. Recent remodelling*
8–15	315		Brown (10YR 4/3) highly calcareous silt loam, psuedomycelium present even in short exposure, common fine fibrous roots, variable (heterogeneous) stone content with a predominantly small subangular (little weathered) chalk pieces, abrupt to sharp boundary A horizon material and spoil *3.2 Downslope colluvial soil (and spoil)*
15–27	316		Dark grey to dark greyish brown (10YR 4/1-2) – the colour reflecting less chalk in the matrix) - silt loam, common fibrous roots, abundant small and medium chalk pieces, medium blocky structure in places abrupt boundary spoil and colluvial soil *3.2 Downslope colluvial soil (and spoil)*
32–42	326	OSL 8	Brown to yellowish brown (10YR 3/3-4) silt loam, variable stone content but generally common to many subrounded chalk stones, weak blocky structure in places Colluvial soil ?B *2 Base of former soil*
42+	Chalk		C chalk

TRENCH A: *Left foot*
TRENCH A: north-facing: upslope (profile 1)

Depth (cm)	Context	Sample	Description
0–9	1		Dark greyish brown (10YR 4/2) stone-free humic silt loam, common fine fleshy roots, massive dense, and seemingly apedal, clear to abrupt boundary Topsoil (Ah)
9–18	7		Dark greyish brown (10YR 4/2) to greyish brown (10YR 5/2) slightly stickier (siltier) calcareous silt loam, abrupt boundary A (or redeposited Ah/A, revetment for chalk *5. Recent remodelling*
18–31	10/19		Dark greyish brown (10YR 4-5/2) silt loam with common to many small chalk pieces, rare medium subrounded chalk pieces, and very rare medium flints Dump or colluvium *4. Upslope colluvium and dumped soil* and *3.1. Upslope colluvium*
31–50	25/27		Grey (10YR 5/1) to greyish brown (10YR 5/2) calcareous silt loam, weak medium structure, essentially stone-free, few to rare medium chalk pieces (27) at base (with sharp contact against chalk figure fill) Colluvial soil (contemporary, 27 probably pre-dating figure) *2. Base of former soil and 3.1 initial colluvium* *1. Base of former soil/pre monument colluvium*
50+	Cw		Weathered Chalk Cw

TRENCH A: north-facing: chalk fill (profile 2)

Depth (cm)	Context	Sample	Description
0–7		2 & 4	Dense clean brilliant white to pale yellow (2.5Y 8/1-2) compacted calcareous marl with rare very small chalk pieces, occasional small pieces of chalk embedded in cement, sharp boundary Marl 3 silt wash (2019–2020) over marl (2019 fill)
7–22		5	Slightly off white (pale yellow 2.5Y 8/2-3) dense cemented chalk marl paste with rare very small chalk pieces (and rarer small chalk pieces), sharp (occasionally abrupt) boundary Marl 2 (?1979–2017, Nancy Grace, pers. comm.)
22–33		6	Abundant small rounded and surrounded (water-worn) chalk pieces (chalk kibble), in little matrix (a fine chalk marl but many small/very small voids), abrupt to sharp boundary Charcoal from chalk rubble and paste on south-face side Kibbled chalk
33–39		20	Very pale brown (10YR 8/3-4), to pale yellow (10YR 8/4) calcareous paste with common very small subangular-subrounded chalk pieces – gritty chalk marl, sharp to abrupt boundary Gritty chalk
39–57		23	Small and medium chalk stones with loose dry chalk silt (capped intermittently with a thin broken chalk marl paste marl 1). Rubbly chalk fill
57–64		28	Yellowish brown (10YR 5/4) silt loam with common to abundant medium subrounded chalk pieces Possibly trampled/mixed A horizon material / and calcareous colluvium – clearly dates the chalking
64+	Chalk		C chalk

TRENCH A: north-facing: downslope (profile 3)

Depth (cm)	Context	Sample	Description
0–4		3	Dark grey (10YR 4/1) to dark greyish brown (10YR 4/2) stone-free silt, with weak small/medium crumb structure, Ah
4–5	rill		Lenses of small and very small surrounded chalk pieces Chalk rill
5–20		9	Greyish brown (10YR 5/2) to grey calcareous silt loam, dense, stone-free fine roots, possibly mixed chalky soil and chalk turves abrupt boundary. Anthropogenic / soil material / dumped soil *5. Recent remodelling*
20–33		14	Greyish brown silt loam, with very small chalk flecks – contains rare discontinuous chalk bands – rills), abrupt boundary Soil wash including chalk washing from figure *5. Recent remodelling* *3.2 Downslope colluvial soil and dumped soil*
33–41(–50)		26	Greyish brown silt loam, with some medium and rare small chalk pieces, weak medium blocky structure, over (28), but in some places over chalk where (28) is not present or clear, abrupt boundary Colluvium and spoil *2. Downslope colluvial soil*

Depth (cm)	Context	Sample	Description
41–50	28		Intermittent and mixed layer of yellowish brown (10YR 5/4) silt loam with common to abundant medium subrounded chalk pieces, clear small to medium blocky/prismatic structure Colluvial soil pre-dating the chalking figure 2. *Base of former soil / pre-monument colluvium* 1. *primary fill of Giant 1 under Giant 2 chalk stack*
50+	Chalk		C chalk

Auger surveys

Michael J. Allen

Three auger surveys were conducted; one of the putative severed head (the knoll), during the excavation phase, a second to examine the nature and depth of the natural rendzina soil profile on the scarp slope (conducted in 2023), and the third of the belt/phallus (2024). This last was conducted as a consequence of the proposal by Martin Papworth of the possible existence of pre-existing belt. He suggested the continuation of the belt through the phallus, which had not been definitely recorded or proven to exist previously, so this was tested. A previous larger-scale auger survey had been conducted below the area under the left arm by Alister Bartlett in association with the 1979 resistivity survey (see Clark *et al.*, above), and is reported here.

The severed head

Three hand auger profiles were conducted through the knoll. The knoll had slightly more tussocky grass than the surrounding hillside, concentrations of old rabbit droppings and at least one rabbit scrap and shallow burrow. Augering through the deepest deposits near the centre of the mound revealed a shallow topsoil about 20 cm thick, beneath which was, in two of the holes, a chalky deposit 6 cm thick sealing two mixed soil deposits each about 10–11 cm thick. These may be colluvium and a dump of upcast mixed soil. Both holes off the centre of the mound encountered soft ant-ridden deposits. It seems, therefore, that there is a thin layer of chalk beneath the turf and topsoil. It is uncertain whether this represents the remnants of chalk dump relating to previous scouring and rechalking of the figure, or an original construction of a soil mound capped with chalk. The chalky layer was only recovered in two of the three auger holes so does not exist, or survive, as a consistent capping. The deposits of the mound generally suggest upcast soil which has been mixed by worms, ants and, locally, rabbits. Whether this knoll is a part of the original design, a later but intended addition, or waste soil from the many scouring and rechalking events could not be confidently discerned. The interesting facial features observed by Castleden in *c.* 1995 (1996, 162–5) could not be seen in the field at least. The auger record was, however, reminiscent of a spoil heap and we may question whether it is anything to do with the Giant and could readily be explained by movement during the numerous scouring (and rechalking) events; the area under the left arm being the largest open blank space around the figure.

Auger: centre of mound ('head' below outstretched hand, directly below fingers – level with second rib up)

Depth (cm)	Context	Description
0–16	A	Dark yellowish brown (10YR 3/4) humic silt loam, stone-free, abrupt boundary A, topsoil
16–20	A	Dark yellowish brown (10YR 3/4) humic silt loam, with common fine chalk flecks-free, abrupt boundary A, soil
20–26	Chalk deposit or rill	Dark yellowish brown (10YR 3/4) silt loam dominated by abundant small and medium rounded chalk pieces, abrupt to sharp boundary Chalk deposit or rill deposit
26–37	Fill/colluvium	Dark yellowish brown (10YR 4/4) silt loam, common fine chalk flecks and very small chalk pieces, abrupt boundary Deposit – fill or colluvium
37–48	Fill/dump	Very pale brown calcareous marl, and soil material Mixed deposit, fill or dump
48–70	Fill or weathered chalk	Light yellowish brown (10YR 6/4) calcareous silt matrix with greyish brown soil inclusions, abrupt boundary. Mixed chalk marl and some soil Origin uncertain: deposit or heavily weathered chalk (not seen elsewhere)
70–80	Fill or weathered chalk	Light yellowish brown (10YR 6/4) calcareous chalk marl becoming more chalky and chalk marl ?weathered marl or primary fill
80+	Chalk	C Chalk

Area beneath the left arm (aka cloak or lion skin)

Alister Bartlett† and Michael J. Allen

The 1979 resistivity (and magnetometer) survey (see above) conducted over the whole figure recorded that a series of higher readings present below the outstretched left arm in the area of the purported lion skin or cloak. In 1980 a programme of corroborative augering in grid 20 m × 20 m was set out beneath the left arm and 125 auger holes recorded over the higher readings. The augering provides useful information for the geophysical survey, but in its own right forms an important set of data. The records principally indicate total depth to the chalk. The soils themselves were not recorded except that in two locations, where a chalky hillwash to below 25 cm and a hillwash to below 50 cm was noted. Not only was moderately deep 'hillwash' recorded, but it was reported 'white chalk and lighter brown earth streaks (darker brown seen in upper layers)' clearly indicating the presence of layers of soil with chalk stones largely absent from the adjacent soils (Allen 2023b).

Soil depths were consistently in excess of the typical natural soil depth of 20–35 cm. Over 40% of the holes were 35 cm, many were at about 40 cm and some up to 75 cm deep. This clearly indicates an area of varying, undulating but generally increased soil depth, and one not typical or expected especially such a steep scarp slope. This general area of enhanced or thicker soil forms the area commonly

FIGURE 3.18. Auger profiles (2024) across the top of the legs showing the chalk outline trench (left), and the phallus showing no corresponding trench (right). Image: Justin Russell 2024

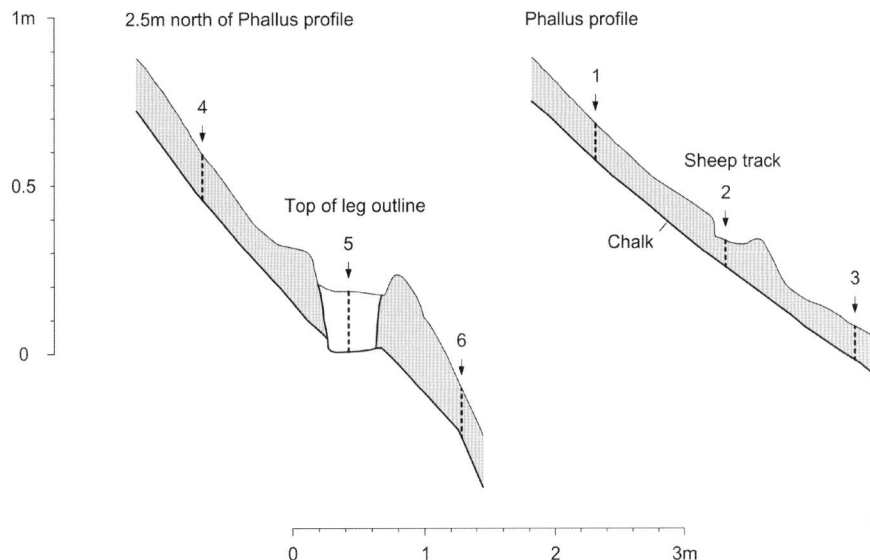

considered to be the cloak or lion skin beneath the empty space under the left arm, and is typical of upcast soil and spoil. Previous geophysical surveys, and the 2020 augering, also confirm its presence. The severed head is also a mixture of soil and chalk, so whether these are part of a deliberate design, or the remnants of other activity is not immediately clear. It does not, however, match the chalk outline of the Giant and this is the most obvious place for chalk and soil to be dumped during scouring and restoration, as discussed later (Chapter 7).

The putative belt

A belt under the phallus was postulated by Martin Papworth based on a line clearly seen on the rendered aerial photogrammetry image (Fig. 3.3), and this suggestion was posted in the National Trust website (Papworth 2021a) and released to the press along with the OSL dates, getting national and global coverage (Booth 2021; Bridge 2021; Brown 2021; Mead 2021; Morrison 2021; Simpson 2021, etc.). Consequently geophysical surveys were conducted over the belt (Cheetham, above) and as this was not wholly conclusive, six points were cored with a 1 cm diameter gouge auger across the belt and across the top of the legs (Fig. 3.18; Allen 2024). This clearly indicated that there is no subsurface feature and that the line was a sheep track evident in the field and present on aerial photographs from at least the 1970s. The concept of a belt is therefore untenable.

Summary of infill events

Michael J. Allen and Martin Papworth

The development of both the chalk stack and deposits developing against the chalk outline are summarised in nine phases of the sections development in Trench B across the right foot (Fig. 3.19). Evidence from three other trenches has been used to comprehend this developmental sequence.

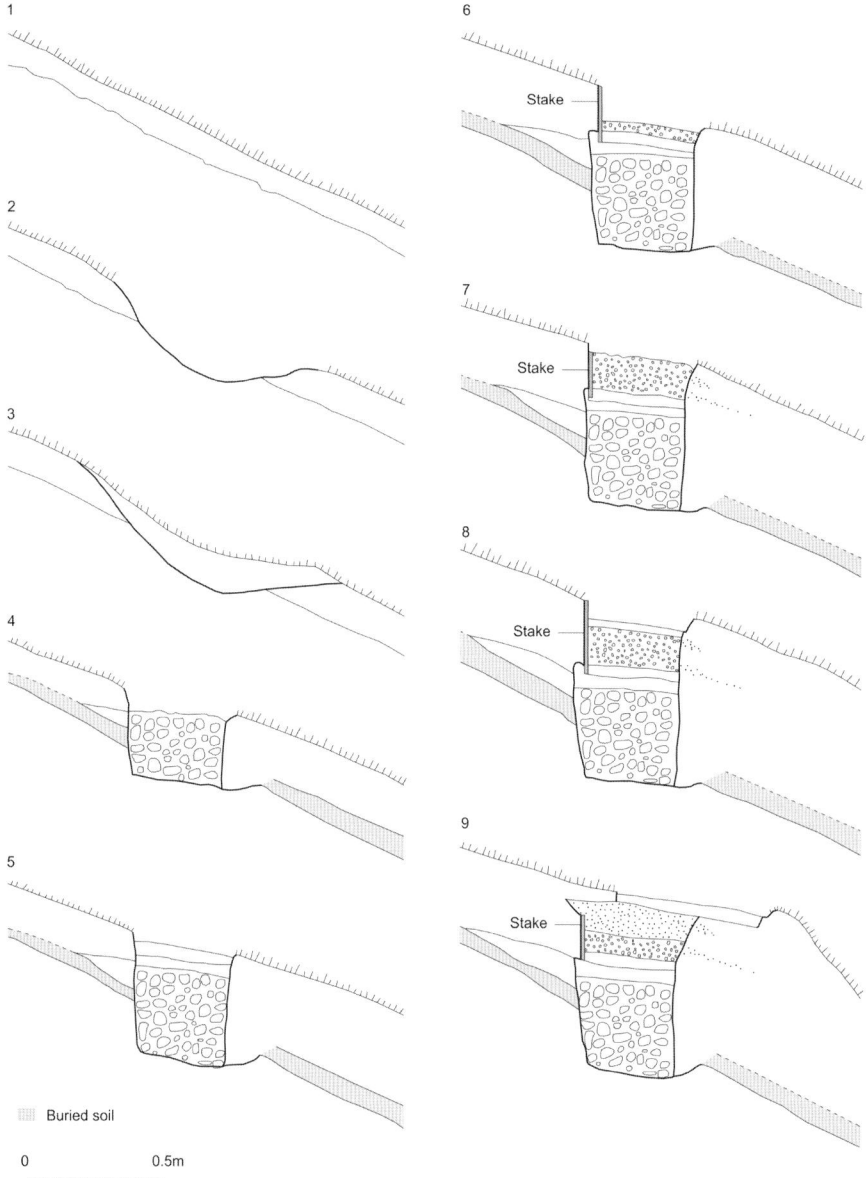

FIGURE 3.19. Interpretation of the development of the Giant's outline (chalk stack), associated soils and accumulating and colluvium. Image: Justin Russell & Mike Allen

1: The pre-Giant shallow hillside rendzina soil developed *in situ* on the steep (45°) slope under permanent pasture.

Giant 1: early medieval

2: An initial broad shallow cut though the soil and creating a shallow hollow (*c.* 60 to 80 cm wide) in the chalk form the original early medieval outline of the first Giant (Giant 1). Any chalk was clearly removed, and soil spread, possibly creating a shallow low 'bank', essentially just thickening the soil locally.

3: The cut for the first Giant infills relatively rapidly with soil wash.

Giant 2a and 2b: pre-AD 1250 to eighteenth/twentieth century
4: A narrower (40 cm wide) well-defined vertical-sided cut though the infilled Giant 1 scoop and into the chalk, re-establishing the figure pre-AD 1250; initially as an *open trench* and latterly in the eighteenth–twentieth centuries as an exposed chalk base, and then chalk-filled, outline. The cut chalk was carefully and totally removed from the immediate vicinity of the outline. The soil was spread on either side of the figure preserving the base of the former soil (buried soil) and this spread acting as a barrier allowing essentially stone-free colluvium to build up against the Giant's outline, and soil development to continue. The chalk rubble seen here is post-eighteenth century at the earliest and probably at pre-1945 (following Martin Papworth's chalk outline succession; Fig. 3.16).

Twentieth century
5: Soils wash down (colluvium) against the outline. Scouring and replenishment of the Giant require additional chalk infill (chalk paste), probably in the twentieth century.
6: Further colluviation and soil build-up on the upslope side. The stake is the remnant of scouring events when boards shored the side during removal and replenishment of the upper part of the chalk infill stack. Soil and colluvium masked the upslope portion of the outline, and replenishment (Trench B) does not extend over all of the former outline. The outline is narrowed at this point, possibly mid-twentieth century.

1956–1979
7: Continued colluviation and increasing soil depth. Scouring and replenishment with kibbled chalk in two episodes of 'industrial' contractor (E.W. Beard & Co. and E.W. Beard Ltd.) in 1956 and 1979. In other locations (e.g., Trench A, left foot and Trench D left elbow) the outline is widened, and in these two periods moves downslope by as much as *c.* 20 cm. The outline is clearly filled to surface at this line (or before) as chalk wash rills can be seen in the section. This is fine chalk rain-washed from the outline into the adjacent grass (as seen in 2019, see Figs 2.2 & 2.3), or spilt chalk from the scouring and replenishment.
8: Continued colluviation and increasing soil depth, regular maintenance (weeding and cleaning) and small scale replenishment.

1979–2020
9: Continued colluviation and increasing soil depth. Scouring and replenishment with further chalk increased the thickness of the chalk stack. Finally three phases of heavily tamped chalk with National Trust volunteers in 1995, 2009 and 2019 result in the outline not covering and former upslope outline, and extends it up to a further 40 cm downslope increasing its width (at this location) to about 70 cm.

108 Michael J. Allen

Optically stimulated luminescence dating

Phillip Toms, Jamie Wood and Michael J. Allen

Archaeological context for the dates

Michael J. Allen

The date of the Cerne Abbas Giant has long been the subject of speculation by archaeologists, historians and locals (Darvill *et al.* 1999) with arguments that he could be prehistoric (Bronze Age/Iron Age) to Roman or historic. Despite this long debate no archaeologist had grasped this nettle; obtaining an absolute scientific date for the Giant, however, was going to be a challenge. The chalk outline is cut through the turf and into the chalk and then infilled with rammed chalk; it was unlikely that archaeological finds would be associated with the outline, nor any organic material present with which to date the chalking by radiocarbon methods. However, accumulations of sediments against the sole of each foot and crook of each elbow (see above), could potentially be dated by optically stimulated luminescence (OSL). This had previously been successful in dating the chalk figures of the Uffington White Horse, Berkshire (Miles *et al.* 2003; Rees-Jones & Tite 2003; Chapter 16), and the Long Man of Wilmington, East Sussex (Bell & Butler 2014; Chapter 15). The dating of the Giant was the basis of this entire fieldwork project.

Dating these colluvial deposits would provide a *terminus ante quem* date (date before) for the establishment of the Giant, although relict former soil at the base of the colluvium might provide a date coeval with him. A series of nine locations were sampled from three trenches by Phil Toms, University of Gloucestershire (Table 3.3; Figs 3.14; 3.20 & 3.21), and gamma spectrometer measurements made from key deposits in Trench B and C (Fig. 3.21). *In-situ* gamma spectroscopy measurements were made from two locations. Nine samples were taken of which five were selected for dating (one of which failed). The dating project was funded jointly by the National Trust and University of Gloucestershire.

The successful dates came from samples taken from soil-rich deposits accumulating against the chalkfill outline, and not from any of the chalk-fills themselves. The measured samples fall into two clear categories i) those associated with the construction of, or soil development associated with,

Sample	Trench	Context	Deposit	Comment
OSL 1	C	226	Downslope, lowest soil/fill	Measured
OSL 2	C	206	Chalking, kibbled chalk	Not progressed
OSL 3	C	216	Downslope colluvial soil	Measured
OSL 4	C	221	Chalking, marl	Not progressed
OSL 5	B	128	Base of colluvial soil downslope	Measured
OSL 6	B	125	Upslope colluvium	Measured
OSL 7	B	121	Chalking, chalk marl	Not progressed
OSL 8	D	326	Downslope, colluvial soil	Measured, no result
OSL 9	D	321	Chalking, chalk marl	Not progressed

TABLE 3.3. The location of the OSL samples

FIGURE 3.20. Phil Toms monitoring the gamma spectrometer during OSL sample of Trench B. Images: M. Papworth, digital manipulation, D. Ingram (Oxbow Books)

the first phase Giant, and ii) colluvium accumulating upslope and development of colluvial soils post second phase Giant. Samples OSL 1 (right elbow, 226), and 5 (left foot, 128) related to the first phase Giant, and OSL samples 6 (right elbow, 125) and 3 (left elbow, 216) related to post second phase Giant deposits (Figs 3.14 & 3.21; Table 3.3). There is, however, some ambiguity over the precise location of OSL 1. The sampled section (south side) was not drawn and the location projected onto the opposite section drawing (Figs 3.14 & 3.21). Although there is a possibility that this came from the chalk rubble this seems unlikely as this was predominantly chalk with virtually no material (0.35%) <0.5 mm, and it was so compacted and rubbly it was not possible to get metal kubiena tins which buckled (see Appendix 4). The result is also chronologically inconsistent if from this deposit (i.e., the downslope colluvial soil).

Optically stimulated luminescence dating

Phillip Toms and Jamie Wood

Nine conventional sediment samples, those located within matrix-supported units composed predominantly of sand and silt, were extracted from Trench B (right foot) Trench C (right elbow) and Trench D (left elbow) on 20 March 2020 and submitted for optical dating (Fig. 3.21; Table 3.3).

Based on the Central Age Model (CAM) and incorporating uncertainties, sediment sequences associated with the Cerne Abbas Giant span 0.79–2.25 ka (AD 1230–230 BC) in the fine sand quartz fraction and 12–72 ka (9980 BC–70,000 BC) in the counterpart fine silt quartz (Table 3.4). For the latter fraction, the age estimates are highly unlikely to be synonymous with the creation of the Cerne Abbas Giant. Instead they most likely reflect a mixture of 'older' fine

Sample (CEAB)	Lab code	Location and deposit	Aliquot type	Grain size (μm)	FMMMin Age (ka)	CAM Age (ka)
OSL 3	GL19103	Downslope Trench C, Colluvium (216)	Single grain	180–250	0.77±0.26	1.28±0.32
			Multi-grain	5–15	–	17.5±2.2
OSL 6	GL19104	Upslope Trench B, Colluvium 125	Single grain	180–250	0.78±0.16	1.06±0.27
			Multi-grain	5–15	–	14.3±1.9
OSL 1	GL19102	Downslope Trench C, Soil or First Fill (226)	Single grain	125–250	1.04±0.33	1.33±0.26
			Multi-grain	5–15	–	64.2±7.8
OSL 5	GL20067	Downslope Trench B, Colluvium 128	Single grain	125–250	1.11±0.21	1.90±0.35
			Multi-grain	5–15	–	41.9±6.9

silt quartz originally held within the chalk and 'younger' aerially/sub-aerially deposited material, producing age overestimation. This has potential implications for previous OSL dating of chalk figures that have drawn on multi-grain aliquot, fine silt quartz OSL dating (e.g., Uffington White Horse; Rees-Jones & Tite 1996; 2003), where the age of construction may be overestimated owing to incorporation of grains whose OSL signal partially relates to the formation age of the chalk.

All sand fractions in the samples exhibit >20% overdispersion (37%–82%: Table 3.4), suggesting inter-grain D_e variation is forced by variables other than counting statistics, spatial variation in D_r and experimental error. Therefore, in estimating the burial age, drivers beyond these variables must be considered. We exclude MAM estimates from further consideration as these can be strongly influenced by outliers (Rodnight *et al.* 2006). The relatively young age estimates across the spectrum of single grain data from each sample suggests that this material is not contaminated by 'older' fine sand quartz that may have been held in chalk and not exposed to sunlight prior to burial at the site. All samples exhibit a positive skew in their inter-grain D_e distribution; this is particularly marked for samples OSL 3 (GL19103; skewness 3.1) and OSL 6 (GL19104; skewness 4.2). This may be symptomatic of partial bleaching, where samples receive insufficient exposure to sunlight prior to burial and thus minimum ages will reflect either fully bleached grains and reliable estimates of burial period, or well-bleached grains and maximum estimates. The relatively high proportion of zero dose grains (26–30%) and shallow depths of the OSL samples might suggest active pedoturbation, where sediment is being constantly reworked to the surface and that minimum age estimates would be indicative of the timing of that reworking rather than creation of the Cerne Abbas Giant. However, the positive skew in the inter-grain D_e values of samples OSL 6 (GL19102) and OSL 5 (GL20067) is far less than that associated with material undergoing active pedoturbation (cf. Gliganic *et al.* 2016). It is, therefore, more likely that the high proportion of grains with D_e values statistically concordant with zero in these samples reflects a short burial period that is synonymous with the creation of the Cerne Abbas Giant (Table 3.5; Papworth 2021b).

The sample from the downslope colluvial soil (326) in Trench D failed to provide a result. In summary, FMM_{Min} OSL age estimates from sample OSL 1

TABLE 3.4. OSL aliquot type, grain fraction and ages derived from the Finite Mixture Model minimum age population (FMM_{Min}) and Central Age Model (CAM) (Toms & Wood 2021)

Position	Location	Deposit	Sample (CEAB)	FMMMin Date
Giant's right elbow	Downslope trench C	Colluvial soil (216)	OSL 3	AD 1250±260 (990–1510)
Giants right foot	Upslope trench B	Colluvium (125)	OSL 6	AD 1240±160 (1080–1400)
Giant's right elbow	Downslope trench C	Soil or first fill (226)	OSL 1	AD 980±330 (650–1310)
Giant's right foot	Downslope trench B	Colluvial soil (128)	OSL 5	AD 900±200 (700–1100)

TABLE 3.5. The OSL results (Toms & Wood 2021)

(GL19102) from the basal soil adjacent to the chalking in Trench C (the right elbow) and OSL 5 (GL20067) from basal sediment in Trench B (immediately adjacent to and downslope of the right foot) suggest that the Cerne Abbas Giant was created no earlier than AD 650–700 (Table 3.5).

A preliminary consideration of the OSL results and their interpretation: what we have dated

Michael J. Allen

The tenth-century results for the creation of the Giant and thirteenth-century evidence of his maintenance (Table 3.5) are a conundrum as historians have no evidence for the figure at Cerne before 1694 (see Hutton 1999b; Edwards 2020; Chapter 18;), but Phil Toms confirms from a scientific perspective that the science and measurements of these results are 'as good as it gets' in such contexts, and the geoarchaeology clearly confirms the deposits are contemporary with, and post-date, the earliest Giant. The dates are cohesive and scientifically robust, and are a stratigraphically consistent group of results. The two sets of dates (Table 3.5) of pre- or early Giant AD 900±300 and AD 980±330 (AD 650–1310) are statistically indistinguishable from each other and the second set being of some of the hillwash against the Giant (AD 1240±160 and AD 1250±260 (AD 990–1510)) are also statistically indistinguishable making this a very robust set of results. They are also consistent with snails in the deposits which were introduced into Britain in the medieval period. The results give a construction date of about AD 900±300 (AD 700–1100) and AD 980±330 (AD 650–1310), and the other confirm that the Giant was in existence and being rechalked in the twelfth to sixteenth centuries (AD 1080–1560).

In order to use the OSL results sensibly and appropriately it is important to understand precisely what we have dated in terms of the event relating to the sampled deposit. That is why so much effort was expended comprehending the deposit taphonomy (formation processes, above). Four OSL results were obtained; two from the cut and fill/soil development associated with Giant 1 (128 and 226) and construction of Giant 2, and two stratigraphically higher and overlying the cut of Giant 1 and are in post Giant 2 colluvium and upcast – one from the upslope colluvium (125) and one downslope colluvium (216).

The early deposits (Giant 1)

The precise location of the two samples associated with Giant 1 is imperative. That from context 128 (right foot) and 226 (right elbow) are the lowest possible location and are a part of a relict well-structured largely *in-situ* rendzina soil, subsequently engulfed in colluviated soil and soil material, in the shallow wider cut of Giant 1. These contexts represent the first fill and soil developing in the former shallow hollow, or a combination of soil material buried in the shallow hollow dug for the outline combined with the very early soil development. These samples are from the right foot and right elbow, but comparable stratigraphic locations, and as such they are the earliest soil deposits associated with the Giant. They both gave late tenth-century dates (AD 900±200 and AD 980±330) respectively. The coincidence of these dates, both from the earliest surviving deposits, is reassuring and strongly suggestive that these date very early events associated with the Giant (if not his creation itself). This suggests an early tenth-century date for *both* of these deposits, and thus by inference for the Giant himself. These deposits can be no earlier than AD 650–700, and represent a date soon *after* the first Giant. We suspect that Giant 1 is not much earlier than the late tenth-century date, he could still be tenth century; it would not take very long for a shallow, 0.65–0.8 m wide the probably only 0.3 m deep, scoop, but he could be slightly earlier.

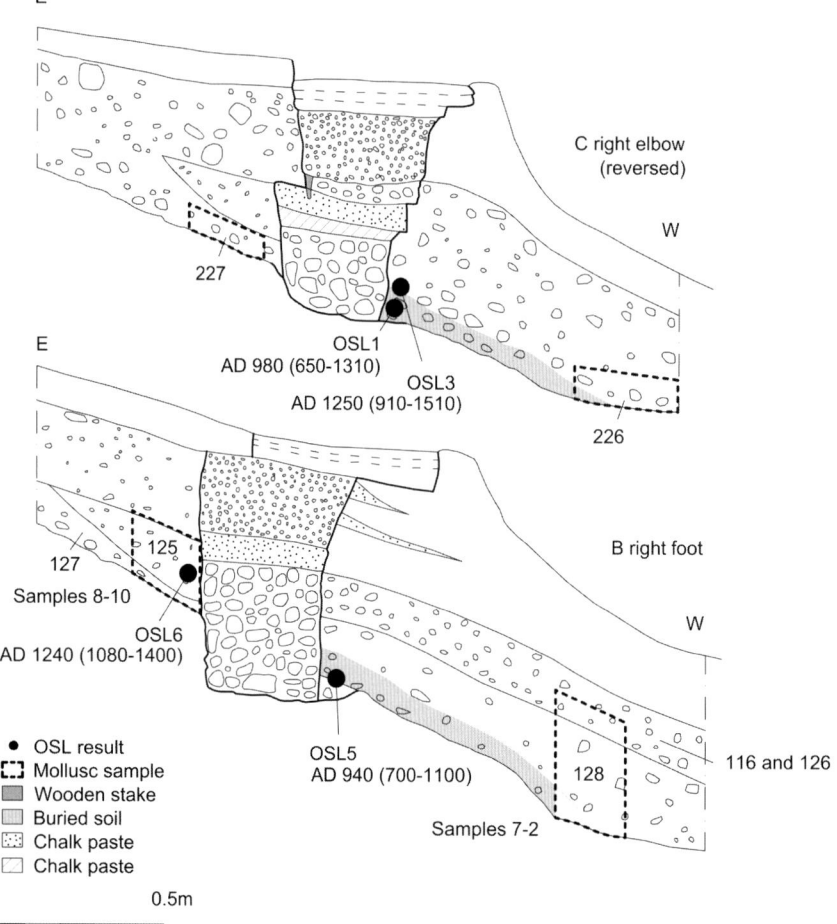

FIGURE 3.21. The two main sampled sections showing the land snail samples and OSL sample locations in Trenches B (south side) and C (north side. reversed). Samples in Trench C were taken from the south side and were transposed on the north side section, and reversed (see Appendix 3). Image: Justin Russell 2023, based on field drawing by M. Papworth

The later, post-Giant, deposits (post Giant 2)

Two other dates were from the downslope and upslope colluvium over the cut for Giant 1 and which post-date Giant 2 (the narrow cut trench and hillwash accumulation), but indicate a time when he was clearly extant. He had clearly been in existence for some period of time to enable colluviation against the chalked line upslope, and soil wash and soil development on the downslope side. On the upslope side (Trench B, 125), the sample was clearly above the lowest context, the relict rendzina (or colluvial) soil (127), and about 10 cm into the hillwash. Similarly, on the downslope side (Trench C, 216) the sample was again above the relict buried rendzina soil (226) and well into the initial colluvial deposit (216). The two samples were at comparable locations, but from the right foot (Trench B) and right elbow (Trench C), and gave remarkably similar dates of AD 1240±160 and AD 1250±260, suggesting a phase of destabilisation on the hillslope above, and of colluviation against the Giant in the mid-thirteenth century. What this represents is unknown, but could indicate tillage on the top of the hill, scrub clearance on the hillside or even increased stocking and grazing levels on the hillslope. It might simply relate to climatic vagary and an episode of increased rainfall and soil wash. It could also represent a phase of scouring clearing out and cleaning of the Giant. What is clear is that this was a hillside event in the mid-thirteenth century, by which time the Giant had been long extant. Historical research covered this time period and that evidence can be seen in Chapter 4 by Barbara Yorke, and Chapter 9 by Katherine Barker.

The land-use history of a hillside: land mollusc evidence

Michael J. Allen

The principal aim of fieldwork was undoubtedly to date the Giant, however, the small excavation trenches also provided the opportunity to examine the land-use history and local environment in which the Giant was carved and survived, via land snail analysis. The figure is carved on a steep scarp slope that we might expect would have remained wooded until relatively late in prehistory, and possibly into early historic times, as its earlier clearance for settlement, and tillage would not have been necessary. If the Giant was, however, of historic date and woodland had been removed from the hillside, then defining subtle changes in the hillside land-use and environment (lightly grazed grassland, meadow, downland etc.) may be very challenging to detect, and separate, via the analysis of land mollusc data.

Land snail analysis could determine whether the scarp slope of the Down was cleared before, or for, the Giant, or whether he was cut into a pre-existing open downland, short-grazed grassy sward or long ungrazed downland, and then examine land-use history of the slope after he was cut. In particular the short stratified sequences might indicate changing pressures of land-use (grazing) on the Giant. A series of samples was taken for land snail analysis following standard field methods (cf. Evans 1972; Allen 2017a; 2017b) to address the following aims and research questions, i.e., to:

- provide the first palaeo-environmental record associated with the monument, and examine the palaeo-environment contemporary with the construction of the Giant;
- determine if it was built in recently cleared, or pre-existing open downland landscape; i.e., was the hillside wooded?, or was the woodland removed specifically for, or well before, the Giant;
- establish if episodes of vegetation regeneration representing abandoned and neglect could be detected; and
- in view of the long hiatus between the creation of the Giant and the first documentary records, can the evidence of land-use help explain its absence for several centuries? i.e., thirteenth to seventeenth centuries.

The locations selected to sample included colluvium on both the upslope side and downslope side of the figure (see geoarchaeology above). It was assumed that the upslope colluvium was the more immediate and earliest, with the downslope colluvium (and soil development as a shallow calcareous colluvial brown earth) following slightly later and then coevally. Sections in two trenches were selected to sample; principally Trench B of the sole of right foot and a few support samples from Trench C, the crook of the right elbow, as the deposits in both these trenches were the best developed and had been selected for OSL sampling. Consequently most of the palaeo-environmental, geoarchaeological investigation and OSL dating was concentrated on these profiles. The best and most pronounced of these was the right foot (Trench B). A series of eight samples were taken in two contiguous columns from this sequence, and augmented by two spot samples taken from the right elbow (Trench C).

Analysis

Samples were taken from the stratified colluvial soil, and ancient spoil/soils (Trench B, samples 3–10), and lowest deposits (Trench C 225/227 and 226); these included possible pre-monument relict soils at the base of Trench B (base of 128) and Trench C (226). Samples of >1500g were taken from cleaned and described section faces (Figs 3.14 & 3.21) following standard methods and taking care not to cross context boundaries (Allen 2017b). These aimed at providing a landscape and land-use context for the monument. Although very short sediment sequences (42 cm and 20 cm) they might indicate changing pressures of land-use (?grazing) on the monument, but more importantly indicate the nature of the pre-monument land-use. Samples of 1500 g of air dried sediment <16 mm were processed for land snails following standard methods outlined by Allen (2017b, 35–6) and Evans (1972, 44–5). Snails were sorted and identified under ×6.1–×55 magnification using a Leica stereo-binocular microscope, and the results are presented in Table 3.6 and as a histogram of relative abundance in Figure 3.22 where nomenclature follows Anderson (2005). The species are discussed by broad ecological groupings (open country, catholic or intermediate and shade-loving) as defined by Evans (1972) and modified by Entwistle and Bowden (1991) to assist in interpretation.

Following analysis in 2021–22 a series of modern assemblages was collected from the steep scarp slope adjacent to the Giant in 2023. Modern fauna were sampled from the short, lightly grazed grassland north and south of the Giant and about his waist/elbow level where the herbaceous grass vegetation had a typical stand height of about 4–6 cm. The second area to the south, at about the same level, was of more tussocky longer grass with a typical stand height of 15+ cm, but with some up to 35 cm. The significance of this was variation in ground shade and moisture content, which might reflect different mollusc faunas. Soil moisture was measured but seemed to reflect the current conditions (rainfall, periods of sunshine), rather than variations across the various vegetation stand heights. Modern fauna were collected in two ways: i) with petrol power leaf blow in reverse with collection bag (see Mags Cousins in *Mollusc World* 48, fig. 4, p. 19) and ii) from auger cores during auger survey of the modern soil depths in September 2023 (Allen 2023a). Samples were processed in the same way as those from the excavation.

The first phase of hillwash accumulated against the upslope side (227, 127 and 125). Soil development occurred on the downslope side and included overspill soil wash, hence the colluvial soil on the downslope side (226, 128 and 126/116) consistently showed better, albeit weak pedological structure, than all the upslope lower colluvium. It is on this basis that the chronology of the land-use history is based, and not the changing mollusc assemblages. Thus we can group the deposits, and order the land-use history, as follows:

Upslope		Successional chalking	Downslope		Deposit Phase Group
Trench B	Trench C		Trench B	Trench C	
113	210	102/202	109		3.1. Upslope colluvium and dumped soil
	213	103/203			
118	218	204	114	214	
		105/205		215	3.2. Downslope colluvial soil
		106/206	**126/116**		
		212	**126**	216	
		121/221	**128 upper**	226	
125	**225**	120		**226**	2. base of former soil
127	**227**	123/223			
			128 base		1. Pre Giant 2 soil

Context numbers in bold are those with land snail analysis

Despite the calcareous nature of the deposits and the large samples (1.5 kg), mollusc numbers were generally quite low (43–134: i.e., 29–89 mollusc/kg), with only that from the very calcareous upslope colluvium (227) in Trench C having 258 shells (172 mollusc/kg). The assemblages are dominated by open

	Cerne Abbas Giant 2020								
	Upslope				Downslope				
Trench	B			C	B				C
Context	127	125	125	227	128	128	128	126	116	226
OSL date AD		1240			910			(1250)		980
Sample	8	9	10	1	7	6	5	4	3	2
Depth (cm)	51–61	41–51	31–41	31–41	70–75	63–70	53–63	43–53	33–43	32–42
Wt (g)	1500	1500	1500	1500	1500	1500	1500	1500	1500	1500
MOLLUSCA										
Pomatias elegans (Müller)	5	6	1	3	2	3	7	17	5	3
Carychium tridentatum (Risso)	–	–	–	–	–	–	4	–	–	–
Cochlicap cf. *lubrica* (Müller)	1	–	–	–	–	–	–	–	–	–
Cochlicopa spp.	–	–	6	–	–	1	2	1	2	–
Vertigo pygmaea (Draparnaud)	8	24	41	15	4	6	7	16	9	4
Pupilla muscorum (Linnaeus)	13	37	49	18	12	14	39	38	13	8
Vallonia costata (Müller)	1	8	30	3	4	–	12	2	2	–
Vallonia cf. *excentrica* Sterki	2	6	16	2	4	1	6	7	8	–
Acanthinula aculeata (Müller)	1	–	1	–	–	1	2	–	–	–
Punctum pygmaeum (Draparnaud)	–	2	6	1	–	1	–	1	3	–
Discus rotundatus (Müller)	2	6	1	2	2	6	8	6	–	18
Vitrina pellucida (Müller)	–	1	–	–	–	–	–	–	–	–
Vitrea contracta (Westerlund)	–	1	1	1	–	1	3	–	–	1
Aegopinella pura (Alder)	1	–	–	–	–	–	3	1	–	–
Aegopinella nitidula (Draparaud)	3	[1]	2	–	2	3	6	3	–	–
Oxychilus cellarius (Müller)	–	–	–	–	2	2	4	–	1	–
Nesovitrea hammonis (Ström)	–	–	–	–	–	–	–	–	1	–
Limacidae	3	3	8	2	–	–	1	2	1	1
Cecilioides acicula (Müller)	(13)	(23)	(87)	(25)	(23)	(68)	(131)	(175)	(58)	(58)
Cochlodina laminata (Montagu) [lost]	–	–	–	–	–	–	1	–	–	–
Clausilia bidentata (Ström)	–	1	–	–	1	1	2	1	–	–
Candidula sp.	–	3	17	–	–	–	–	6	3	–
Cernuella virgata (da Costa)	2	2	12	–	–	–	–	–	5	11
Helicella itala (Linnaeus)	6	7	6	6	6	6	6	5	14	2
Candidula / Helicellids	5	18	47	6	5	11	18	6	6	6
Monacha cantiana (Montagu)	–	–	1	–	–	–	1	–	2	1
Trochulus hispidus (Linnaeus)	1	2	10	3	–	1	1	3	2	–
Cepaea spp.	2	–	3	–	–	1	–	–	+	3
Cepaea / *Arianta* spp.	–	–	–	–	–	–	2	+	–	–
Taxa	15	13	17	11	9	15	20	14	15	10
TOTAL	**56**	**127**	**258**	**62**	**43**	**59**	**134**	**116**	**78**	**58**

TABLE 3.6. Land snails from the Cerne Giant Trenches B and C (OSL date in parentheses is from Trench C)

country species indicating a post-woodland environment, and one in which open conditions prevailed and were long-established. The sporadic presence of *Monacha cantiana* tends to suggest a post Roman occurrence for these deposits (Kerney 1999, 189); and although it was still rare until the medieval period (Kerney 1970), the consistent significant presence of Introduced Helicellids

(*Cernuella virgata, Candidula* spp., principally *intersecta*) suggested a medieval or later date (cf. Kerney 1966). When the OSL dating programme was held up by Covid, this provided the first evidence that giant was not prehistoric and was post Roman, and probably medieval, in date.

The giant was constructed in an open, post woodland clearance environment and the hillside had long been maintained by grazing (226, 227, 127 and base of 128); as might be expected at this date. What is particularly important is that this includes assemblages from the earliest deposits from the first Giant (Giant 1) at the base of context 128. Although no buried soils survive, there were hints at the bottom of the sequence of localised and sporadic preservation of the base of a former rendzina or shallow brown earth soil, which may relate to the land-use contemporary with the cutting of Giant 2. This is suggested by clear but weak and intermittent blocky (or prismatic) at the base of downslope colluvial soil (context 128) and base of the upslope colluvium (227) and indicate the remnants of a former soil. In both cases shell numbers were very low (43 and 56 respectively), nevertheless open country species represent over 80% of the assemblage with *Pupilla muscorum*, *Helicalla itala* and *Vertigo pygamea* being dominant elements; these are species typical of dry calcareous grassland, especially short-turved, sheep-grazed grassland (Evans 1972, 143; Kerney 1999, 94, 103 etc.). Both the *Vallonia* species present (*V. excentrica* and *V. costata*) are typical of old calcareous short grassland habitats. None of the catholic or intermediate species present would be out of place in this habitat. Most of the shade-loving species (e.g., Zonitidae: *Aegopinella nitidula* and *A. pura*), can be found in grassland, only *Discus rotundatus* is rare in open downland (Evans 1972, 185), but can be found in clumps of longer grass and readily in loose bare soil under open hawthorn bush (Allen pers. obs.; it is also present today in the light woodland at the foot of the slope). Overall, although small assemblages, these tentatively suggest long established open grazed or trampled grassland (Chappell *et al.* 1971). The only slight contradiction is the assemblage from the base of the upslope colluvium (context 127), which suggests open, but slightly more mesic (longer grass) habitats, as do those from the basal remnant soils on the downslope side (226). Are these remnants of an earlier longer ungrazed grassland that existed just before the Giant was inscribed on the hillside?

The assemblages from the main colluvium upslope, which we largely take to be coeval with the land-use soon after the Giant was constructed, show continued open dry, short grazed grassland, suggesting the hillside was grazed by sheep and the Giant readily observable in the Dorset countryside as we might expect. This hillwash is unsorted and fine-grained with few stones typical of low-energy soilwash (rather than soils disrupted under tillage). Although generally we see hillwash as predominantly a result of tillage (Allen 1988; 1992; Bell 1983; see *Geoarchaeology of the two hillwash deposits*), here we assume that it is soil washed up against the chalked figure under stable grassland in which patches of loose soil exist in terracettes and with patches of loose soil under small isolated bushes.

The most significant, although subtle, development is seen in the assemblages from the downslope hillwash and soil development. Although *Pupilla* remains the dominant species (Fig. 3.22), increases in *Pomatias elegans* suggests areas of loose broken soils (possibly spill and soil wash from the rechalking events). If the assemblages are considered to represent a more mesic long grassland then the absence of *Carchyium tridentatum*, which is common in this habitat (Cameron & Morgan-Huws 1975), needs considering. Auger surveys on the downland slope immediately adjacent to the Giant in Autumn 2023 showed thin soil depth (typically 15–20 cm), but also enabled the recovery of soil and of snail assemblages from the upper 8 cm (Ah) horizons as 'modern; comparators' from short grazed, trampled grass (4–8 cm; 3 samples) and from longer less-well grazed grassland (8–35 cm; 3 samples).

These all produced shells of predominantly open county species in all six samples. Although sample sizes were small (670–885 g) and shell numbers varied from 46 to 109 (66–154 shells/kg) no *C. tridentatum* apices or fragments were present in any sample. Those from the longer grass on the steeper slope tended to contain more *A. nitidula* and *Nesovitrea hammonis* than those in the short grass (Allen in prep/archive). This tends to confirm the tentative interpretation from the subtle variation on the subfossil assemblage of slightly longer grass with less grazing pressure versus a lightly grazed short grassland (Fig. 3.22).

The presence in the subfossil assemblage of other shade-loving species such as the typically rupestral species in the Clausiidae, as well as the increase in the Zonitidae (*A. nitidula*, *A. pura* and *Oxychilus cellarius*) do not detract from the overall open country conditions but these changes in species composition are seen in grassland succession environments (Cameron & Morgan-Huws 1975) and grassland hawthorn sere (Evans 1972, 136 & 186; 1984), and here suggest more mesic conditions such as the base of longer tufts of grass that

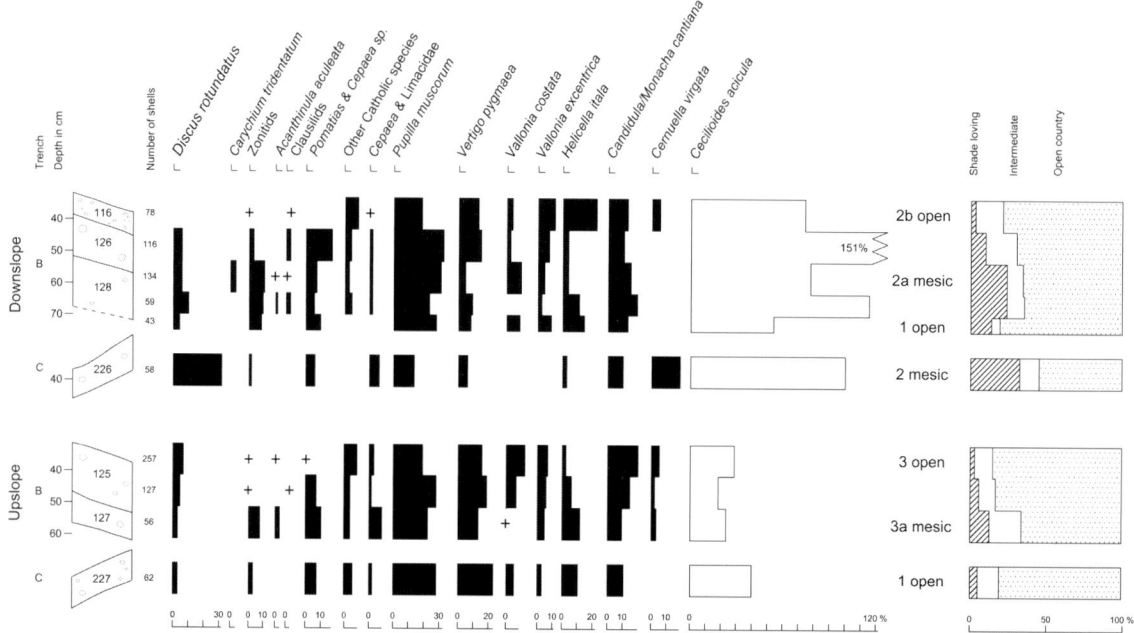

FIGURE 3.22. The land snail histograms from Trench B and C. Graphic: Justin Russell 2023

develop in ungrazed conditions where the grasses are allowed to grow longer, and other longer herbaceous, and occasional woody, species also occur. We are not suggesting woodland or even closed scrub, but examination of open steep ungrazed scarp downland slopes with occasional hawthorn bushes and other light scrub (in Sussex) and ungrazed winter grass heights reach about 350 mm reducing to 200–250 mm in the summer had similar open country assemblages. Even well-grazed grassland (30–50 mm) will recover to about 80–100 mm and under-grazed chalk grassland tends to be 250 mm and ungrazed can reach 500 mm (P. Hawes pers. comm. 5/4/23).

A fragment of unidentifiable charred material (Lisa Gray pers. comm.) was embedded in chalk from the kibbled chalk (6) of the chalk outline in Trench A (11051-20).

The two mollusc sequences from the right foot (Trench B) and right elbow (Trench C), comprising just 10 samples (Fig. 3.22; Table 3.6), provide a subtle, yet potentially highly significant local land-use history, especially when placed with the OSL dates. The Giant seems to have been cut into a pre-existing long-established open grazed downland. The scarp slope was lightly grazed, and although the evidence is scant and limited, it can be strongly inferred that there is no indication of, nor necessity, for major preparation of the hillside for the Giant in the ninth to tenth century AD. There are, however, tentative hints that a longer lightly grazed grassland sward (226 and 127) was replaced by a more tightly grazed short grassland (base of 128 and context 227). The giant seems then to have existed in this open grazed downland, allowing the figure to seen on the hillside.

The land-use succession seems clear if we accept the geoarchaeological arguments summarised here; that is that the upslope colluvium in general occurred before the downslope colluvium, although they must be in part coeval. How these relate precisely to the OSL dates is, however, more difficult, especially the upslope and downslope colluvium show subtly different land-use regimes, yet produced essentially the same OSL dates (AD 1250 and 1240 respectively). If we base the land-use history on the stratigraphic sequence as argued above then we can suggest that, not surprisingly, the Giant existed in an open short-turfed grassland allowing him to be shown off to the Dorset countryside. Moreover, there then followed a period during which the grassland grew to higher stand heights, possibly with the occasional bush (?hawthorn) and base loose soil. If left ungrazed and unmanaged it would rapidly grow over with rank grass and ultimately bushes and hawthorn (cf. Hawes 2015; Hawes *et al.* 2018; Ridding *et al.* 2020) as indeed was reported in 1868 when the Giant had 'a shabby appearance on account of the trenches being choked up with weeds and rubbish' (Keithley *et al.* 1999, 18). Similarly a lessening of rabbit grazing will result in the same effect, as occurred on the downland at the onset of myxomatosis (cf. Thomas 1960; 1963). If either occurred in the mid-thirteenth century as one of the OSL dates might suggest, did this correspond, for instance, with changes in land management and of the economic fortunes of the abbey; a reduction in sheep grazing or changes in farming or even increase in rabbits? These might account for long ungrazed downland on the scarp slope at least. The snail assemblages hint at a phase of longer more mesic, perhaps ungrazed downland rough pastors, possibly with the occasional bush or hawthorn. This may have led to bryophytes and lichens, and latterly light vegetation colonising the chalk figure. More significantly, was this enough to obscure the Giant from immediate view and from remembered consciousness? Could this, in part, explain the disappearance for the Giant until the mid-seventeenth century, perhaps when (more intensive) sheep grazing came back to the scarp slope? Certainly the land snails suggest a return to more open grazed downland, but we have no chronological indicator for these changing events.

Discussion and conclusions: putting the Giant in his place in the landscape

Michael J. Allen

The geoarchaeology and land-use history gleaned from the land snails now provide a dialogue to accompany the Giant (and the OSL results). Whatever the date of the first historical reference in the seventeenth century, we have an absence, as far as we know, of any written comment or sketch for over 700 years … and in view of his striking imposition on the Dorset landscape that seems inexplicable and improbable if he was there in full and plain sight. So the Giant must have faded away physically and disappeared, but also gone to sleep in terms of local memory. As there is no record, nor evidence, of him being physically removed from the hillside, nor covered up (turfed over), we can only take the lead from the land snail evidence and suggest that he became subsumed and lost on the downland hillside no longer maintained by grazing.

If grazing regimes are reduced and chalk grassland allowed to develop, changes will occur very quickly, over the decadal scale (Hawes 2015). Taller grasses will suppress many short-turved species and change the ecological composition of the grassland, rapidly leading to scrub (<40 years) with hawthorn and dogwood (Pywell *et al.* 1995; Hawes 2015; Hawes *et al.* 2018). Large areas of open bare chalk will colonise within 40 years or so (Hawes *et al.* 2020; 2024). Those of the Giant are only very narrow chalk rubble zones being <0.5 m wide, packed with loose chalk which promotes and encourages bryophytes, the first stage of grassland colonisation (Hawes *et al.* 2018; Ridding *et al.* 2024). Where there is a grazing regime vegetation growth is very controlled. On sites or slopes that are not grazed growth it is variable and generally if the hillside comprises a mix of moderate grass, tussocks, and brush, stand heights may typically be 0.3–1 m high. Regardless of the actual grazing regime, only regular maintenance of both the site (i.e., the local hillside) *and* the figure itself (weeding, edging, chalking) will ensure that the figure survives. Certainly the character of the hill figure will definitely change as a consequence of varying degrees of maintenance (as we have seen at the Giant), unless great and fastidious care is taken.

This leaves two unanswered questions. Why was the Giant allowed to be lost? And how (and when) was he refound?

In physical and ecological terms we can see that reduction of, or change in, grazing and browsing regimes, may allow long grasses to grow around and obscure the Giant. Although this could be seen as an incidental consequence of changing farming regimes or even economic fortunes, the fact that any figure requires regular maintenance (cleaning, edging, weeding, rechalking) suggests that a cessation of this care and management was a conscious decision. The narrow chalk lines of the figure would rapidly be colonised by bryophytes and invaded by the existing grasses (Ridding *et al.* 2024) and the sward height increased so that with one or two generations he would only be a shadow if anything on the hillside and after three generations (*c.* 75 years) would have been lost from memory of the living. Whether this was religious, cultural, political or financial we cannot yet tell, but it was via a deliberate or wilful act of neglect in the medieval period. Further historical and documentary research, possibly at the abbey and local sources, rather than the Society of Antiquaries of London may, in this case, find some clues. We revisit this point in Chapter 7.

And then, as if by magic, in the seventeenth century the Giant is back on the Dorset hillside, yet there is no record of his rediscovery nor even of his re-awakening by what must have been a large endeavour of clearing, cleaning, weeding, recutting and rechalking, unless this is what was happening in 1694. Why then? What prompted or enabled this? Such a large scale of works may have led to inadvertent subtle changes in the outline, and loss of some elements, and possibly additions of others. After all this, research has clearly proven the date of the outline of the Giant, but not of any of the internal features and embellishments.

Was he refound actively through community of societal memory and deliberately searched out and brought back to life? Or did a change and increase in grazing regimes and browsing (or even just the rabbit population, which is more likely in the seventeenth century) enable a shadowy image of the Giant to reappear, which re-awakened local interest in the figure? Whichever it was, there is still a surprise that there is no record of the inevitable clean-up and restoration unless it still resides hidden in local, county or Abbey records.

Primary fieldwork conclusions

Excavation has revealed the presence of not just one, but two Giants on Giant Hill, and the research and excavation results have shown that he was not always the white chalk-filled outline we now recognise. For centuries he existed as deep cut trenches on the hillside, as represented by Eric Ravilious's painting of 1939 (Fig. 7.4), and the 1920s postcard of the Giant (Fig. 7.3a).

The OSL dating has now enabled us to discount theories of the Cerne Abbas Giant as a prehistoric representation of a Celtic god or a Romano-British figure of Hercules. It is now clear that the origins of Giant 1 at least belong to the earlier medieval period, perhaps centring on the mid-tenth century and linked to the foundation of Cerne Abbey itself. Though the true reason for his creation may never be known, this new dating enables new theories to be considered.

CHAPTER FOUR

The Giant and the early medieval history of Cerne

Barbara Yorke

The knowledge that the Cerne giant was probably created during the early Middle Ages means that in order to try to understand the circumstances in which he was produced we need to review what is known of Cerne against the background of the history of the West Saxon kingdom. Unfortunately, there are relatively few documents that record what was happening in Cerne in the pre-Conquest period. Many of the records we do have come from later in the Middle Ages, and though one would like to think that they are based on genuine records or traditions of Anglo-Saxon origin, their interpretation is often not straightforward. Even what might seem to be fixed points in Cerne's history, such as the date of 987 often cited as the foundation date of the monastery of Cerne Abbas, are not quite what they seem and need some decoding. Hutchins recorded that he was able to see cut faintly into the chalk between the Giant's legs the date of 748 (though on the engraving the '4' looks more like a '9') (Hutchins 1774, II, 293). Whatever it may mean, the date cannot have been carved in the early Middle Ages as it is in Arabic numerals not used in England until the later Middle Ages, but it does, coincidentally or not, point us to what has turned out to be the correct era.

Cerne and the West Saxon kingdom

Cerne was incorporated into the kingdom of the West Saxons at some point in the seventh century AD. Its history before then, following the ending of Roman rule in Britain in the fifth century, is even more obscure than its early Anglo-Saxon history. However, Gildas, writing somewhere in the west of Britain *c.* 500, records an ordered Christian society where power had seemingly devolved to local leaders (Winterbottom 1978). This society is usually referred to as 'British' to distinguish it from areas to the east that had adopted Anglo-Saxon culture, and by which it was increasingly menaced. By 600 a dynasty of West Saxon kings had become a dominant power in central southern England. We cannot plot the exact circumstances in which areas of Dorset came under their control, but evidence from land-grants suggests that West Saxon kings were in control of western Dorset during the second half of the seventh century; the diocese of Sherborne was created in 705 (Yorke 1995, 60–79; Hinton 1998, 38–46). The new order was in part the result of West Saxon aggression. Victorious battles over the British are recorded in the annals of the ninth-century *Anglo-Saxon Chronicle* though none of these are in the vicinity of Cerne. Local leaders presumably surrendered on behalf of their communities. The results are recorded in the lawcode of King Ine (688–725) where a British land-owning and ranked society is recognised, but with half the wergild (amount payable in compensation if killed) of their Anglo-Saxon counterparts (Attenborough 1922, 36–61; Grimmer 2007). However, British who were in royal service were treated more favourably, and this may mark the way forward for the indigenous inhabitants of the west country. They are likely to have assimilated themselves into Anglo-Saxon society, taking service with West Saxon kings and other lords, and adopting Old English names and language. The results are implied by the next known West Saxon lawcode, that of King Alfred (871–99), where all his subjects are treated equally and assumed to be West Saxons (Attenborough 1922, 62–93). This type of blending of incomers and natives had already happened further east in Wessex in the fifth and sixth centuries. If the Cerne Giant has been correctly interpreted as Hercules, a Roman hero who seems to have had a particular resonance for Anglo-Saxons as well (see below), he could be seen as an appropriate symbol for the blended British/Anglo-Saxon society of Cerne and the south-west.

The West Saxon royal family owned much land in Dorset and had probably assumed ownership of key estates when areas were taken over. Cerne may have been one of these. The original estate seems to have stretched either side of the River Cerne or Char from Dogbury Camp and High Stoy to Charminster where it joined the River Frome (Barker 1988b). Such river valley estates are typical of western Wessex and may have existed for some centuries before the Anglo-Saxon period. Any part of the Cerne estate could be referred to as 'Cerne' or 'Cernel', and in Domesday Book there are a dozen or so places with these names, including what would be later distinguished as Cerne Abbas, with only Up Cerne and Charminster having identifying qualifiers (Thorn & Thorn 1983). The fact that anywhere on the wider estate might be referred to as 'Cerne' during the Anglo-Saxon period must be borne in mind when trying to disentangle which references relate specifically to the settlement later known as Cerne Abbas.

There are a number of pointers that suggest that the estate of Cerne was at one time in royal hands. A fourteenth-century list of estates of Sherborne Abbey records grants of land in Cernel from two West Saxon kings, Ecgbert (the grandfather of King Alfred; 802–39) and Æthelbert (brother of King Alfred; 860–65) (Finberg 1964, nos 563 and 570; O'Donovan 1988, 81–2). These two estates were Charminster and Up Cerne, which in Domesday Book are recorded as possessions of the bishop of Salisbury (successor to the bishop of Sherborne) (Thorn & Thorn 1983, 2.1 & 2.3). Remnants of royal ownership appear in other Dorset Domesday Book entries where a number of small estates are recorded as being held from the king by men in royal service (Thorn & Thorn 1983, 24.5; 26. 8–12). The largest Cerne Domesday Book estate is the 26 hides of Cerne Abbey (Thorn & Thorn 1983, 11), which, as considered more fully below, seems once to have been a possession of the family of Æthelmær. As Æthelmær's father Æthelweard explains in his Latin version of the *Anglo-Saxon Chronicle*, they were descendants of King Æthelred I (865–71), another brother of King Alfred (Campbell 1962, 2) and so they may have inherited a royal estate.

Alternatively, one might see that central Cerne Abbey estate as one that had been assigned for the support of ealdormen of Dorset and their successors, the ealdormen of the western shires, a position held by both Æthelweard and, more briefly, by Æthelmær (Cubitt 2009). One of the duties of the ealdormen was to supervise the raising of the shire levies (fyrd) to fight as part of the West Saxon royal army (Yorke 1995, 84–103). In the ninth and tenth centuries the importance of military service increased because of the threats posed by Viking attacks. The Dorset fyrd would not only have been in action when Viking armies were in Dorset, as in 876 when they occupied Wareham, but might be called out to serve further afield as in 893 when contingents from western Wessex were called out to support King Alfred's son-in-law and ally, Æthelred, lord of the Mercians (Whitelock 1961, 48–58). Even more damage seems to have been wreaked by Viking armies in the reign of Æthelred II Unræd (978–1016) with 998 being mentioned as a particularly disastrous year for Dorset. Æthelweard, the father of Æthelmær was one of the senior ealdormen of Æthelred's reign, for instance, taking a leading role in the peace negotiations with Olaf Trygvasson in 994, though his role in battles is not specifically referred to in the annals of the *Anglo-Saxon Chronicle* (Whitelock 1961, 83; Roach 2016, 176–8). A hero with a reputation as a doughty fighter such as Hercules would have been a resonant symbol for either of these periods of hard fighting. The suggestion that the Cerne Giant might have been connected with a mustering site for the Dorset fyrd is one that seems quite feasible (Morcom & Gittos 2024; and see Brookes Chapter 19).

The foundation of Cerne Abbey and the family of Æthelweard and Æthelmær

None of the events of West Saxon history so far discussed can be related specifically to Cerne, but it undoubtedly did have an abbey founded in the late tenth century as part of a wider movement of church reform and monastic foundation often referred to as the 'Tenth-Century Reformation'. Although led by churchmen, the movement received much support from the West Saxon kings and their leading nobles such as Ealdormen Æthelweard and Æthelmær (sometimes referred to with the by-name 'the Fat') (Yorke 1995, 210–25). What is less certain is exactly when the abbey of Cerne was founded and by whom. The key document is a record dated 987 that is often referred to as Cerne Abbey's foundation charter, but it is not actually a charter nor concerned with the foundation of the abbey as such (Sawyer no. 1217; Squibb 1988). The document is known only from its entry in a thirteenth-century *Chartae Antiquae* roll and, as it stands, does not seem to make complete sense (Kemble 1839–48, no. 656). Æthelmær addresses the king, bishops and other notables and records that he had given the place called Cernel to God in honour of the Virgin Mary, St Peter and St Benedict, and that after a few years he had arranged that the vills of Cernel and *Æscere* would go to the abbey after his death. Now, he is announcing additional gifts that will go to the abbey for his lifetime and after his death, and further arrangements of estates from three individuals, one of whom is a relative. On the face of it, it seems rather strange that Æthelmær should have founded a monastery at Cerne and kept the main revenues of the place for himself, but it is, of course, possible that the references are to different parts of the Cerne estate. Then, given that Æthelmær was still probably quite a young man in 987, how much earlier had he founded the monastery? There is in fact a record of a grant from King Edgar (959–75) to Abbot John of Cerne of land at Muston in Piddlehinton (adjoining the Cerne estate) in the account of an inquisition held at Dorchester in 1440 (Squibb 1988, 13). This brief, late record is far from an ideal source, but it does raise the possibility that the abbey of Cerne had been founded some years earlier by someone other than Æthelmær (who would probably have been too young to do so before the death of King Edgar in 975). William of Malmesbury, writing in the twelfth century believed that Æthelweard had been the founder of Cerne (Winterbottom 2007, 292–3), which is a possibility, but, as Æthelweard is thought to have died in 998, he was still alive in 987, and so it does not explain why it was his son apparently making new arrangements for the abbey then. Perhaps the founder was another relative and Æthelmær was his heir. Leland recorded from his tour on the eve of the dissolution of the monasteries in the sixteenth century that Cerne had been founded in the reign of King Edgar by Æthelmær '*comes* of Cornwall' (Hearne 1774, IV, 67). Was this an earlier Æthelmær after whom Æthelmær, the son of Æthelweard was named?

Æthelmær's 987 document recalls those records, often in the form of leases, whereby individuals sought to vary the grants made by a relative to a religious house by negotiating a return of the disputed land for life, or a fixed series of lives, with reversion to the religious community. The document also resembles the declaration made, in Old English, that is included at the end of the foundation charter of Æthelmær's new monastery at Eynsham in 1005 (Sawyer 1968, no. 911; Roach 2018, 48–50). That appears to be a first-person address to a royal assembly informing them of Æthelmær's intention to be leader of the Eynsham community during his lifetime, but granting the monks the right to choose their own leader after his death. One can suggest that there might have been a comparable declaration to an assembly behind Æthelmær's 987 document concerning Cerne. A putative explanation might be that a kinsman of Æthelmær had founded a monastery of Cerne in the reign of King Edgar, but, perhaps on the death of this kinsman, Æthelmær (and some others) sought to reclaim some of the donated lands. The document of 987 would then represent a compromise brokered between Æthelmær and the no doubt aggrieved abbot of Cerne which would see the lost lands eventually returned (as well as additional gifts to take effect straightaway). What was enrolled in the thirteenth century was not necessarily exactly what was in the original on which the enrolment was based especially if Æthelmær's declaration had been in Old English like that included with the Eynsham charter. The monks of Cerne in the thirteenth century might have thought that this Æthelmær was their original founder and assumed that that was what his declaration must have said.

This leaves open the question of who was living at the site of Cerne Abbas in 987. Did Æthelmær and the monks all reside there, or had Æthelmær banished the monks to a different part of the Cerne estate? Æthelmær had at least three children so even if he was often absent at the court, there is the issue of where his family was based. It is quite likely that before the monastery was founded the Cerne estate had a private chapel to meet the needs of its royal and ealdormanic lords. Leland believed that there had been a small foundation with three 'monks' before its refoundation in the tenth century (Hearne 1774, IV, 67). This would have been distinct from the minster church of Charminster, which had probably once been part of the Cerne estate but had been granted by King Ecgbert to the bishop of Sherborne in the early ninth century, and would have carried out parochial functions for lesser inhabitants (Hall 2000, 90–3). So it is an interesting question exactly where the homilist Ælfric resided when he was sent by his bishop Ælfheah of Winchester (and so after 984) to the *minster* at Cernel at the request of the thegn Æthelmær (Clemoes 1997, I, 3). It is usually assumed that Ælfric joined Cerne Abbey, but it is possible he was the private chaplain of his patrons Æthelweard and Æthelmær who are closely linked to a number of works that Ælfric produced during his time at Cerne. His *Lives of the Saints* was commissioned by Æthelweard and Æthelmær jointly, a translation of part of *Genesis* was made specifically for Æthelweard and the two series of *Catholic Homilies*, while dedicated by Archbishop Sigeric, were also intended for the use of Æthelweard (Jones 2004, 1–18; Cubitt 2009). When Æthelmær was obliged to move to Eynsham in 1005, he took Ælfric with him and installed him as abbot of the new monastery.

When Æthelmær made his declaration in 987 he was a close associate of King Æthelred II and held the position of *discthegn* at the royal court, which his father had once held at the court of King Edgar (the father of King Æthelred) (Roach 2016, 159). But in 1005 Æthelred had one of his periodic fallings-out with erstwhile supporters; Æthelmær and another royal relative were obliged to withdraw from the royal court to live in monasteries (Roach 2016, 200–16). This was the occasion when Æthelmær took-over the minster of Eynsham, having apparently acquired it by an exchange of land with his son-in-law (Jones 2004, 16–18; Hardy *et al.* 2003). Æthelmær retired there with Ælfric, and one wonders why he did not stay at Cerne. Perhaps the king had been unwilling for him to go to Cerne, which was an established powerbase of his family. Alternatively, could the abbot of Cerne have complained successfully of Æthelmær's apparently somewhat cavalier treatment of the monastic endowment? The retirement of laymen into monasteries could have penitential connotations and King Æthelred II was particularly attuned to the need for penitential behaviour to win God's support against the Vikings (Cubitt 2009; Roach 2016).

Æthelred's war with the Vikings continued not to go well, and in 1012 or 1013 Æthelmær resumed his public life and took up his father's former position of ealdorman of the western shires, that had apparently been vacant since his death in 998 (Roach 2016, 287). In 1013 Æthelmær headed the delegation of western thegns who submitted to King Swein of Denmark at Bath, which was his last known appearance, and it is assumed that he died soon after (Whitelock 1961, 92–3). His son Æthelweard was put to death on the orders of King Cnut (son of Swein) in 1017, and his son-in-law, another Æthelweard, who had succeeded to the position of ealdorman of the western shires, was outlawed in 1020 and not heard of again (Keynes 1994, 67–70). He had been implicated in a putative rebellion against Cnut while the king was absent from the country in 1019–20. However, the family did not lose all influence as it appears that Archbishop Æthelnoth of Canterbury whom Cnut appointed in 1020 was also a son of Æthelmær (Bolton 2017, 120–1). Æthelnoth seems to have enjoyed particularly close associations with Cnut, and perhaps it was due to his influence that Cerne flourished as a monastery and came to believe that Æthelnoth's father, Æthelmær was its founder. In Domesday Book the abbey of St Peter's Cerne owned a substantial 120 hides including several estates that had been acquired in addition to the majority of those that Æthelmær had promised them (Thorn & Thorn 1983, no. 11; Barker 1988b). It had also acquired a house-saint, Eadwold, the brother of the martyred East Anglian king Edmund (d. 879) who was said to have lived as a hermit, but Eadwold did not necessarily have any prior link with Cerne (Licence 2006; and see further below).

The family of Æthelmær were evidently central to the history of Cerne and the foundation of its abbey in the tenth and eleventh century, and were major political figures as well. Æthelmær and his father Æthelweard were exceptionally learned laymen, and Æthelweard who produced a Latin version of the *Anglo-Saxon Chronicle* is particularly impressive and appears as the embodiment of the learned layman whom King Alfred had apparently hoped his campaign of learning would produce (Ashley 2007; Gretsch 2013). One possible explanation for the production of the Cerne Giant would be that it was a tribute made by Æthelmær to his father for whom a rendering as Hercules might seem appropriate. Unfortunately, we do not know where Ealdorman Æthelweard was buried, but Cerne is a possible contender. At the beginning of his Latin translation Æthelweard styled himself as *Patricius Consul Fabius Quaestor* and he may have seen himself as embodying Roman and Saxon virtues as a political and military leader (Campbell 1962, 1). Ælfric did not approve of Hercules (Irvine 2003, 181–2), and one can assume the abbot of Cerne did not either, but perhaps Æthelmær wished to assert his own control of the family estate. Or should we see the Giant as an embodiment of Æthelmær himself? It is also possible that the Giant was created in one context and subsequently reinterpreted in others (as is explored more fully below).

Giant–heroes in the Anglo-Saxon world

The Cerne Giant as Hercules

There is a long tradition of referring to the Cerne Abbas chalk-figure as a giant and since at least the eighteenth century he has been interpreted as Hercules (Stukeley 1764; Piggott 1938). The clincher is the large, corrugated club that he wields, which is Hercules's iconographic identifier (Ogden 2021). Although the Cerne figure seems to reference the classical iconography of Hercules, the style with its round head, distinctive tear-drop depiction of features and nakedness apart from a possible belt is typically Anglo-Saxon (Morcom & Gittos 2024). The classical iconography of Hercules could have been known to Anglo-Saxons and other early medieval peoples from surviving statues, manuscripts, mosaics and other artworks including gemstones (Piggott 1938). A prestigious early medieval artwork on which the Labours of Hercules was depicted is the throne of the Frankish king Charles the Bald, that was probably made for his coronation in 875 in St Peter's, Rome (where it is still) (Nees 1991). A daughter of Charles the Bald was married to King Æthelwulf of Wessex, the father of King Alfred, and then to one of Alfred's brothers. In their recent study Helen Gittos and Tom Morcom have made a strong case for the Cerne Giant as an early medieval depiction of Hercules, and have also shown that any identification with a god called Helith is a chimera, a misunderstanding by a medieval author of the name 'Helias', the Latin version of the name of the prophet Elijah (Morcom & Gittos 2024).

Hercules is described as 'giant' an (*ent*) in the Old English version of Orosius' *Seven Books of History Against the Pagans*, written in the late ninth or early tenth century, probably in Wessex (Godden 2016, 192–3), and in Ælfric's *Lives of the Saints*, probably written in Cerne in the 990s (Skeat 1885, 35, lines 111–15). Such giants (*ental eotenas*) were among a considerable range of supernatural beings that are associated with places in the Anglo-Saxon countryside (Semple 2013, 179–80). Literate Anglo-Saxons might also recognise two additional categories of giant: *orcneas* from the classical world and *gigantes* from the Bible (Hall 2007, 73). Biblical condemnation of giants as descendants of Cain came to be applied to the two other categories by some early medieval Christian writers who saw them as monstrosities with no place in a properly ordered Christian society (Orchard 1995, 58–85). Such negative opinions were not shared by all, and stories about classical and native giant-heroes seem to have been enjoyed by many and they might even be presented as positive role-models for their courage and ingenuity (Irvine 2003). The carving of the Cerne Giant can be interpreted as belonging to that more positive tradition.

Hercules and the Old English giant–heroes

The conversion of the Anglo-Saxons to Christianity led to the issue of what to do about supernatural beings who might have roles that were significant to the ruling elite, for instance, in the generation of royal houses. Fathers of the Church such as Augustine and Isidore had set out the two possibilities; either such beings were really men and should be treated as such, or they had to be assigned to the monster/demon category of beings and seen as enemies of God and man (Johnson 1995). Even a hardliner such as Bede seems to have had no trouble in treating the chief Anglo-Saxon god Woden as a human ancestor 'from whose stock the royal families of many kingdoms claimed their descent' (Colgrave & Mynors 1969, 50–1). Such a clear-cut binary divide was not always easy to achieve in practice as, although gods and heroes could be presented as men in certain respects, supernatural elements and otherworldly powers were often an integral part of the stories that circulated about them. Thus, the great Anglo-Saxon hero Beowulf appears in the first part of the epic poem as behaving like a regular human hero when he arrives at the court of King Hrothgar, but when he battles with the two *eotenas*, Grendel and his mother, he shows comparable supernatural strength and abilities, for instance, being able to stay submerged and to fight underwater for several hours (Orchard 1995, 27–85). It would seem that at that point in the poem Beowulf was also, like Grendel and his mother, superhuman in size as he could wield their giant-size weapons. In other words, Beowulf was not really a heroic ordinary human, but what could be termed a giant–hero, though one who could be seen as an admirable figure and be distinguished from the evil giants, the monstrous Grendel and his mother, who menaced mankind.

The origins of giant–heroes such as Hercules can be found in the classical period when Roman temples might display the bones of extinct megafauna as those of deified heroes. Such practices arguably influenced the Germanic world as well (Orchard 1995, 105; Yorke 2015). The Anglo-Saxon compilation the *Liber monstrorum de diversis generibus* ('Book of monsters of various kinds'), which may have been produced in western Wessex in the late seventh or early eighth century, is mostly concerned with classical monsters and giants (Lapidge 1982; Orchard 1995, 86–115). It is a tribute to the depth of classical knowledge available in Wessex at the time, and also to a fascination with classical legends, which in theory are being condemned (Herren 1998). The only Germanic entry in the *Liber monstrorum* is for King Hygelac of the Swedish Geats who is Beowulf's apparently normal-sized uncle in the Old English poem. In death Hygelac is presented, like his nephew in his battle with Grendel and his mother, as being giant-sized:

> And there are monsters of an amazing size, like King Hygelac, who ruled the Geats and was killed by the Franks, whom no horse could carry from the age of twelve. His bones are preserved on an island in the river Rhine where it breaks into the ocean, and they are shown as a wonder to travellers from afar (Orchard 1995, 105).

This is all the more remarkable as Hygelac may have been an actual ruler in sixth-century Sweden and is mentioned by Gregory of Tours (Chambers 1967, 381–87; Leneghan 2020, 121–6).

Hercules appears in all three sections of the *Liber monstrorum*, first among the monstrous men, then as the slayer of monstrous beasts and serpents. However, in spite of Hercules' apparent monstrosity, and condemnation, the author declares:

> Who does not admire the courage and weaponry of Hercules, who, at the western entrance to the Mediterranean, erected pillars of an amazing size as a spectacle for the human race, and who constructed trophies of his wars in the East by the Indian Ocean, as a memorial for posterity, and afterwards travelled in battles through almost the entire world, and spattered the earth with so much blood, and at the point of death wrapped himself in flames to be consumed? (Orchard 1995, 264–7).

The *Liber monstrorum* not only shows a knowledge of Hercules, the favoured candidate for the identity of the Cerne Giant, in early medieval Wessex, but that he could be presented as an admirable figure. An even more enthusiastic notice is to be found a century or so later in the Old English translation of the *Consolation of Philosophy* by Boethius, which is one of the works associated with King Alfred's campaign to translate key works from Latin to Old English (Irvine 2003). Although doubt has been expressed about the degree of involvement of King Alfred, and whether the work actually dates to the early tenth century (Godden 2007; Godden & Irvine 2009, I, 135–46), that does not affect its use as evidence for attitudes in late ninth- and early tenth-century Wessex. For what it is worth, Ealdorman Æthelweard believed the Old English Boethius translation was rightly associated with King Alfred and that 'for any who might hear it read, the tearful passion of the book of Boethius would be in a measure brought to life' (Campbell 1962, 51). Boethius praised the fame and courage of Hercules, and this is enthusiastically developed in the Old English translation. Hercules' killing of the homicidal King Bosiris is judged as being done 'very rightly in God's judgement (*dom*)' (Irvine 2003, 172–4; Godden & Irvine 2009, I, 274–5), while admiration for Hercules' *cræft* in devising a way to kill the hydra by fire leads to the aphorism: 'if someone embarks on a [task] then he leaves it with difficulty; he never comes to a clear conclusion unless he has an understanding as keen as fire' (Irvine 2003, 172–4; Godden & Irvine 2009, 360–1). *Cræft* is a skill much admired in the Alfredian canon (Pratt 2007, 287–95). As Susan Irvine has commented, '[Alfred] is presenting Hercules as a kind of prototype of himself … the concept of a task which requires a combination of *cræft* and the appropriate tools for its successful execution is exactly how Alfred envisages his own role as king elsewhere' (Irvine 2003, 175). The destruction of the hydra by fire is not found in any other known version of the Hercules legend and so could perhaps be an indication that popular stories about Hercules were circulating in Wessex where, as discussed above, Hercules would have been a particularly appropriate cultural figure for the intermingled British and Anglo-Saxon population of western Wessex. The Old English Boethius also contains a passage on how the stories of heroes of the past who strove for honour and reputation in this world can provide an inspiring example for those who come after (Godden & Irvine 2009, I, 372).

The positive praise for Hercules in the Old English Boethius is perhaps all the more surprising because he was often viewed negatively in contemporary Francia to which the Alfredian court circle looked for inspiration and from which two of Alfred's advisers came (Nees 1991; Morcom & Gittos 2024). In this the Frankish commentators followed those patristic authors who condemned all classical mythology. Ælfric the homilist echoes such opinions when he wrote of 'hateful Hercules' 'the immense giant who killed all his neighbours' (Skeat 1885, 35, lines 111–15; Orchard 1995, 124–5). As Ælfric was living in Cerne at the time he wrote one has to wonder whether his antipathy was sharpened by the image of the Cerne Giant on his doorstep. Whoever had created the Cerne Giant had presumably done so in the more positive spirit of legends in which Hercules and other giant–heroes were considered admirable and their bravery and martial prowess worthy of emulation.

Giants in the Anglo-Saxon countryside

Hercules does not appear in Old English place-names, but *enta* as a general category do, though they are not that common. They are usually associated with manmade structures such as dykes or burial mounds (barrows) and the latter are referred to in *Beowulf* as well. In other poems *enta* are associated with Roman-period structures, such as roads and forts (Semple 2013, 179–80). The relation of the Cerne Giant to the Trendle earthwork that lies just above its head is therefore of potential interest, as is the existence of other earthworks in their vicinity (Piggott 1938, 327–8; Putnam 1999). Also associated with burial mounds and earthworks were various gods and semi-mythical kings, many of whom may have moved into the giant–hero category in the early Middle Ages. Woden, the former god who became the founder of Anglo-Saxon dynasties has given his name to Wansdyke in northern Wessex and several associated places especially in Wiltshire as well as further afield. One of Woden's by-names 'Grim' is even more widespread and frequently associated with large earthwork structures such as dykes (Semple 2013, 172–6).

One place that may be particularly helpful for understanding the Cerne Giant in its landscape setting is the prehistoric long barrow in Oxfordshire (formerly Berkshire) known as Wayland's Smithy; the name is first recorded in the boundary clause of a charter dated to 955 (Sawyer 1968, no. 564). Wayland, or Weland, belongs to the same quasi-mythological/human world as Beowulf. The armour that protected Beowulf from Grendel and his mother had been made by Weland's father, and Weland was also a smith who was captured and lamed by a king who wanted to make him work for him (Orchard 1997, 389–91; Hall 2007, 39–47). The story of how Weland took his revenge and flew away by magical means was apparently well known to both English and Scandinavians in the early Middle Ages (Davidson 1958), and was represented on the Franks

Casket that was probably made in Anglo-Saxon England in the late seventh or eighth century (Webster 2012a). Weland is the only mythological figure besides Hercules to be mentioned and praised in the Old English translation of Boethius where he has been substituted for the Roman Fabricius (from *faber* 'a smith') in a section on the emptiness of fame and death as a great leveller:

> Where now are the bones of the famous and wise goldsmith Weland? (I said wise because the craftsman (*cræftegan*) can never lose his skill (*cræft*) nor can it be easily taken from him any more than the sun can be moved from its place.) Where are now Weland's bones, or who knows now where they were? (Godden & Irvine 2009, I, 283 and II, 30).

Weland is clearly being praised in this passage; he exercised the same *cræft* for which Hercules was commended and has the quality of wisdom that is much desired in the translations associated with King Alfred (Yorke 2017, 50–5).

Of potential interest for understanding the Cerne Giant is the fact that Wayland's Smithy lies only about a mile along the Ridgeway from the Uffington White Horse, the celebrated prehistoric chalk-carving (Miles 1999). Various legends exist for these two sites, and in one the White Horse is said to leave its hill every hundred years to be reshod by Weland in his smithy, possibly implying a tradition of a connection between them (Grinsell 1939, 16–20). Both sites are also in the vicinity of Ashdown where in 871 a celebrated battle was fought by King Æthelred I and his brother Alfred against the Vikings (Whitelock 1961, 46–7). The Uffington White Horse is the only known precedent in central Wessex for a giant-sized carving in chalk before the creation of the Cerne Giant, and Tom Morcom and Helen Gittos have suggested that it could have been the inspiration for the carving of the Cerne figure (Morcom & Gittos 2024). There are similarities in the positions of the two sites on prominent positions on hillsides close to significant routeways, and they also draw attention to the existence of prominent stones that had the capacity to be used as blowing-stones in proximity to the two sites. Both chalk-carvings may have marked the way to places of assembly and mustering sites for the shire army, as is discussed more fully by Stuart Brookes (Chapter 19).

Another place of potential relevance for understanding the role of the Cerne Giant is the site of *Cwichelmeshlæw* also on the Berkshire/Oxfordshire Ridgeway and known today as Scutchamer Knob. The *hlæw* is a large prehistoric barrow, which it may be presumed was seen as the burial place of Cwichelm, one of the early West Saxon kings who is said to have died in 593 (Whitelock 1961, 14; Williams 2006, 208–11; Baker 2019, 37–9). A number of the sixth-century West Saxon kings are claimed in similar place-names to have been buried at prehistoric sites in Wessex, but it is unlikely that they all were in reality (Semple 2013, 1–2, 159–92). Although the activities of these early kings were recorded in the annals of the *Anglo-Saxon Chronicle*, some of them may have come to straddle the real and mythical worlds in a similar way to King Hygelac or the supposed founders of the Kentish royal house Hengist and Horsa. We only have tantalising glimpses of the legends that may have circulated about them in the early medieval period. There is a hint of one concerning Cwichelm in the 1006 annal of the *Anglo-Saxon Chronicle* when a Viking raiding army:

> Turned along Ashdown to *Cwichelmeshlæw*, and waited there for what had been proudly threatened, for it had often been said that if they went to *Cwichelmeshlæw*, they would never get to the sea (Whitelock 1961, 88).

Perhaps the implication was that Cwichelm would rise up against any hostile army foolish enough to come to his burial place (though, if so, he failed to do his stuff for the West Saxons in 1006). *Cwichelmeshlæw* is twice recorded as a meeting place of the late Saxon shire assembly and, as has been suggested for the vicinities of Cerne and the White Horse of Uffington, may have served as a mustering place for the shire armies (Williams 2006, 208–11; Brookes Chapter 19). The 1006 annal gives a rare indication of the type of boast made as armies were rallied under the inspiring protection of their presiding heroes.

The Giant–Hero reinterpreted: the case for St Eadwold

There is a long tradition of reading the iconography of the Cerne Giant as Hercules and this identification fits well with the positive attitude to Hercules in some West Saxon sources between the seventh and the tenth centuries when the Giant was probably carved. However, that need not preclude the Giant being given other identities as well. It has often been suggested that Old English heroic figures may have been equated with prominent laymen of the period in which they were recorded thus giving an added contemporary relevance to the material. The poem *Beowulf* in particular has received this type of interpretation aided by the fact that some of the characters such as Offa and Wiglaf have the same names as known Anglo-Saxon rulers, in this instance of Mercia in the eighth and ninth centuries (North 2006). Beowulf in the form Beow appears alongside other kinsmen from the poem in some versions of the West Saxon royal pedigree, and this has led, for instance, to theories that the poem was commissioned in the context of King Alfred extending his overlordship to Viking-occupied areas (Chase 1981; Neidorf 2014). Other poems too have been seen as coded references to matters of contemporary concern, and in the eleventh century in particular there was a vogue in England and Normandy for referencing classical figures in this sort of way (Van Houts 1992; Licence 2020, 282–97). It is therefore quite plausible that the Cerne Giant could have been iconographically depicting Hercules, but intended, or subsequently interpreted, as prominent West Saxon leaders such as King Æthelred I or his brother Alfred, or the ealdormen Æthelweard and Æthelmær, descendants of Æthelred I and patrons of Cerne. Æthelweard claimed a 'Roman' identity when he styled himself 'patricius consul Fabius quaestor Ethelwerdus' (Campbell 1962, 1). One might also reference the precedent of the Roman emperors, especially Commodus, identifying themselves with Hercules (Piggott 1938, 325–6).

The Giant could also have been radically reinterpreted in later centuries, as arguably happened in the eleventh century when he may have been repurposed as St Eadwold (Morcom & Gittos 2024). Eadwold was Cerne Abbey's medieval saint, and was said to have been an East Anglian prince who retired to live as a hermit after his brother King Edmund was martyred by pagan Vikings in 869. Eadwold may not in fact have had any connection with Cerne until his remains were transferred to the abbey in the 1020s, as Tom Licence who has made a detailed study of the available texts suggests (Licence 2006). This was the time when there was a revival of the cult of Eadwold's supposed brother at Bury St Edmunds. The translation of saintly bodies from sites that had become insignificant to newly founded monasteries such as Cerne that were in need of a saintly patron was rife in the late tenth and eleventh centuries. Sometime between *c.* 1060 and *c.* 1080 the abbey of Cerne seems to have commissioned a Life of their saint from the professional hagiographer Goscelin, though it survives only in part as a set of lessons in a later compendium (Licence 2006). Specific references to any earlier connection with Cerne are conspicuous by their absence in what survives of Goscelin's text, which contains only hagiographical commonplaces. However, Tom Morcom and Helen Gittos have suggested that Goscelin visited Cerne while he was composing Eadwold's Life (as was his usual practice) and used the Giant Hill as the basis of his description of Eadwold arriving at a steep hill, where his staff miraculously blossomed into a tree, while a silver spring appeared at its foot where he could found his hermitage, thus reimagining the Giant as a representation of Eadwold himself and his club as Eadwold's blossoming staff (Morcom & Gittos 2024).

It's an attractive suggestion and would have been a convenient way for the abbey to deal with the large chalk figure, which even if partially grassed-over is unlikely to have completely disappeared from view. An apparent recutting of the figure in the thirteenth century could then be interpreted as connected with promotion of the cult of the saint, perhaps at a time when the abbey was trying to retrieve its lost estate of *Æscere* promised to them originally by Æthelmær (Squibb 1988, 13). Although Goscelin seems to have stopped short of explicitly claiming that Eadwold had lived as a hermit at Cerne, in the following century William of Malmesbury in his *Gesta Pontificum* (*c.* 1125), after some prevarication, implies that the monastery of Cerne had been founded in the place where Eadwold had had his hermitage (Winterbottom 2007, 292–3). Unfortunately, William does not refer to the Giant figure and we shall never know for certain whether an identification as Eadwold was generally accepted. Giant saintly figures were not unknown in the Middle Ages and St Christopher is a notable example (Orchard 1995, 12–18). Brave, resourceful giants were also to be found in the literature of the time and remained popular with all classes of society (Boyer 2016), and are still intriguing people today.

Conclusion

Unfortunately, no clear statement exists of exactly when or why the Giant was created in Cerne in, as it now appears, the early Middle Ages so that the best that can be done is to review likely scenarios against what can be recovered of the history of Anglo-Saxon Cerne and of early medieval attitudes to giants. The Cerne Giant is most likely to belong to the positive appreciation of giant–heroes to be found in certain elite circles in Anglo-Saxon England in the eighth to tenth centuries. Such figures were valued as exemplars of brave and successful warriors in a period when elites were much involved in warfare, and Hercules, whom the Cerne Giant is likely to represent, is the main such classical figure referred to approvingly in certain texts (even if condemned in others). The ealdormen Æthelweard and Æthelmær of Cerne were exceptionally learned laymen and it is tempting to associate them with the creation, or possibly a reinterpretation, of the Giant. St Eadwold of Cerne may have been a further stage in his reimagining. Will we ever know more? The rediscovery of additional early medieval written records for Cerne is probably unlikely, but further scientific studies could refine knowledge of the date of the original carving and of significant periods of recutting. The excavations of the abbey site may reveal more of its origins and possibly, as at Eynsham, how the abbey related to earlier use of the site (Hardy *et al.* 2003). Further understanding of earthworks on and around Giant Hill might also throw light on the context in which the Giant was constructed. There may be further surprises to come in unravelling the fascinating history of the Cerne Giant, and room for many more hypotheses that try to explain his existence.

CHAPTER FIVE

Hide-and-Seek on a Dorset hillside

Brian Edwards

If an early medieval date range has proved surprising, the stratigraphy promotes a potentially cogent biography for this hill figure: distinct waves of restoration followed a revival that punctuated long periods of inactivity. In other words, a not so attentively managed early medieval hill figure at Cerne Abbas has been displaced by a very well-cared for Giant. Reviewing reports about the Giant's condition may then prove informative.

Hide-and-seek in the age of photography

Now you see it – now you don't. Beckoning those travelling through the landscape, a form of hide-and-seek is inherent in the creation of hill figures. This game is compounded as the seasons lay siege, so nature's attempts to reclaim turf monuments demands an annual strategy.

As landowners managing the monument, the National Trust's plan for the Cerne Giant in October 2023 is to update the website to ensure visitors' expectations match what they will encounter.[1] This statement was in response to criticism about the poor visibility of the Cerne Giant, which had been the catalyst for the condemnation of heritage management in online exchanges (*Daily Telegraph* 2023; Lumb 2023). Endorsing this outlook, the condition and visibility of the Giant has proved a regular theme among 426 reviews posted on Tripadvisor between 2006 and 2023. One was posted only 14 months after the major restoration of 2019:

> Instead of a crisp white image that you see in the media, it was barely visible, greyed out and not in keeping with its, and England's, heritage (Tripadvisor 2020).

It was much the same in 1993 when Ivan Smith, the Land Agent for the National Trust, found that 'virtually every photograph' published was an aerial image. In Smith's opinion these 'inevitably lead many visitors to feel that the view of the figure will be equally as good on the ground, when sadly it never will be' (NT CGa, letter I. Smith to P. Tompson 19/10/93). Returning to the present time, commonplace drone photography compounds the visualisation issues, with the fresh chalk and manicured edges of a newly restored Giant in 2019 proving particularly popular in the media (see Fry, Chapter 13). Indeed, whilst social media pondered news of the Cerne Giant's early medieval beginnings, on 12 May 2021 (Morcom & Gittos 2021), many media headlines were accompanied by aerial photographs of a restored Giant (Milmo 2021). Herein lies a further difficulty with interpretation, for beyond living memory the Giant's deep and wide lines were never filled to the brim with brilliant white chalk. An artist's impression of the original is then clearly needed, and how the Giant appeared at various stages of history might prove revolutionary.

Despite aerial photography fuelling visibility complaints, its popularity is evident and concerns about discernability became a regular theme long before aerial photography found favour with the media. Complaints of this nature are manifold in National Trust files, having been passed on by the parish council, the local post office, cafés and shops (NT CGa, I. Smith to K.S. Griffin 30/11/21). Concerns about the Giant's visibility will be found from every decade in National Trust files and widely reported even before that (e.g., *Dorset County Chronicle* 1864, 1 September 1864, 106, col 2; *Southern Times* 1864; Fitter 1947; Anon 1956; Rickman 1967, etc.).

Hide-and-seek 1920–45

The Trust first cleaned the Giant in 1920, the year it was gifted. Three years on the *Observer* offered advice on 'The Cerne Giant: Ancient Turf-Cuttings and How to Preserve Them', detailing 'the ever recurring danger' and explaining that these 'turf monuments' 'get obscured by the process of time and the growth of herbage' (*The Observer* 1923). The Trust's second cleaning operation followed in 1924, costing £5.00, the outlay coinciding with the Giant being listed as a Scheduled Monument. 'The grass has been carefully cut by the sides of the trenches and the trenches themselves have been weeded,' reported *Country Life* (1924). Their advice on future management, was that there should be 'no general cutting of the grass; only that growing by the side of ditches should be clipped and cut back', and there should be 'no disturbance of the surface, only periodic weeding' (*op. cit.* 1924). The further suggestion that a 'small fund' would 'secure scouring at regular intervals', gained a response from Sir Henry Hoare, who gifted £150 from The Captain Henry Cole Arthur Hoare's, 1st Dorset Yeomanry Memorial Fund (BNL). In contrast to these plans for careful management of any further restorations, it seems, were on hold to a background of accusations of obscenity combined with attempts to have the Giant internally censored. With a potential threat of vandalism mounting in 1932, a village rota maintained a watch on the Giant overnight (Wilcox 1988). The official recognition and care of the Cerne Giant in the early 1920s had created an intense public focus, which continues to reactively impact on the management of the monument to the present day.

In June 1940 it was decided to camouflage hill figures to prevent enemy aircraft from using them as a navigational aid (Robertson & Schofield 2000). The Giant's trenches were filled with vegetation by the Home Guard (Dorset History Centre RON/2/2.149) and to prevent injuries to and from horses a fence was introduced or moved in 1942 (NT CGa). When the brushwood was removed in 1945, the Giant's outline was overgrown and sections of the trench sides had been damaged (*The Tatler* 1945). Fencing the monument against cattle but removing/moving a lower strand to allow sheep access, appeared to be the answer (NT CGa).

No sooner had the war ended, than the parish council had brought the poor visibility of the Giant to the attention of the Trust. As we have seen (Chapter 1) the charity persuaded the experienced archaeologist Stuart Piggott to oversee restoration work, and on 16 August 1945 Piggott listed the 'difficult' areas he had to tackle as the Giant's 'face, hand holding the club, and genitals' (NT CGa; Chapter 1). We might then note that a great deal of turf had to be imported for repairs, and in 1946, the landowner eventually instructed the wardens not to take turf from outside the enclosure (NT CGa). Most memorable of all, Piggott informed the National Trust's agent, Eardley Knollys, on 25 August 1945, 'The men have made a splendid repair job of the Giant's testicles with pegged down turf.' The following June, less than a year after Piggott's extensive restoration, Wells Natural History and Archaeological Society expressed disappointment that the Giant was overgrown and only parts could be seen.

Early twentieth-century hide-and-seek

On Monday 21 August 1905, Dorchester born surgeon Sir Frederick Treves and his wife Ann travelled to Cerne Abbas after a stay on the Dorset coast at Swanage. The following day, the Giant's annual tendency to vanish beneath overgrowth was dramatised into news (*Daily Mail* 1905; *Western Chronicle* 1905a; 1905b). Treves had been catapulted into public consciousness in 1902, after performing an emergency appendectomy that saved the life of King Edward VII and the coronation just days later. Thereafter, the national press frequently reported the surgeon's whereabouts and activities. With the celebrity surgeon's visit to Cerne Abbas in 1905, whilst researching and writing a book about Dorset, the disappearing Giant story was serialised soon enough. 'Cerne Giant Threatened' was among the headlines, where others preferred 'Vanishing Giant', or 'Renovation Wanted' with an estimate of £12 for a restoration featuring prominently (*Barbados Agricultural Reporter* 1905; *The Belfast News-Letter* 1905, 25 August; *Western Chronicle* 1905b). *The Bystander* meanwhile reported the Giant was 'barely to be seen'. The accompanying photograph, by H.G. Archer, confirmed the Giant was 'hard to trace, although still recognisable' as was his belly button (*The Bystander* 1905, 46). The Giant's penis would appear to be overgrown but quite separate from his navel, which was detectable in early Edwardian photographs and garishly enhanced postcards (Archer 1903; Fig. 5.1).

Published in 1906, Treves' enormously popular *Highways and Byways of Dorset* reported from his visit that the Giant was 'grown over with grass and is nearly invisible to the eye' (Treves 1906). Although accurate at the time, media interest in the vanishing Giant had stimulated a response within weeks. In October 1905 a party of 'ladies and gentlemen, armed with spades, hoes, etc.'

FIGURE 5.1. Detail from an early Edwardian postcard (*c.* 1902–7), which garishly highlighted the Cerne Giant's penis and navel; just the lower half of the navel is discernible

took 'Cerne by storm', and with the long grass removed the Giant was 'now more easily discerned' (*Western Gazette* 1905, 12). This guerilla gardening exercise, performed by outsiders it would seem, potentially stimulated a Giant clean-up by local 'ladies' in 1907 (Vale & Vale 2000, 13). A formalised restoration of the hill figure followed in 1908, for which a subscription was raised, and a committee appointed, a response perhaps to Treves blaming 'sleepy and indifferent Cerne' for the length of the grass.

Prior to National Trust ownership, those openly commenting would never dare blame the landowning lord of the manor when nature overtook the Giant, so the village bore the brunt of criticism. Extending this preconception into the interwar period, another author stated the 'Dorset rustic cares very little about the matter' (Thurston-Hopkins 1922). This prejudice hardly fits with Victorian indications that the Giant was ousting the traditional name of Trendle Hill as its location. Some 20 years before references to 'Giant's Hill' were being characteristically popularised by Murray's *Handbook for Travellers*, it featured in connection with cricket in 1835 (*Dorset County Chronicle* 1835; Murray 1856, 105). Local sports clubs and teams playing matches on Giant's Hill, started competing under the adopted name of the 'Cerne Giants', with this name becoming particularly noticeable in media reports between the 1880s and 1920s. It was soon more widely propagated by footballers, cricketers, and tug-of-war teams, that those born in Cerne were identified as 'giants'. In turn, those estranged through work or marriage may on visiting be dubbed 'returning giants' (*Bridport News* 1899, 5; *Dorset County Chronicle* 1920, 2; *Southern Times and Dorset County Herald* 1900, 7). The evidence of the nineteenth century and since suggests residents have maintained a fond connection with the Giant but, of course, at no point did the people of Cerne Abbas own the site.

Hide-and-seek in the nineteenth century

Following an Ancient Order of Forester's fete held at Cerne Abbas in July 1867, which saw the entire village splendidly decorated for the occasion, it was remarked in sharp contrast of the Giant that 'his mightiness didn't appear to us in good form, and we commend him to the attention of the good folks of Cerne against another festival' (*Dorset County Express and Agricultural Gazette* 1867, 2) The following year it was announced that the Giant was to be restored. Only the landowner could be held responsible for the condition and upkeep of the Giant of course, and yet in a masterclass in propaganda the lord of the manor was spun into a hero giving 'orders'. It was detailed that the Giant had 'for many years presented a shabby appearance on account of the trenches being choked up with weeds and rubbish, and the outlines being otherwise defaced'. The orders were 'to have his "mightiness" cleaned and restored, as near as possible, to his original state and condition' (*Bridport News* 1868a). The following week the *Bridport News* announced the serialising of a 'History of the Cerne Giant' (*Bridport News* 1868b), regional interest had been aroused and when that happens more visitors arrive.

In July 1886, 62-year-old Charles Way, a master blacksmith of Long Street, Cerne Abbas, fashioned and installed the Giant's railings soon dubbed the 'coffin' enclosure (*Bridport News* 1886). Within this coffin, the Giant's location was readily picked out, so the condition of the outline and annual overgrowth was more noticeable than it had ever been. In 1897 it was recalled the Giant was restored 10 years earlier by Jonathan Hardy, a shoemaker living in Long Street that served as church sexton (March 1899). The year was misremembered, for in June 1888 General Pitt Rivers refused the Dorset Field Club access to the Giant on account of ongoing work and the completion of the restoration was announced in October. The Giant was therefore restored after, and not for, Queen Victoria's Golden Jubilee (*Western Chronicle* 1888; *Bridport News* 1888). The year was not the only thing the sexton misremembered; Hardy was put in charge of the restoration in 1908 when the Giant's penis was extended to incorporate the previously independent navel (*Southern Times and Dorset County Herald* 1908).

Some antiquaries play hide-and-seek

In addition to the earliest known reference to the Giant being a 'repair' in 1694, which highlights a late seventeenth-century want to make the Giant whole, in 1742 the Oxford archivist and antiquary Francis Wise usefully highlighted that the Giant was 'not to be seen at any great distance' (Wise 1742, 48). Visiting the Giant in October 1754, Richard Pococke provides a description of the Giant we would recognise today, before adding that 'The lord of the manor gives something once in seven or eight years to have the lines clear'd and kept open' (Cartwright 1888). In a letter sent in 1763 to the antiquary William Stukeley, the Rev John Hutchins, the rector of Wareham, affords us the further details that the Giant's 'furrows are about a foot or more deep' by two wide, and that 'The people sometimes cleanse the furrows and fill them up with fresh chalk' (letter from Hutchins to Stukeley 1763; Lukis 1883). Even were this exercise entirely financed by the lord of the manor it would seem unlikely furrows that deep were being filled up with fresh chalk in this period.

Telescopes were in use but not mentioned when an anonymous letter was published in the *Royal Magazine* of September 1763, describing the Giant when 'viewed from the opposite Hill'. The Giant's periodic revival potentially explains both a relationship with this viewing spot and the disappearance of three figures between the hill figure's legs, first mentioned by this anonymous writer. Despite being illegible, these figures were interpreted as representing a date (Morgan Evans 1999, 114). We may conclude that these figures, which were shallow and not of the same construction as the outline (Allen, Chapters 3 & 7), were not maintained, having perhaps faded before the remainder of the outline had become overgrown. This conclusion is further suggested by a later published account, reporting that during an earlier period the three figures were accompanied by a line of three letters or characters cut into the turf below the numbers (Hutchins 1860–1, 871–3). The overall impression from these very few episodic details between 1694 and 1763, is that if the eighteenth-century outline of the Giant was revived by the local community, it was when from a distance the figure had all but disappeared. This would undoubtedly have an impact on the impression formed on rare visits by those unable to climb the steep hill, with the timing relative to both the season and any previous restorations affecting what could and could not be readily recognised.

The primitive and the industrial

When the earliest known drawings started circulating during the 1760s, the Giant gained a neoteric emphasis that reshaped the vision of a centuries old hill figure. The earliest known images could hardly be more dissimilar in outline, for they project two very different versions of the same Giant (Morgan Evans 1999, 108). One descends from a tradition of primitive homemade sketches and copies of field drawings: for the price of postage these portable drawings were requested and swapped for various reasons, including the difficulties of travel, the survey of known monuments and to illustrate newly discovered sites. Of much wider circulation and impact than those primitive sketches, the alternative published versions could also be copied and shared of course, but they could only be expensively purchased, they were not particularly portable and therefore largely unsuitable for fieldwork.

The earliest sketch thus far mentioned, was created in the autumn of 1763 for John Hutchins, who in his words was no draughtsman. Long after visiting Cerne Abbas in the 1730s, Hutchins had been encouraged to investigate Dorset's past. With his interest further stimulated by the widespread community of antiquaries, some of whom were clergymen, it would seem probable that in writing to a friend, Hutchins had approached a local clergyman to provide a sketch of the Giant. The timing of the request suggests this sketch was created in a relatively short window in autumn. Despite this being a time of year when the Giant's lines were probably at their most difficult to discern, Hutchins received the sketch on 28 November 1763. The following day, when despatching this sketch to Stukeley, Hutchins appears to indicate he was sending the original (Lukis 1883). This in turn, alongside subsequent events, suggests Hutchins had retained at least one copy. Despite what Hutchins had said about being no draughtsman, that copy was created overnight.

Within a few months Hutchins sent an ink sketch of the Giant to Charles Lyttleton, the president of the Society of Antiquaries who was serving at that time as the bishop of Carlisle (Fig. 5.2a). After Lyttleton received this letter, dated 12 January 1764, the sketch was bound into the *Minute Book* of the Society of Antiquaries alongside the entry for 9 February 1764 (Morgan Evans 1999, 111). Examination of that sketch reveals an underlying pencil drawing, which is perhaps most noticeable in a gap in the ink sketch, between the upper part of the club and the fingers of the Giant's right hand. Pencil can also be readily detected around the eyebrows, whilst absent from lines interestingly added below the eyes (Society of Antiquaries 1762–1765a).

Familiar with the appearance of the Giant's outline but not perhaps recalling it as he might until receiving the sketch he had commissioned, Hutchins added no caveats in respect of accuracy of the sketches when writing to either Stukeley or Lyttleton. Despite what can be interpreted today as inelegance, the lack of any qualifying remarks by Hutchins suggests the earliest surviving

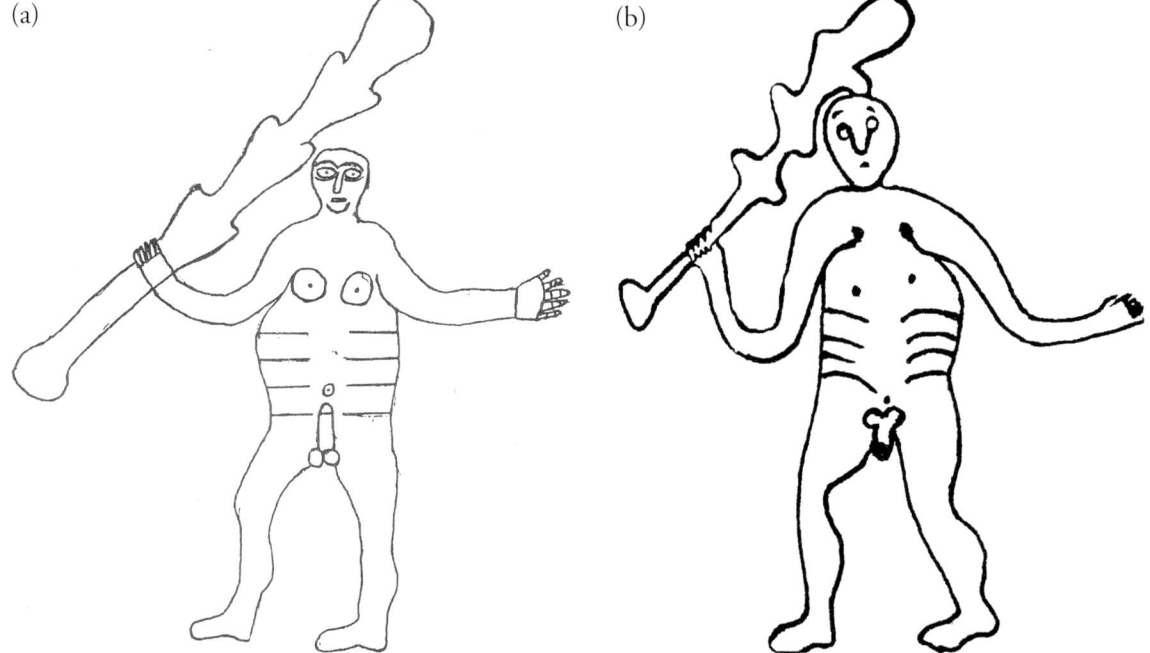

FIGURE 5.2. a) The 1763 sketch forwarded by John Hutchins in January 1764, now in the Minute Book of the Society of Antiquaries; b) A 1763 miniature of the Giant included on an estate map. After Benjamin Pryce 1768. Illustration: B. Edwards 2013

sketch of the Giant offers a relatively representative contemporary impression of the original on the hillside. For endorsement we may turn to Robert Lumley Kingston (1712–1773), a Dorchester solicitor, who on 28 June 1794 wrote one of his many missives to the lawyer, librarian and archivist Dr Andrew Coltée Ducarel (1713–1785) (Nichols 1818, 624–5):

> I have got into my hands a drawing of the Giant at Cerne, very rudely done, and indeed I may say the more like the original by that means, a copy of which shall attend this letter.

Dr Ducarel had been one of those present when William Stukeley read a note 'some Observations on Mr Hutchins's Account of the Giant' to the Society of Antiquaries on 16 February 1764. Ducarel's sketch was not then merely similar, but most probably had been copied from the earliest known image at that time (Society of Antiquaries 1762–1765b).

These early sketches and copies have other relatives. On creating an estate map of Cerne Abbas in 1768, Benjamin Pryce included a miniature of the Giant (redrawn here as Fig. 5.2b). It echoes the earliest known sketch in all but one remarkable detail, the Giant's penis is not erect. In the 1790s another estate map was produced, based on Pryce's earlier survey and a new miniature image of the Giant was included once again. Sadly, the condition of this map is not as stable as its forerunner, so this miniature Giant disappointingly keeps it secrets. The final contributor to this Georgian sequence is the charming sketch of the Giant as witnessed on the spot by Samuel Hieronymous Grimm in 1790 (British Library Add.15537 f.122; see Fig. 5.3). The outline of Grimm's Giant is similar to the earliest known sketch produced more than a quarter of a century before,

FIGURE 5.3. The Cerne Giant depicted by Samuel Hieronymus Grimm in 1790; 'The Giant, Cerne Abbas', ink wash on paper. © British Library Board Add. 15537 f.122

144 *Brian Edwards*

and also the several miniature illustrations of the Giant that featured in estate maps (see Dorset History Centre 1768a [Giant]; 1768b [Giant]; 1884 [Giant, no genitalia]). It would appear Dr Maton witnessed similar when travelling to Cerne in this period, for he opined that the Giant was 'cut out with as little meaning, perhaps, as shepherds' boys strip off the turf on the Wiltshire plains' (Maton 1797, 17–18).

Turning to the sophisticated eighteenth-century versions of the Giant, the first to appear in print (and fourth of the sequence mentioned here) made its debut in the *Gentleman's Magazine* in August 1764 (Fig. 5.4, left). Dr Maton surely wouldn't consider this the work of shepherds' boys, were this Giant to resemble what he had witnessed at Cerne (Maton 1797, 17–18). To the sci-fi generations this Giant resembles a humanoid robot, with a face reminiscent of a Dr Who Cyberman or the *Star Wars* droid C-3PO (Minden *et al.* 2002; Grazier & Cass 2017, 198). As a product of its own time, this interpretation of the Giant descends from the early years of the Industrial Revolution, an observational *académie* study creatively influenced perhaps by Edme Bouchardon's 1741 anatomy book, *L'anatomie nécessaire pour l'usage du dessein* (Bouchardon 1741). To those with bound copies of the *Gentleman's Magazine* or encountering the later amended versions in expensive tomes in the library of a country house, this Georgian Giant was quite a statement. Presented as a modernised monument, which was as far from the ruggedly wild figure in antiquarian field sketches as one might imagine, was probably the point.

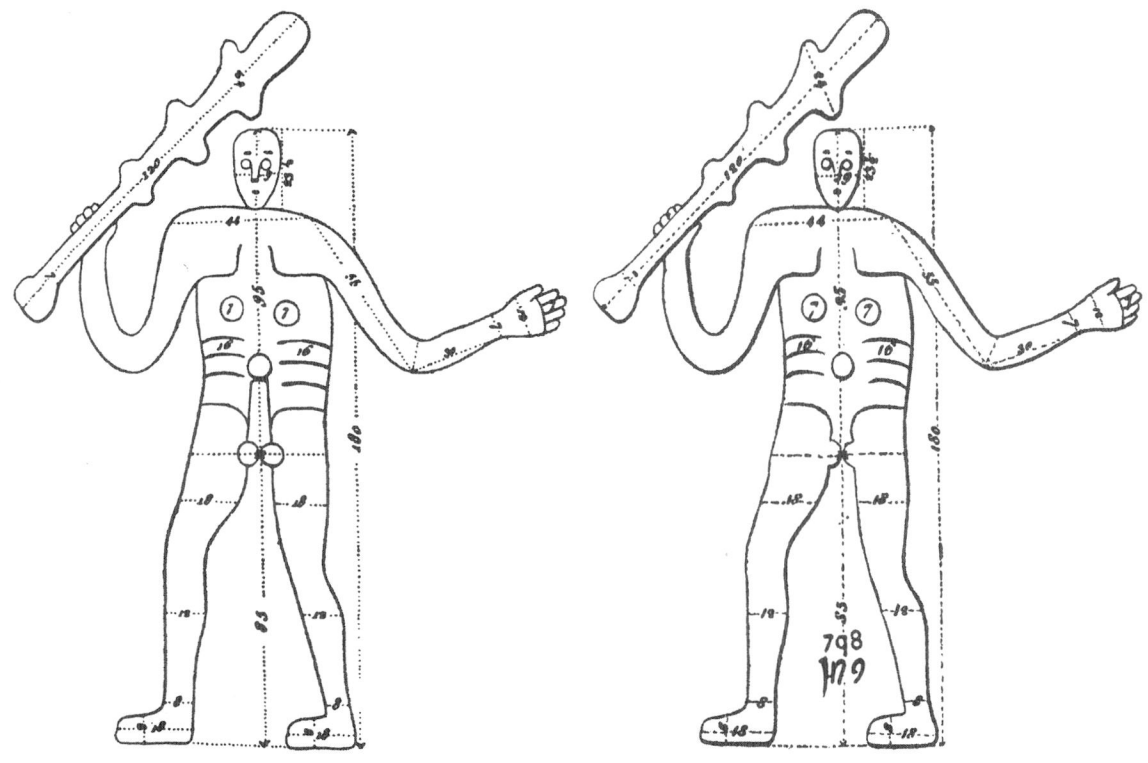

FIGURE 5.4. (left) The Cerne Giant's debut as a published drawing, *Gentleman's Magazine* in August 1764; (right) The Cerne Giant as he appeared in John Hutchins' posthumously published *History and Antiquities of the County of Dorset*, 1774

5. Hide-and-Seek on a Dorset hillside

Having circulated again and again in an amended version, it may be argued that the robotic Giant in the *Gentleman's Magazine* had an impact that foreshadowed the experience of our own time. The amended version was first perpetuated in a popularly published Dorset county history associated with the late John Hutchins in 1774 (Fig. 5.4, right). The same image then featured on a sumptuously engraved plate in Richard Gough's 1789 edition of the enormously popular *Camden's Britannia* (Camden 1789b; Gough 1809, vol. 1, 50 pl. II, and vol. 1, 68, pl. II, fig. 2). Thus the sophisticated illusion of the robotic Giant heightened expectations, whilst offering little resemblance to what was encountered in Cerne Abbas most of the time. In other words, this Giant created a misleading impression with an outcome similar one imagines to the frequently repeated aerial photographs of today.

Hide-and-seek in the age of pornography

'It would be a disgrace if such a primitive work of Art should be permitted to fade away,' read a report of the Giant in 1883 (*Southern Times and County Herald* 1883): the opinion being stressed was that a stranger would not have spotted the figure on the hillside, which was only found by those that knew where to look. This instance of inconspicuousness among so many similar experiences is of particular interest. Firstly, it arose between two known Victorian restorations, when the Giant might have been expected to be seen. Secondly, it was the year before A.H. Green sketched a miniature of the Giant when creating an estate map. Of Green's forerunners that included a miniature Giant with genitals in estate maps, each had surveyed Cerne and yet none of those surveyors included an erection. Green didn't either, but whether the erection was visible or not during these times is a moot point.

A routine trap awaits discussion of the Giant's genitalia/erection, the supposition being that explicit images were censored, which on an individual basis may or may not have been the case. The visibility and condition of the Giant throughout the entire span of these centuries had been poor, however, a fact somewhat eclipsed by the impression projected by the perfected images published in the eighteenth century. When the Giant's lines were choked and overgrown, the genitalia were difficult to spot let alone record. Any impression that an erection was in evidence would have been further obscured by overgrowth on the raised bank that once created the impression of a tip of the penis. The penis was seen and drawn in 1897, but by using a field telescope (Fig. 5.5).

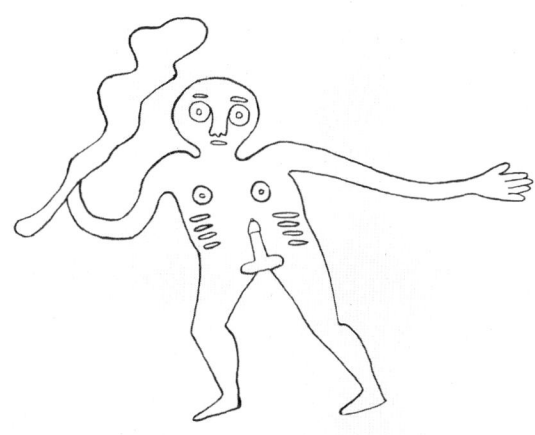

FIGURE 5.5. 'This Figure of the Cerne Giant was drawn in 1897, with the aid of a Field Glass, and afterwards corrected on the spot; but the effect of foreshortening, caused by the slope of the hill, has not been altered' (March 1902, 8)

Confusion created by overgrowth is perhaps how the separate navel, traceable in early Edwardian photographs and postcards, became joined to the shaft of the penis (Castlelden 1996, 27). The Giant's penis was inadvertently lengthened at some point between the guerilla gardening of 1905 and the publication of Petrie's drawing in 1926 (Fig. 1.5). If Leslie Grinsell made the correct evaluation of the Rev C.W.H. Dicker's photograph of Easter 1911 to confirm the navel has been integrated with the penis, the restoration of either 1905 or 1908 would appear to be responsible. The latter being the more thorough restoration, was thus the more likely cause (Grinsell 1980a, 30; Fig. 1.7). That this cojoining took place in this early twentieth-century period, when reference works were available and those involved were as or more informed than anyone in history, it is surely indication enough that the details between the upper part of the Giant's navel and the inner leg area was always mystifying.

It is particularly ironic that the charge of censoring the Giant should be cast in the direction of Victorians and Georgians from the twentieth century. In addition to multiple attempts to internally censor the Giant, only censored postcards of the Giant were available between the World Wars, and the postwar National Trust took the decision not to produce postcards for fear of prosecution for obscenity.

Follow the science

Given the surprise arising from the announcement of the OSL date range of AD 700–1100 for the Cerne Giant, we have mostly forgotten that in 1872 the Dorset born physician and antiquary Dr Thomas William Wake Smart (1805–1894), explained why a medieval origin was entirely feasible.

> The Benedictine Abbey of Cerne, founded by Æthelmær or Aimar, AD 987, being richly endowed with lands, became the seat of learning, and, no doubt, of much dissipation also: hence it is quite within the bounds of probability that, … in conjunction, perhaps with some of the townsfolk, may have occupied some of their vacant hours in portraying the lineaments of this legendary personage. Nor might this ascription be deemed at all derogatory to their artistic taste or skill, for we have abundant proof, in the carvings and sculpture of medieval age, that the principles of aesthetic taste engendered within the cloister were not essentially of that refined, pure, and chaste style which prevails in modern works of art (Smart 1872).

Smart summarised all the essential ingredients: the creation of the Giant meant a high cost and considerable effort for something innovatively imaginative but ephemeral that couldn't be seen at any great distance. To be considered likely as an origin for the Cerne Giant, any hypothesis simply must fit this criteria. An association with a mustering point has recently been presented, but seems highly unlikely, when it would be more effective to fly a flag, build a beacon fire, or simply dig a hole or a trench to expose some chalk (Brookes, Chapter 19; Morcom & Gittos 2024, 23–4). With academic discussion retreating towards antiquarian interpretations of hill figures, argued in the eighteenth century as marking Saxon battles and conversions to Christianity, it is as if the identity recently cast off by the Uffington White Horse may well lay siege to the Cerne Giant in the media (Marples 1949, 31–2, 69, 138–145; Schwyzer 1999). Our oldest equine hill figure is also thought to have survived through regular scouring of course, and perhaps that wasn't the case.

Whatever the Giant represented in origin, one stand-out will remain among the surprises: the long silence before this extraordinary monument is first mentioned in 1694 (see Darton 1935; Bettey 1981; Hutton 1999b; Edwards 2020). If the figure was as routinely overgrown as it has been since the seventeenth century, however, we may have a feasible explanation.

Whilst we hear nothing of the Giant during the 500+ years of Cerne Abbey's existence (Hutton 1999b), the stratigraphy reveals the Giant was yet to be re-awakened. If overgrown, a turf figure is clearly irrelevant to contemporary identities and otherwise preoccupied lives. Where there is little or no concern about the condition of the Giant, there is no reason for the figure to be mentioned and as evidenced by many witnesses, strangers passing-by probably wouldn't spot it anyway.

Once the Dissolution of the monasteries had led to the closure of Cerne Abbey in 1539, what remained on that hillside retained meanings for contemporary residents of course. If the Giant could occasionally be detected and believed to be of Roman Catholic origin, any concern would have likely been complex, but any fear of it being interpreted as idolatrous or evidence of continuing worship would surely have drawn some response (Walsham 2011; Davis 2015). That no such intervention was in evidence when the stratigraphy was examined, makes it probable that any earlier hill figure wasn't visible until the Giant was restored in the seventeenth century. However, as Harriet Lyon has closely argued that monastic remains were also adopted and recycled in moves that tortuously complicated engagement with the past, if the Giant was not entirely invisible but too indistinct to fathom during the sixteenth century, then perhaps it was subject to the suppression of past associations hence absences from topographical reporting (Lyon 2022, 125–90). We are after all, yet to discover what notice anyone took of the Cerne Giant between the Reformation and the arrival of William of Orange in 1688. The fascination for visible relics of antiquity had continued to enthuse generation after generation of scholars and travellers, but we hear nothing of the Giant on the hillside at Cerne Abbas. The silence engulfing the Cerne Giant is even more puzzling in view of the national celebrity of giants across the same centuries; but not if the figure was almost unnoticeable (e.g., Fairholt 1859; Massingham 1926; Barker 1997; Cohen 1999, 26–61).

Surveyors of the standard of the father and son named John Norden, would surely though have encountered the Giant when carrying out their survey of Cerne in 1617 (Bettey 1981, 119). Perhaps it was the Nordens that rediscovered the Giant, but if so, one can only surmise as to why their survey of Cerne in 1617 made no reference to the hill figure whatsoever.

The senior of this Norden partnership was a more complex character than the royal surveyor hitherto discussed in the context of the Giant. Faced with some unspecified charge taken seriously by Elizabeth I, Norden senior had felt conflicted. Having argued he had been confused with someone else, the royal surveyor appears to have invented a second John Norden, and claiming mistaken identity with this contrived individual enabled Norden the surveyor to distance himself from his own devotional writings (Kitchen 1997). These earlier circumstances would appear potentially haunting: if the local land agent at Cerne Abbas didn't suggest the surveyor should forget all about the hardly discernible impression on the hillside, and Norden signalled the hill figure's presence, when the Giant would have been the property of the reigning monarch's heir, the future Charles I, made Prince of Wales only the previous November. If uncertain what the figure represented, and discovery might somehow lead to the surveyor's unwelcome past experiences being revisited, Norden may have been convinced that the Giant was so far from obvious it need not be identified.

A forthcoming centenary

> All through those centuries there must have been folk to remember him, and to prevent his obliteration on that steep hill by siltage, drainage, desiccation, and strong vegetation – forces that only toil and constant effort can keep at bay for valuable crops on far less perilous surfaces (Darton 1935, 330–1).

Despite many twists, turns and complexities, the adoption of hill figures by communities and popular culture has clearly played an important role in the survival of these monuments that are vulnerably exposed in respect of their lonely locations. It would then appear that an instinctive response may have played a key role in the survival of these hill figures. On taking over this responsibility, the National Trust has not only battled waves of vandalism in over a century of ownership and care, but thankfully resisted pressure to censor the Giant's explicit details. Calls for more frequent restoration and grass management should equally be resisted, for it is when the Giant is recently chalked that its appearance is at its most uncharacteristic. As a Scheduled Monument, a balance must be considered between, on the one hand filling the trenches with chalk to make the Giant a visible tourist pleaser, and on the other increased deterioration and inauthenticity. In desiring picture perfect hill figures modern taste is misinformed, not least in respect of the Giant. Himself has remained barely visible if not unnoticeable throughout most of his history.

Acknowledgements and thanks

Many thanks to my ASAHRG colleagues: Martin Papworth for welcoming my input and for arranging access to the National Trust's Giant files, and Mike Allen for his patience during my commitment to a series of demanding projects. Many thanks are also due to Gordon Bishop, whose contributions and assistance has been invaluable, and to George Mortimer for assisting with source material. To the Cerne Historical Society also, for being a focus for interest from near and afar, and not least for producing a most entertaining and informative magazine. My thanks to Peter Lush and the Society of Dorset Men for informative exchanges. Multiple thanks are also due to the staff of Dorset History Centre, the Wiltshire History Centre, and Jane Schon, archivist at Wiltshire Museum. The author also wishes to thank and acknowledge Garry Gibbons who has contributed in various ways to the thinking behind this chapter; as to some extent has everyone that contributed to the trial initiative, the subsequent volume, or has ever written about hill figures.

Note

1 Responses shared on Facebook from a National Trust, Business Services Co-ordinator for West Dorset (part time), 6 October 2023.

CHAPTER SIX

Know your Giant

Brian Edwards

From when first encountered, to being seen from time to time, hill figures are inevitably reimagined and reinvented. This chapter is about impressions such as these in history: particularly those impressions of the Cerne Giant that have an immediately strong contemporary presence, only to unexpectedly fade and when superseded may be misremembered or even very quickly forgotten.

Scholarly interest in recent times led to the discovery that an Edwardian postcard manufacturer had garishly highlighted the Cerne Giant's erect penis. In mundanely applying thickened white lines to photographs of indistinct hill figures, the Giant's penis was so well defined it was quite plainly set apart from the adjacent navel (see Fig. 5.1). With the Giant's navel subsequently vanishing from the hillside, it was concluded that the penis had been accidently extended to incorporate the navel during an over-zealous restoration (Pitman 1978, 27; Legg 1999, 133; NTf 1). Where the Giant's navel and the garish postcards had all but been forgotten, we now recall them both, but we also tend to forget that by 1935 only 'carefully censored' postcards of the Giant were on sale in the village shop in Cerne Abbas, and that a decade further on the National Trust decided not to sell postcards of the Giant for fear of prosecution (*The Bystander* 14 August 1935; NTf 2). These incisive episodes of censorship have been eclipsed by more relaxed times: now the Giant features 'intact' on not only postcards but on the side of milk bottles in the village shop (Fig. 6.1), and the National Trust is comfortable posting an early morning joke about 'members' on social media (NT Twitter, 2019). If we recognise from these Giant related episodes that forgetting has repeatedly happened in the age of photography, who then could imagine

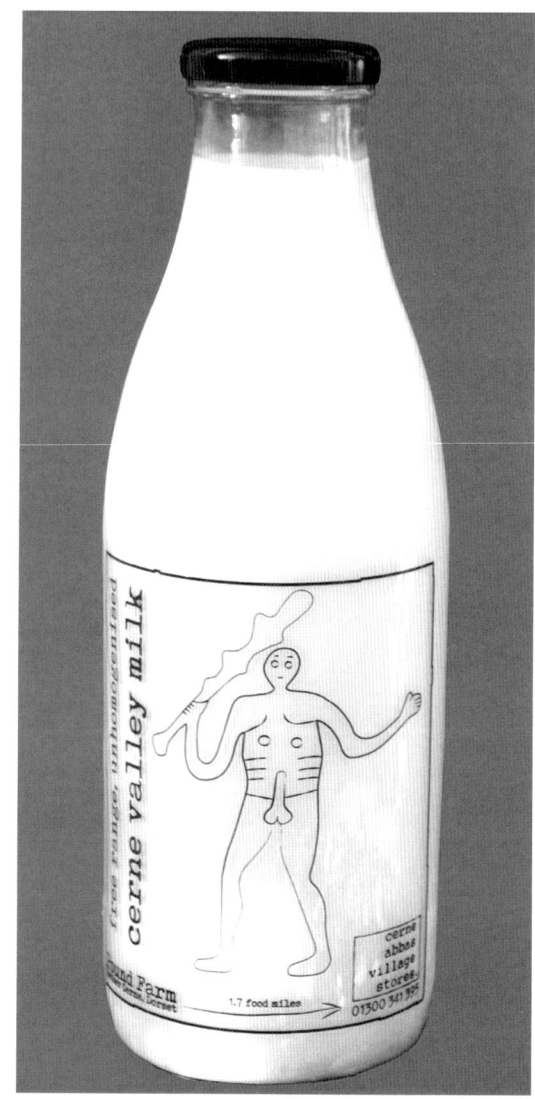

FIGURE 6.1. The Cerne Giant 'intact' on a rechargeable milk bottle. Image: B. Edwards 2024

the Giant of Cerne Abbas hasn't changed in appearance or identity, let alone in presentation and celebration in over 1000 years?

Headline acts

'Woman sits on penis and sings Abba' is, at the time of writing, just the latest example of the Cerne Giant being celebrated in media headlines and straplines (*Bournemouth Daily Echo* 24 November 2023). Track back through 'Would you put a huge penis on a packet of cheese?' (*The Guardian* 11 June 2023) to 'Huge penis to be polished by hand' (*The Daily Star* 29 August 2019), and beyond lies such as 'Fine figure of a man may not be the real thing' (*Western Daily Press* 11 March 1996).

Offering more than a snapshot of the British obsession with sexual innuendo (Porter 1998), it takes remarkably few headlines to illustrate a range of parallels through which the Cerne Giant may be adoptively associated. The context and yet more memorable examples have been explored in this volume by Sarah Fry (Fry, Chapter 13). From film promotion stunts to good causes and health related prompts, several groupings are evident. Yet more will be found in a list of onsite occurrences at the Cerne Giant since the 1960s, which has been added to the Cerne Giant's Timeline (Chapter 7). The timeline makes readily obvious how quickly a change overtakes public consciousness of the Giant. When a monument is used as a billboard it is equally obvious, even in the cyber age, that some of these occurrences can just as quickly be superseded, misremembered and in some cases forgotten.

Hill figure innovation

Posing a rich vein of history in connection with the Giant, many creative links have been formed through music, poetry, and artworks right up to and including the present century. Of those not making media headlines, not all will be widely remembered but each were memorable to someone.

On the eve of the second lockdown in 2020, having not been out for a stroll on May Day as originally planned, a 5 m tall 'intact' Giant puppet paraded through the village of Cerne Abbas at Halloween (Mead 2021; Sasha Constable, pers. comm.). Social media may prompt recall of this and other recent examples, such as the handheld Giant puppet from 2020 with wobbling genitalia mounted on a spring (Trevor Sproston, pers. comm.). This creative line can be traced back to the earliest known Giant replicas made by Mr J.A. Thwaites, an elderly disabled resident of Cerne Abbas, during the interwar period. As the first widely recognised maker of 'intact' wooden models of the Giant, sold to visitors from all over the world, Mr Thwaites also pioneered the now extraordinary range of commercially available Giant inspired products (*The Bystander* 14 August 1935; Vale & Vale 2000, 13).

Of the projects linking hill figures with education (Mical Sorga, pers. comm.), local schools have a long history of engagement with the Cerne Giant, evidenced by such as local talks, guided walks and assisting with restorations. The Cerne Giant has also been fashioned into one form or another by at least three schools remotely located from Cerne Abbas (*The People* 3 July 1994; *News of the World* 20 August 2000; *Plymouth Herald* 5 November 2008). Education should not then be overlooked as a link with the origin of the Cerne Giant, especially having influenced engagement with hill figures elsewhere. An undated photograph of the newly completed Penleigh White Horse in Wiltshire, suggests schools have been involved in designing and creating equine hill figures since at least the Victorian period (Ivan Clarke Collection). Extending this back even further, Garry Gibbons has established that under adult leadership, restorations of equine hill figures by schools and children of school age stretches back to the 1780s (Gibbons 2017; 2021; see also Jennings 1894; Reeves & Morrison 1989, xxxiii, footnote; Sadler 1989, 7).

From slogans to flags and insignia, additions to hill figures are a regular occurrence, the most common of which is the addition of genitalia. A penis was added to the Wilmington Long Man on more than one occasion (e.g., *The Times* 25 May 1939; *Daily Mirror* 31 March 1959; *The People* 28 March 1976), and equine hill figures have similarly been made into stallions with regularity (*Bath Chronicle* 19 April 1930; *Wiltshire Times* 5 November 1955). In contrast, the design of the Cerne Giant's penis does not appear to be a later embellishment or prank. The earliest identifiable additions could be the shallowly cut symbols and numbers recorded between the Giant's legs in the eighteenth century (Hutchins 1774; Morgan Evans 1999). Reinvention is after all, as inherently central to hill figure associations as it is to passing clouds, and it is highly likely that the reuse and recycling of imagery and sculpture has always been commonplace (Ng & Swetnam-Burland 2018).

His Highness's Nose in 1993

In addition to the above unrecorded twentieth-century penis extension of 1908, and having his postwar testicles returfed in 1945, the Cerne Giant got a new nose in March 1993 (NT archive, Edwards, Chapter 5). An identifiable feature of every extant drawing from 1763 to 1897, the Giant's nose survived as a slightly raised mound which had no outlying trench. The importance attached to the nose was later explained: 'the Giant was not just a two-dimensional engraving in the chalk, but a figure with three-dimensional relief, of not inconsiderable sculptural sophistication' (Keithley *et al.* 1999, 24).

The roots of this nasal exercise of 1993 can be traced to a remark about the 'phallic nose (now grassed over)', in a caption to an illustration in the *Dorset Yearbook* in 1986. This remark struck a chord with Gerald Pitman, a Sherborne resident and local historian who began a letter writing campaign in 1990 in a bid to persuade the National Trust to reinstate the Giant's nose (e.g., letter Gerald Pitman to David Thackray, 25 February 1991). Oral testimony from a long-standing National Trust warden suggested there had been no change in the nose since 1981, but it was suspected from photographs that the Giant's nose had been more pronounced in 1974 (NT archive). The renovation of the nose received Scheduled Monument Consent in 1993 and was undertaken that year. The timing was by coincidence such that the Giant's nasal exercise of 1993 followed the earliest reference to the Cerne Giant being unveiled, by historian and Cerne Abbas resident Vivian Vale, as a record from November 1694 (Vale 1999).

The double celebration

When Queen Mary II died of smallpox aged just 32 in December 1694, there would have been much to reflect upon in the month following the repairing of the Cerne Giant. With the churchwardens sanctioning that repair, there is no doubt that the Cerne Giant was recognised as Hercules at this point. Through both artworks and the written word, an association with Hercules can be traced for every state leader known to the population in England during the seventeenth century. Were the number of artworks for all others added together, the total would not come close to the number of designs projecting William III as Hercules (Fig. 6.2). The correlation was amplified in everything from broadsides to etchings in pictorial publications, from medals to coins, and architecture to fine art. This predominance means that whilst making comparisons with Hercules were commonplace, the only leader the population at large would familiarly associate with Hercules was William III (Edwards 2020). The significance of the date on which the Herculean hill figure at Cerne Abbas was being repaired would not then be misunderstood or underrated in 1694, for 4 November was William III's birthday and the anniversary of the joint monarchs' (William and Mary) marriage (Edwards 2005).

Upon close inspection of the churchwardens' accounts an ampersand

FIGURE 6.2. William III as Hercules from a Dutch broadside of 1689 by Aert Dircksz. Illustration: B. Edwards 2023

clearly follows the 4th and below, attached to the N for November and in the same hand, is a distinctive 5th (Edwards 2020). Resembling a toppling pi symbol, this characteristic 5 is repeated elsewhere in the churchwardens' accounts (Fig. 6.3): there is then no doubt that 4 and 5 November was indicated as linked in this entry in the Cerne Abbas churchwardens' accounts (Dorset History Centre: ref: PE-CEA/CW/1-2). When the date column appears alongside the respective entries the importance of the combination is apparent.

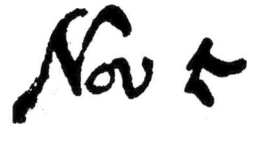
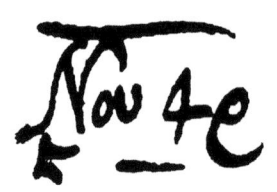

FIGURE 6.3. The twinned anniversary dates of the 'repairing of the Giant' recorded in the Cerne Abbas churchwardens' accounts. Top: 5 November as recorded by the same hand elsewhere in the churchwardens' accounts. Bottom: 4&5 November are clearly linked. © Dorset History Centre: PE-CEA/CW/2. Cerne Abbas Churchwardens' Account Book, 1687–1694

| Nov 4& | gave ye Ringers | 00 | 7 | 6 |
| 5 | for repairing of ye Giant | 00 | 3 | 00 |

This joint entry made in the churchwardens' accounts in November 1694 was highly unusual, it signals that the Giant's repair related to the then recent tradition of a national joint celebration staged across consecutive days. Between 1689 and 1701 the anniversaries celebrating William III's birthday and marriage to Queen Mary on 4 November, was combined with the anniversary of his landing at Torbay on 5 November twinned with the existing observance of deliverance from Gunpower Treason on that same day in 1605 (Edwards 2020). Following the coronation on 11 April 1689, when William and Mary were crowned joint monarchs, the new double celebration radically recharged and reshaped the annual observance for those upholding the traditions of 5 November (Cressy 1992; Hutton 1994).

Directly connected to the biggest political anniversary of the late seventeenth-century calendar, Vivian Vale's discovery of the earliest reference to the Cerne Giant has proved a game changer.

Heroes and villains

A complex background of mythmaking awaited the birth of the future William III (Enenkel 2018; Stern 2009). This upbringing is detectable in a portrait of the 17-year-old Prince of Orange by Jan de Baen *c.* 1667, in the Royal Collection Trust (Fig. 6.4). Theatrically posed in ancient armour, William's index finger gestures towards an ornate plumed helmet, beyond him at the water's edge stands a statue of Hercules wrestling with the Nemean Lion in his First Labour. William was associated with Hercules much like the Trojan hero Aeneas in Virgil's *Aeneid*, who was to be instrumental in the founding of Rome.

In political poems found at Wrest Park, which date from both before and after 1688, Linda Halpern has highlighted 'veiled references' likening William to Aeneas (Halpern 2002). It is then a very short imaginative hop from William of Orange landing at Torbay on 5 November 1688, with its traditional celebration of deliverance on Gunpowder Treason day, to conjuring a parallel with Aeneas's arrival during the annual feast commemorating Hercules defeating Cacus, the cattle stealing cave dwelling giant that had been terrorising the population

FIGURE 6.4. Jan de Baen, William III when Prince of Orange c. 1667, oil on canvas, 180.3 × 133.1 cm. Royal Collection Trust. © His Majesty King Charles III 2024

(*Aeneid* 8). William was identified with both Aeneas and Hercules, so the fight with Cacus presented a role into which James II was immediately cast (Baxter 1992; Sellers 2001). In contrast to the characterisation of James II as a monstrous terrorising thief, William of Orange's Hercules was projected as a symbol of peace that would bring unification through the restoration of the Protestant faith in Britain (Kalinsky 1972, 133; Garrison 1975, 200). There is then perhaps no mystery behind a parish church funding the restoration of the Cerne Giant as a depiction of Hercules on the double anniversary, for once James II had been deposed the churchwardens' accounts recorded opposition with certainty:

1689
Feb 20
Pd for a boke of Thanksgiving being preserved from Popery and arbitrary power. 1s. 0d.
(Dorset History Centre: ref: PE-CEA/CW/1-2).

Among the ongoing examples that emphasised the importance with which the new order was regarded in Cerne Abbas between 1689 and 1694, was the regularity with which the bells of St Mary's were rung in connection with William III. Albeit not always consistently maintained, bell ringing for royal occasions and anniversaries were existing conventions, and the traditions before 1688 included regular charitable donations to the poor, seafarers and soldiers. What did change was the detailed descriptions within the accounts, which after 1688 reflected sectarian empathy (Dorset History Centre: ref: PE-CEA/CW/1-2). Added to the bell ringing for events which ranged from military victories to safe landings upon the King's return, the close connection with William III's campaign is unmistakeable. This inventory of support for the new regime is the context for the unique entry that mentions the Cerne Giant in connection with the double anniversary. Where once it was thought there was only one reference relevant to the Giant in the churchwardens' accounts, there are a great many of relevant interest.

The parish church funding the restoration of the Cerne Giant as Hercules, and the support shown for William III, explains the non-involvement of the landowner (Vale 1999, 72–3). Thomas Freke had owned the Giant since 1666 but was not known as a supporter of William III, it was noted that he played no part in the Revolution and was 'conspicuously absent from the warrants for raising money for William of Orange' (Henning 1983, 365–6; see also Hayton *et al.* 2002). What he was known for was the unkind nickname the 'great Freke', which, even were it not a direct reference, would surely put paid to any financial contribution towards the Giant's restoration (Henning 1983).

Forgetting Hercules

In many of the instances in which coins and medal, and popular illustrated chronicles, promoted the new regime, it was the inclusion of William's characteristic nose that closely identified him with either Hercules or Aneas. Famously, Dryden's 1697 edition of Virgil's *Aeneid*, saw the publisher Jacob Tonson pay an engraver to alter the face of Aeneas on existing plates to resemble William III (Barnard 2015). It is not then beyond the bounds of possibility that the Giant's nose featured in the repair for the dual anniversary in 1694. How then, with the Cerne Giant so readily resembling Hercules, was the Williamite context overlooked by everyone from eighteenth-century antiquaries through interwar archaeologists to the National Trust's nose restorers of 1993?

This chapter has repeatedly stressed how easily impressions come and go. Memory is then slippery enough without traditions being so amenable to resisting accurate recollection. Traditions do not form and grow in isolation, they merge with anything that matches or can be adapted, even if originating in a rival form (Laidlaw 2014). Let us recall Cacus, and the context that plunged James II into the role of the cave dwelling giant that terrified locals and stole their cattle. Add the widespread and sensational impact of Jonathan Swift's *Gulliver's Travels* being published in October 1726 (Anon [Swift, J.] 1726; Just 2002), and by 1735 the curate of Cerne Abbas is sharing a local tale of villagers pinning down a giant that had fallen asleep after eating a flock of sheep (Morgan Evans 1999). The giant is nowhere to be seen of course; the imprint of the Cerne Giant is the only evidence that he ever existed.

Make of the Cerne Giant what you will. No one knows the Giant like you know him, many will not know your Giant at all.

Acknowledgements and thanks

The author would like to thank Barbara Yorke and Garry Gibbons for reading and commenting on early extracts from this chapter. The author is also most grateful to Sasha Constable, art teacher and Cerne Abbas resident, for information about the Giant puppet, to Trevor and Margaret Sproston for discussions about both puppet making and memories of Cerne Abbas, and to Mical Sorga, Teacher of Music, and Mical's pupils at Matravers School, in connection with a song about the Westbury White Horse.

CHAPTER SEVEN

The Giant's story: the archaeological results considered

Michael J. Allen

Several clear statements can now be made as a result of the excavation and subsequent analyses and research (see below). Based on, and leading from the primary and secondary research aims (date and environment), these analyses allow us to address aspects of the two (or more) Giants, the nature of his form, and outline and manufacture, and finally review the nature and history of the scouring, maintenance and its record.

Main conclusions

1. The Giant is early medieval, and the fact that this broadly encompasses the period of both incorporation (seventh century) and the founding of and refounding of the abbey (AD 987) is probably not a coincidence. The significance of this is explored by Yorke (Chapter 4) and Morcom and Gittos (2024).

    There seems to be some activity around AD 1250 as two OSL results (from early colluvial soils, contexts 135 and 216 Trenches B and C) both produced dates centring around this period, but for which we can't identify any associated activity here, nor in the local landscape. This probably relates to undocumented changes in farming and land management (see Barker, Chapter 9).
2. There are at least two, if not three Giants, or phases of the Giant.
3. The Giant in his present white chalk figure form seems to be largely a twentieth-century manifestation.
4. His outline has not radically changed in over 1000 years. The most obvious change is the incorporation of the penis into the navel in 1908, but there are other subtle and potentially significant minor changes.
5. The largest conundrum was outlined by Ronald Hutton at the Trial in 1996 and still exists, which is the absence of any historical records of the Giant from the tenth century until the mention in the churchwardens' accounts for repairing the Giant in 1694 (Vale 1999). We argue that this is in part due his former form as just a trenched, not chalked outline,

and changes in grazing pressure and regimes; the longer grass obscures the figure, as evidenced by the National Trust who regularly receive reports from visitors complaining that they could not see the Giant, or who were disappointed in his condition.

The date of the Giant

Despite the fact that the early medieval OSL dates were a surprise to both the majority of the archaeological and historian audience, and even to those who study the early Middle Ages, an early medieval context for the Giant's creation makes good sense. An early medieval date certainly fits with the form of his face. The distinctive tear-drop shape is akin to other human faces in Anglo-Saxon art, such as the ones on the Sutton Hoo sceptre, the Finglesham buckle, and figurines from Kent and East Anglia (Morcom & Gittos 2024, fig. 6). There are parallels for this shape from Scandinavia but it is quite different from classical models and it is less common from the eighth century onwards (see illustrations of faces in Withers & Wilcox 2003; Webster 2012b; Brundle 2019, 118–26; 2020). An early medieval date also fits with recent work on the depiction of human forms in this period. Images of naked human bodies began to be produced from the seventh century onwards and males tended not to be depicted completely naked, but often as wearing a belt (which the Giant doesn't). It is clear that the lines beneath the Giant's ribs indicate the base of his torso/top of his leg, as was often shown on classical images, and not the line of a belt, as proven by geophysical survey (GPR and earth resistance tomography, see Cheetham, Chapter 3) and targeted augering (Chapter 3; Allen 2024).

Some archaeologists and historians have, however, found the early medieval date for the Giant difficult to acknowledge (see Hutton, Chapter 18, for instance). In order to pay due respect to them, the archaeological science and the monument itself, we need to examine our data carefully and critically before proclaiming him a Saxon giant and discussing him as belonging to that period with his contemporaneous landscape. Although there are some well-argued, and potentially justifiable, criticisms levelled of the OSL technique itself, especially when applied to hill figures (see Hutton, Chapter 18, for instance), the main concern is not the OSL results themselves, but the deposits or event that has been dated, and its relationship to the construction of the Giant (i.e., deposit taphonomy or geoarchaeology).

The key issue, therefore, is that the deposits around the Giant post-date his creation – albeit by a few decades, or couple of centuries at most – rather than pre-date him, as a cursory examination of the section drawing or photographs might suggest. Simplistically, the section looks as though a trench has been excavated through 0.6 to 0.8 m of soily deposits and infilled with chalk rubble. The Giant's outline, however, was not *cut through* half a metre of existing deposits residing the steep hillside, but that the Giant was cut through the thin humic renzdina soil as a shallow trench exposing the chalk leaving small soil banks against which subsequent eroding soil came to rest. Hillwash *accumulated against* the trenched outline increasing soil depth locally. The adjacent sediments and the trench (which was later infilled with chalk rubble infill – chalk 'stack') are coeval. The Giant was cut out with low shallow almost imperceptible banks of soil against the outline (see below) against which hillwash accumulated. Recleaning and emptying the trench re-established the low banks and recreated a barrier to hillwash. Consequently, the trenched outline, especially at the feet, elbows and base of the club, became deeper as sediments built up around their edges. This trench was ultimately infilled with chalk (see outline and form, below).

Military assembly points

The idea of hill figures being associated with military assembly is an old one (e.g., Wise 1738) and commonly reiterated. It is not, however, one that can be universally or ubiquitously applied to every hill figure. Certainly it is obvious that as hill figures are sited in well recognised and identifiable points in the landscape, they lend themselves to places of congregation, meeting, assembly and muster points. That is not to say, however, that this applies to, nor that, every hill figure was a military muster point. Despite arguments by Morcom and Gittos (2024), and even Brookes (Chapter 19), that a location at the Giant was a Saxon military mustering point, this seems unlikely at Cerne. Although the proximity of Cerne Abbey is clear, the routeway over the Giant Hill does not help to lead to, or from, any major nor significant (military) location (compare with, for instance, Baker and Brookes 2015a; Baker 2019; Williams 2006). Although the Cerne valley to the south is a direct route to and from Charminster and Dorchester, to the north, towards Sherborne, both the valley and any clear topographical route fade away just 5 km beyond Cerne, at about Minterne Magna and before the Blackmore Vale some 16 km (10 miles) south of Sherborne.

A Saxon Giant and identity

The possible identity of the Giant has been widely discussed; antiquarians and archaeologists have debated his identity for over 300 years. One of the best reviews is provided by Rodney Castleden (1996; 1999) where he considers Celtic gods (e.g., Mars, Sucellos, Nodens) and Helis/Helith, Roman Hercules, a sheep-gobbling giant and a seventeenth-century lampoon of Cromwell, to ridiculing the vociferous local seventeenth-century MP for Dorchester, Lord Holles (1599–1680), while Newman also allies him to the Dorset Ooser (1999, 33; Fig. 1.16). The early medieval/Saxon dating for the Giant now allows us to revisit and reconsider this. On examination, a number of elements of the form of the Giant fit well with other aspects of Anglo-Saxon art, and was created at a time when the Abbey of Cerne was established. It seems inconceivable that these were two entirely independent projects. Morcum and Gittos (2024) clearly argue the Giant could be Hercules; a well-known figure throughout the medieval period (Eppinger 2021; Ogden 2021; Stafford 2021). They argue there is a peak of interest in Hercules in the ninth century and this may go some way to explaining him as the figure in the tenth century.

The lion skin and cloak

The outstretched left arm has for centuries invited the interpretation of it draped with a lion skin or cloak and reaffirming a Herculean identity for the Giant. This was further fuelled by the first archaeological science on the Giant in that Tony Clark's 1970s resistivity survey seemingly showing an area of geophysical anomaly below the left arm exactly where one would expect, and describing a form that could easily be seen as hanging from the arm. The fact Rodney Castleden produced similar results over 10 years later enshrined this interpretation in numerous discourses and publications. However, Alister Bartlett's unpublished relatively large (125 holes) auger survey in 1980, and re-examination of both sets of geophysical results, now allow us to question this. The auger survey clearly shows a spread of soil and stones creating thicker deposits than the natural thin rendzina soil on this very steep slope. Rodney's survey clearly shows a geophysical anomaly, and although the two-probe array was only sensitive to the very near surface (probably 10–15 cm), it nevertheless importantly confirms the auger results. The 2023 GPR survey by Paul Cheetham again suggests thicker depths of soil and stones, providing consistency in all surveys and interpretations. Returning to Tony Clark's survey, the new analysis of this data by Andrew David, Paul Linford and Megan Clements now shows that there is no real indication of the cloak anomaly in the 'raw' data, or even after the new contrast enhancement of the dataset (Linford pers. comm.). The cloak area was originally reprocessed specifically and differently to the rest of the survey to examine the possibility of variation here, and these processed results then incorporated into a draft preliminary plot (Fig. 1.18), which was not taken further.

Given the degree of correlation between Tony's and Rodney's surveys, and Alister's auger survey, it is clear that the minor variations Alister was enhancing were near-surface, rather than deeper, features and tends to confirm the presence of thicker soils here. Whatever this feature is, it's not defined in outline in the same way as the rest of the Giant. Is this a cloak or lion skin but depicted differently, or even added later? Or is this just remaining spoil from the hundreds of years or scouring, cleaning and restorations as I have suggested. Alister's auger survey recorded soil depths to inform the geophysical survey; it unfortunately did not need to record the character of nature of the soils or deposits, so these now crucial descriptions were not made. If there are deeper deposit depths here as his augering shows, then surely this must also represent an, albeit slight, physical earthwork that to date no one has noticed, despite aerial records and other investigations of, and on, the Giant as a whole. Defining the micro-topography of this feature and the character of deposits could answer this question once and for all ... and not only indicate whether that is an integral part of the Giant's design, or an inconsequential element resulting from his upkeep but also help resolve the identity of the Giant. Indeed this is the basis of a new project 'cloaked in mystery' being undertaken in 2024, as this book goes to press, with the intention of publishing the results and following on from discussions by Stuart Piggott published in *Antiquity* in 1932 and 1938.

The Giant as Eadwold

During a reinvention of the Giant, which may occur when people engage with him during frequent scouring events necessary to maintain the figure, Morcom and Gittos argue he was Hercules reimagined as St Eadwold at Cerne, and that the eleventh-century devotion to Eadwold by the monk Goscelin provides the first record of the Giant. They argue that the monk Goscelin wrote the Life of Eadwold in the eleventh century, probably while at Sherborne Abbey. In one lesson he writes of Eadwold on a sloping cliff [Giant hillside] fixing his staff which then turned green and sprouted [knobbly club], from where he could see a fountain of water [St Augustine's well] and ran down the hillside [to Cerne Abbey]. Morcom and Gittos take this as proof that the Giant existed on the hillside and that this is the first reference to him. Being that as it may, we must, however, be cautious, for you cannot see the fountain (well) from the Giant, nor indeed from the hilltop, and you cannot even see the site of the abbey itself from these vantage points (see Fig. 1.2, for instance). Despite that, this record may well reference Cerne, and they do provide a strong argument for the Giant being carved in the image of Hercules and reimagined as St Eadwold. The figure Hercules reimagined as St Eadwold could, therefore, be the Giant, and via this argument they provide the first strong link between the Giant and Cerne Abbey, whereas before there was just speculation. They also provide a reason *why* the Giant may exist on this hillside, but not that he did. In fact the other side of Giant Hill describes the topography recalled by St Eadwold much better, as that slope is just as steep and directly overlooks St Augustine's well and is in plain sight of the abbey, yet there is no Giant there. Morcom and Gittos also suggest that as Hercules he was marker for a muster station for West Saxon troops; the idea of this being a muster station may, however, be topographically or militarily questionable. Eadwold, however, is said to have lived as a hermit on a steep hill in the Cerne valley. So perhaps the green and sprouting staff is the knobbly club, and although he could not see, he did run down the hillside to the fountain of water that is St Augustine's well, and that he ran to the sanctuary of the abbey? The Giant now stands proudly and authoritatively on the hillside, and possibly we should really see him primarily as the monks' portrayal of St Eadwold after all; his outstretched left arm not designed to sport a draped lion skin of Hercules but politely showing the way for travellers towards the refuge and safe haven of the abbey just around the corner and out of sight. Only later was he to become reimagined as Hercules when he became a popular figure in early Christianity.

In the end, whether or not he is a Saxon or post-medieval Giant, he is a Giant of his time (Bender 1999). There are episodes when he comes to the forefront of local communities' and the population-at-large's consciousness, as well as that of antiquarians and archaeologists. We can see this in the mid-seventeenth century (leading to many scholars believing he was inscribed in the hill at this time) and recently, for instance, in the 1990s as result of the academic debate (Darvill – the Trial) and publication, and possibly again even now as a result of our endeavours and the excavations of the abbey (by Willmott and Gittos). In this respect he is a Giant of these times too, and for the people who engage with him.

The sleeping Giant

Although the primary aim of the fieldwork was to provide a date for the Giant, a secondary aim was to examine the land-use history associated with him (Allen 2019; 2020a). He was clearly built in an open downland, long cleared of woodland and that existing as grazed grassland on thin rendzina soils, very much like today. There are, however, hints in the environmental evidence (land snails) of the development of longer more mesic grassland; it does not take much of an increase in sward height and to obscure the giant from view of everyone bar the most perceptive (or those air borne). Most public images are aerial photographs, yet the public view is of a squatter Cerne Giant who is regularly obscured by just the slight growth of grass, as a consequence of late or reduced grazing. He only truly shines to the spectator on the ground when just newly rewhitened and the surrounding grass is closely sheep-grazed, and enhanced by footfall of the team scouring and rechalking, flattening grass around the outline.

The lack of historical records of the Giant from the tenth century to 1694 still remains a conundrum as Ronald Hutton eloquently pointed out at the trial in 1996 (see Hutton 1999b); why did the two John Nordens (father and son) not mention the Giant in their survey of 1617? They describe all the sites and features around the Giant but not the Giant himself. They mention prehistoric sites, the well (in Cerne Abbey), and Trendle earthwork immediate above the Giant, but do not mention the Giant who later gave his name to the whole hillside. Similarly the great antiquarian John Aubrey (1626–97), who studied in Blandford, never mentions the giant; John Leland, a Tudor writer of itineraries when writing about this area stayed in Melbury only 10 miles away, yet doesn't mention the Giant, and nor does Thomas Gerrard when writing his survey of the county (1732). It seems inconceivable, argued Ronald Hutton, that he was lost for hundreds of years of changes, soil development and land-use and then was refound and recut as the same shape on exactly the same line (BBC TV 1996).

This lack of documentary reference to the Giant for about six centuries is perplexing. The slightly more mesic-representing land snail communities might suggest lower grazing pressure and longer sward heights, hiding the Giant from obvious view. However, even when less evident (and almost invisible), the knowledge, or oral record, of the Giant's existence must have encouraged people to take a closer look. If travellers were not sure precisely which part of the hillside at Cerne he was on, their gaze may not have been on the right point; after all, he was not fenced off until 1886. The environmental evidence suggests periods of longer grass possibly obscuring the Giant, but cannot provide us with any indication of the date, duration or timescale. Careful hand excavation of the colluvial soils and deposits associated with the outline produced, not surprisingly, no artefacts. When John Hutchins visited, he said the grass was so long that the lines were scarcely visible (Hutchins 1774, 292), and this has been a common occurrence as the National Trust have regularly admitted, with him becoming overgrown in 1934 (Keithley *et al.* 1999) and as illustrated by Brian Edwards (Chapter 5).

At least two Giants

Limited keyhole excavations have indicated that there are at least two, if not three, Giants of Cerne. The first (Giant 1) is probably that of the shallow wide cut through the thin soil of the scarp slope in the seventh–tenth centuries AD (as seen in Trenches B, C and D; i.e., feet and elbows). The infill of at least one was clearly dated to the tenth century (right foot, Trench B OSL 5; AD 910 (700–1110)).

The main cut Giant (Giant 2) was in existence by the thirteenth century and was outlined by a narrower trench (*c.* 35–45 cm wide) originally only soil depth (i.e., *c.* 20–35 cm), with shallow, deliberately flattened banks of upcast soil. Soil depth adjacent to the outline increased principally by slow, gradual and progressive accumulation of largely stone-free hillwash against these shallow soil banks, especially at downslope locations, and episodic scouring refreshing the shallow banks. It was only during the National Trust's ownership in the early to mid-twentieth century did he take on the more splendid form that we see today as a permanently white outlined 'king' of Cerne hillside (Giant 3). The presence of the trench-outlined, unwhitened, Giant is well attested in many of the later nineteenth- and early twentieth-century photographs and postcards, yet has been overlooked as poor photographs, low angle shots (Figs 7.2, 7.3 & 7.5), or an overgrown Giant. Indeed the state of the Giant has been a public consternation for over a century; from Alda (and Henry) Hoare's complaint to Florence and Thomas Hardy of his condition in 1925, to a disappointed member of public writing a letter to the *Dorset Echo* nearly a century later (6 October 2023; Lumb 2023), see Edwards Chapter 5.

Outline and form

As we have seen (Chapter 1, previous archaeological work) the Giant has only been fully surveyed twice prior to our 2020 survey, but that six main images of him exist from 1763 to 1925 and 2020 (Fig. 1.15). The 1763 image, the first image on record of the Giant, was a mere crude sketch. The most important are the

1764 image first published in the *Gentleman's Magazine*, and Petrie's survey in the 1920s (published 1926). Most others subsequently are largely derived from Petrie's 1926 image.

1763 (Society of Antiquaries)	Sketch; first recorded depiction of the Giant
1764 *Gentleman's Magazine*	Survey 1 (and recorded measurements)
1774 Hutchins	Same, but emasculated, image; some new / revised measurements
1885 & 1892 Plenderleath	Bowdlerised, but derived from 1764/1794 image (emasculated)
1892 Sydenham	Derived from 1764/1794 images (emasculated)
1926 Petrie	Survey 2: new survey (plate III)
2020 National Trust	Survey 3, Downland Survey & Measurement, aerial photogrammetry image, and Figure 3.5.

In view of the impressive, important and notorious nature of the Giant it is surprising that there have been so few surveys in the past 260 years. Perhaps more so that the National Trust had not undertaken any ground plan survey in their 100 years of ownership until 2020, nor even English Heritage, especially since the RCHME, now English Heritage, were undertaking detailed planned survey of pillow mounds and garden features in Cerne Abbas at the foot of the hill in about 1999 (Riley & Wilson-North 1999, fig. 28; Wilson-North & Riley 2003). Nevertheless, detailed examination and comparison of the surveys and images is both worthwhile and informative, if not part of the original research. No invasive fieldwork was conducted, nor observations made, specifically to examine the nature of the figure itself, other than our hand auger through the knoll under the left hand, Alister Bartlett's of the area under the left arm in 1979–80, and our examination of the belt in 2024 (Chapter 3).

Outline of the Giant

The nature of the chalk outline as exposed in the excavation is striking (Fig. 7.1); it consists of a column or stack of stratified chalk rubble and chalk marl layers, which form the foundation of the chalk outline. Why so deep? We have argued on geoarchaeological grounds that this was eroding soil accumulating against the chalk outline (Chapter 3), however, historically the Giant was not a white chalk figure, but a trenched one, as seen in Rev E.V. Tanner's photograph of the 1920s–1940s (Fig 7.2), and another photograph of 1925 (Fig. 7.3a).

In recent times it seems that the Giant existed as a chalk-cut figure, not a chalk-filled one; Rev John Hutchins records that when he visited the site in 1772 'the outlines are two feet broad and as many deep', and that they were regularly cleaned out (Hutchins 1774, 292). This concurs with the depths recorded in the excavation, though the width may be an exaggeration, as none of the chalk fill stacks are as wide as 0.6 m (2 ft), and we suspect that the Giant's outline trenches were at least 0.4–0.5 m deep only in specific and localised places.

7. *The Giant's story: the archaeological results considered*   165

FIGURE 7.1. Photograph of chalk stack (Trench A). Image: Martin Papworth 2020

FIGURE 7.2. Detail of photograph of the Giant from the road by Rev E.V. Tanner *c.* 1920s–1940s © DCM

This trenched form of the Giant seems to have existed into the first half of the twentieth century. A note from Thomas Hardy written from Max Gate, Dorchester (21 September 1925) to Alda and Henry Hoare at Stourhead describes the Giant as being defined by trenches 'fairly deep and all that can be done to make his shape clear is to keep the trenches cleaned out & spread white chalk over the bottom of them. This will remain white many years if raked over and weeded now and then' (National Trust archive WRO 383/954/97). Indeed in 1919 when Petrie undertook his survey of the Giant he also records that the trenches were 2 ft (0.61 m) deep (1926, 9). Early twentieth-century postcards show the Giant as a brown outline (Fig. 7.3a) and notably when Eric Ravilious painted several chalk hill figures in 1939 (for a proposed Puffin children's book on chalk figures, thwarted by the war and his death in 1942), the Westbury White Horse, Uffington White Horse and Long Man of Wilmington are all shown in white, in contrast to the Cerne Giant depicted as a brown outline (Figs 7.4 top & bottom). As recently as the 1940s the RCHME stated that 'The figure is outlined by cuttings now about 2ft deep' (1974, 82). Though their inventory of Dorset was first published in 1952, it was based on fieldwork largely carried out over a decade earlier. We suspect, however, that the trenches are shallower on the vertical outline element, than the horizontal lines where hillwash is more likely to accumulate.

Continual comments about the depth of the 'trenches' from the nineteenth century makes us realise that he existed, in part, as deeply carved trenches in the hillside until relatively recently in the (mid-) twentieth century. Indeed in a number of images he does not look like a white figure and in September 2023, only four years after a major rechalking, he was again poorly visible (Fig. 7.5). Based on these observations, it would seem likely that the chalking in the early part of the twentieth century (1925–56) was responsible for filling these trenches with chalk to bring this chalk to near turf-surface and make the giant the white figure we see today. Possibly Piggott and Eardley Knollys' restoration of 1945 was responsible for the chalk infilling.

Defining elements of the Giant: lion skin, cloak, severed head, symbols and belt

Stukeley allied the Giant to Hercules (*c.* 1693) and the suggestion of a lion skin draped over the outstretched left arm may have originated from there. Piggott published this idea in 1938 and mentions a lion skin (p. 327), but not a cloak. Although first hailed as the missing lion skin, Castleden seems to be the first to suggest that because of the line's smoothness and the 'soft-focus' of Tony Clark's then unpublished geophysical survey it could be considered more as a non-Herculean draped or folded cloak (Castleden 1996, 156). The cloak/lion skin and severed head have always been known to be a different form to the Giant's outline. Augering and geophysical survey in 1979 have provided a clearer and more defined outline than known previously. If they were part of the original design, it does seem strange that the cloak/skin (and severed head) were not created in the same form as the rest of the Giant, instead of being left as much more ephemeral features. Augering as a part of this project and previously in 1979 now show that they both comprise of a mixture of soil and chalk, and geophysical survey (GPR) in 2023 confirm the heterogeneity of the deposits while this and resistivity provide a more consistent outline than Castleden was able to do in 1989/90 and 1995 (1996, 156–61). We postulate that this could just be upcast and dumped spoil from various restorations that had not been totally removed from the hillside, and was not any part of the formal former figure.

7. The Giant's story: the archaeological results considered

FIGURE 7.3. a) 1925 aerial photograph postcard showing the Giant as a brown outline (CCC8716_823), and b) Aerial photograph 1971 showing Giant as a bright white outline (Aerofilms, 22 February 1971). Source © Historic England Archive, with permission

FIGURE 7.4. Ravilious' paintings of hill figures in 1939; (top) Cerne Giant; (bottom) Long Man of Wilmington

Andrew David, Paul Linford and Megan Clements' reanalysis of the 1979 geophysical survey data also no longer see an archaeological feature here.

If this spoil from early rechalkings possibly from as early as 1694 (or before), as the new evidence suggests, then the former downslope rain wash may have been responsible for giving rise to the illusions of folds in a cloak that Castleden alludes to. This chalk and soil wash may be what was encountered in the 2020

7. *The Giant's story: the archaeological results considered*   169

FIGURE 7.5. Photograph of the Giant from the road September 2023. Image © J. Gardiner 2023

augering (see Chapter 3). On balance, however, we think that the existence of this remains unlikely and at best ambiguous; the evidence is a not as strong as some portray. We are, nevertheless still wary of the apposite comments that Tony Clark made in 1983 about 'the area of the Giant's conspicuously unoccupied left arm' and that cloak or lion skin draped over it 'gives meaning to the position of the left arm, and in the writer's opinion improves the compositional balance of the figure' (Clark 1983, 30), but overall feel this is the remnants of spoil from various scouring and replenishing events.

Between the Giant's shins a set of numbers and symbols have only been noted by Hutchins (1774); all records and comment on them are derived from his depiction (Fig 1.9). Newman says they were cut into the turf (1987, 81), however Hutchins himself only says 'between the legs are certain rude letters scare legible' (1774, 292), nevertheless based on their absence in the very good results from the radar survey (Cheetham, Chapter 3) it is likely that Newman was correct in his supposition and that they might be an eighteenth-century addition later grassed over.

Although it had been postulated that he sported a pre-existing belt, geophysical survey and augering (Chapter 3) shows that this is unfounded, untenable and is almost certainly a sheep track, that can be seen on the hillside on photographs since the later 1970s if not before.

Nose

The nose has always been shown as lighter lines (i.e., not outlined as other facial features), indicating it took a different form to the outline and other features. In the 1970s it was noted as a physical mound of soil, but 20 years later had deteriorated; photographs from *c.* 1980 seem to suggest it was (almost) absent. The National Trust controversially reinstated it in 1993, with the aid of a wire mesh to improve stability and longevity, but without any real reference to its original form. The size and height of the nose was never known, so was created by the Trust to form a feature that they felt balanced with the face (see Edwards, Chapter 6).

Changes in the outline

The loss of the navel is well known and documented (Grinsell 1980a; Edwards, Chapter 5), nevertheless, the rendered photogrammetry images produced in association with the excavations, and the wider research conducted by primarily Brian Edwards (Chapters 5 & 6), allows us to briefly summarise some other more subtle changes in his outline than we discussed in Chapter 1. Other changes are that more of his anatomy was depicted as complete circles or roundels than exist now (head and testes) as Castleden pointed out previously (1996, 27; Chapter 1).

Giant on the move

The most significant change has been from the Giant outlined by shallow trenches to at least 1920/1925 (Fig. 7.3), to one with chalk filled trenches by 1935 when an aerial photograph seems, for the first time, to show him as a truly

FIGURE 7.6. Aerial photograph with a white Giant, taken 8 September 1935. © Ashmolean Museum, University of Oxford

white figure (Fig. 7.6). However, the National Trust research excavations have indicated that the outline of the Giant has slipped down the slope, especially since 1995 and coincides with their use of numbers of National Trust volunteers to assist in scouring, rewhitening and giving him a make-over. He has drifted downslope by about 20 cm since 1995 (Trench C; Fig. 3.14), and as much as 45–40 cm since 1945 (Trench B; Fig. 3.16). Some parts of the figure's outline are now considerably (nearly 50%) wider than only 50–100 years ago.

Creating the Giant and maintaining the Giant

The original Giant would have been cut through the shallow soil to expose the clean fresh white chalk. These trenches would, however, have been susceptible to rapid infilling with soil and soon support downland vegetation. The chalk base, if not regularly cleaned and maintained, would become weathered, grey, frost-shattered, and invaded by grassland flora (both vascular plants and bryophytes, especially *Festuca rubra* (red fescue), *Poa annua* (annual bluegrass) *Agrostis* (bent grass) and *Dactylis glomerata* (cat grass); see Ridding *et al.* 2020; 2024). Soil would not just have fallen into the shallow trenches, but would have washed down the torso, legs, and down the arms infilling the trenches at the elbows and the soles of the feet. It would need clearing out and the chalk refreshed by cutting away the dirty loose weathered surface. The weathered chalk rubble was always wholly and diligently removed from the hillside, but low banks of soil left around the outline. But, at the feet and elbows thin layers of soil spread adjacent to the outline combined with considerably more soil eroding down the steep grassland slope (under grazing herds, from burrowing herbivores, and even maintenance activities), flattened out the slope. The deeper soils, here in particular, would require holding back during any emptying of the chalk infill, scouring and maintenance activity, but only when the trench was to be neatly cut and was destined to be infilled with chalk (rather than left as open trenches). The presence, therefore, of stakes (presumably formerly holding and boards or planks) may relate to episodes of chalk infilling and scouring, as opposed to that of creating and maintaining a trenched outline.

The outline, therefore, for the majority of the Giant's existence, would have a shallow trench being deepened by deliberate cutting to expose fresh chalk and subsequently by repeated scouring and the removal of eroded soil and chalk weathered into the trenches, and eroded in by livestock. Hutchins, however, described 'cleaning the furrows and filling them with chalk' in the eighteenth century (1774, 219). This may be so, but there are a number of references then and after that date that refer to trenches that are 2 ft. When Sir Flinders Petrie surveyed the Giant in *c.* 1919 he does not report a white Giant, but trenches 'worn in the denuded chalk to two feet in depth' (1926, 9), and Lord Digby is said to have run up and down the trenches in the 1920–30s as a young child (Barker, Chapter 9) suggesting some depth to them (Barker pers. comm.). Thomas Hardy in 1925 also records trenches that are 'fairly deep' (Hardy 1925). These all concur with an aerial photograph of the same date (Fig. 7.3a) showing a dark trenched outline, as does Ravilious' painting of 1939 (Fig. 7.4 upper). So where had all the chalk infill gone? This leads us to two suppositions: firstly that the infilling with chalk (as reported in 1774) may not have filled the trenches to the brim as we see today, but raised their base with white chalk so that the outline could be (better) seen. Secondly, when the trenches were reported to be 2 ft or fairly deep, that does not necessarily refer to the entirety of the trenches around the figure, but probably just to some areas that were noteworthy and memorable because of their (localised) depth (i.e., the feet, elbow and possibly base of the club).

Wooden stakes

The remains of the wooden stakes found in the excavation were hammered into the edges of the trenched outline to support wooden boards during excavation, scouring, and rechalking as indicated, for instance, in the National Trust schedule of works (Appendix 3, B3) and their 1978 specification for works (carried out by E.W. Beard Ltd. in 1979). The presence of wooden stakes indicate a chalk-filled Giant. If the preserved stakes in the excavation had hand cut or axed points, rather than sawn or machined, it might indicate older stakes, perhaps pre-1950s. The wooden posts (all three of them) were not recorded in any detail on excavation other than their accurate recording on the 1:10 section drawings; they were after all modern and did not relate to the primary aim of dating the Giant. Two were retained in fragments and one left *in situ*. Although identified as spruce/larch (Barnett, Chapter 3) consistent with a recent date, records were only made of them three years after excavation (Grace, Chapter 3) where they had suffered further deterioration and desiccation. Nevertheless they are recorded as squared (but not square), but little information of the points could be discerned.

The point is (no pun intended) to define when from 1925 and up to 1956 the Giant was fully chalked-up. A trenched Giant is evident in 1925 (Fig. 7.3a), but fully chalked possibly as early as 1935 if the Ashmolean date ascription for Figure 7.6 is correct, and certainly by 1947 as seen on aerial photographs after Piggott's 1945 restoration. Correspondence between Piggott and Knollys regarding the 1945 restoration suggest that he was chalked by then (if not by them). The question is, did Piggott create the white chalk-filled Giant, or was this done in earlier scourings in 1932?

The trenched unwhitened Giant is evidenced, not from the excavation itself, but from the photographic and documentary evidence. Until the end of the nineteenth century (c. 1870s) and early twentieth century (1925) the Giant seems to have been a trenched outline (Figs 7.2 & 7.3), and by 1947 at least it is a white chalked figure.

Recording scouring and maintenance activities

No detailed records were made, or seem to survive, in the National Trust (or English Heritage/Historic England) records of any of the twentieth nor even twenty-first-century scouring. No reports of depths cut out, the nature of the chalk deposits removed, whether this digging out impinged on *in-situ* archaeological deposits (it did), or the methods and practice on site were made; nor any detailed photographic record, save those odd few photographs such as published by Keithley (Keighley) (e.g., Keithley *et al.* 1999, fig. 12), which are now apparently lost. This is true even of the more recent major scourings of 1979, 1995, 2008 and 2019. The current National Trust archaeologist, for instance, does not seem to have been required to write or prepare any such report; nor it seems did his predecessor. There is no formal or casual photographic record of the more recent scourings other those in the public domain (e.g., https://www.nationaltrust.org.uk/visit/dorset/cerne-giant/our-work-caring-for-the-cerne-giant), which largely comprise photographs of the National Trust volunteers (e.g., Fig. 13.5) rather than a record of the work, despite being undertaken under Scheduled Monument Consent. No formal reports were made, nor required of the scouring and replenishing of the chalk of the Cerne Giant, and no records of what was cut out and removed, despite the fact it involved digging away (recent) 'archaeological deposits and inadvertently commonly cutting into older deposit and stratigraphic relationships'.

Similarly, other sites, such as the Long Man of Wilmington, East Sussex (Sussex Archaeological Society) and the white horses of Uffington, Berkshire (National Trust) and Westbury, Wiltshire (Secretary of State/English Heritage), all scheduled or guardianship monuments, seem to have no reports formally documenting scouring, reconstruction and restoration works, and demonstrating that they were undertaken in accordance with an (agreed) method statement/schedule of works and completed satisfactorily with no archaeological impact to furnish either their owners' archives, or English Heritage/Historic England. For the Long Man this was just replacing or repainting the bricks/blocks that are only earth-fast (Bell & Butler, Chapter 15), but well-stratified colluvium associated with the figures lies at the footslope just outside the scheduled area. In many cases the scouring, rechalking and restoration of these scheduled monuments seems to be dealt with more as historic buildings, rather than ancient monuments and archaeological sites, or even just artefacts lying on the hillside without accompanying stratigraphy. Consequently consideration of an archaeological impact seems not to have been considered.

Because the hill figures often involve excavation and removal of previous (sometimes recent) deposits but potentially also of *in-situ* archaeological deposits they should be considered more as archaeological excavations and less as building renovation. This does not apply to all sites, as the Long Man of Wilmington restoration only required replacement or replenishment of bricks and blocks placed on and in the soil, and their repainting, and at the Westbury White Horse, the removal and replacement of the concrete skin over the bare chalk base (but even that could mask an earlier horse).

Perhaps a brief but formal archaeological record of these works is required? Although they are very specific in their location, they can interfere with the stratigraphic record of the history of scouring and rechalking events, but also potentially too, of *in-situ* archaeological deposits adjacent to the monuments form; i.e., deposits that are directly and inextricably related to the figure. Research excavations at the Giant and Uffington have shown the complexity and difficulty in understanding even relatively recent events, and this has not been helped by the lack of a simple record of scouring and replenishing activities – the type of records we expect, for instance, from any and every piece of archaeological fieldwork (geophysical survey, excavation etc.). This is not necessarily a criticism of the current, nor former, National Trust archaeologists and area rangers, but that this had not been required of them, and they had not felt that it necessary to do so. The research here, and at the Uffington White Horse (Miles *et al.* 2003), clearly show how complex the scouring and rechalking records are, and although some may be historically recent, others clearly have an antiquity and are important to the historic story of the figures.

Conclusion

The lack of any written records before 1694 is still perplexing even in view of the fact the Giant was a trenched non-white outlined figure and potentially in long grass for the majority of this time. As Ronald Hutton points out, it is still astonishing that he's not mentioned in any of the major antiquarian records. Despite the research and dating set within clear geoarchaeological reasoning, I still find myself still wondering about that gap of six centuries and asking why was he not mentioned anywhere over that time period?

7. The Giant's story: the archaeological results considered

Whilst this research has provided a clear set of scientifically based arguments and even a research agenda of further work to enlarge our understanding of the Giant, his fate, or origin, and purpose, can still all be questioned. The fear that the scientific analysis would provide all the answers and remove the mystery of the Giant is clearly not true. This research has shifted the focus of the Giant from a prehistoric (and Roman) and seventeenth-century one, to an early medieval one, that only summarily, if ever, was considered previously, and which provides another plank in the research of this ever-fascinating archaeological monument.

The Giant, created in the tenth century, seems to be essentially what we see today with no significant difference, and any sixteenth/seventeenth-century change is largely changes in cultural and social perception of him, as we see mapped to a certain extent in the Giant's timeline (below) and exemplified in part by the additions to and alterations of the Giant (see for instance Fry, Chapter 13). Patterns of change and repeated activity can be discerned. It seems, therefore, appropriate to use the new OSL dates to create a timeline of the Giant's creation and history.

The Giant timeline

Brian Edwards and Michael J. Allen

The principal aim of the Cerne Giant research project, as instigated and led by Martin Papworth, was to provide a scientific date for the Giant (Allen 2019). Both background research conducted to place the Giant into its historical context (Chapter 1), and more focused historical research (Edwards) has shown the Giant has a fascinating, if not complicated, association with the local community and wider world (see for instance Fry, Chapter 13). Patterns of change and repeated activity can be discerned. It seems, therefore, appropriate to place the new OSL dates into a timeline of the Giant's creation and history. The Giant has stood on the hillside at Cerne Abbas for over a millennium, during which he has been changed from a broad outline to a deeply cut trenched outlined and then the white chalk-filled outline we see today. He survived though 1000 years of changing ownership, politics, farming regimes and landscapes, and has endured periods of care, neglect, notoriety and ribaldry. The timelines below summarise some of the key events relating or pertaining specifically to the Giant.

The early medieval Giant

910–980	Creation of the Giant 1 (wide shallow trenched outline)
987	Founding or refounding of Cerne Abbey (Æthlemaer)
980–1200	Creation of Giant 2a (narrow precise trenched outline)
c. 1250	One of many phases of hillwash, suggesting an as yet undefined change in land-use
1250–	Phases of pasture and lessened grazing with long rough downland grass
c. 1300	Cerne Abbas St Mary's parish church built

A public history of the modern Giant

In summarising the steps through which the public have come to know and engage with the Cerne Giant, this timeline is not exhaustive and is only a guide. A particular motive was a timeline that would enable readers to reflect on groupings and patterns developing through on-site and select artistic responses to the Giant within living memory. Victorian concerns about the Giant's condition had first made this hill figure newsworthy, with subsequent restorations receiving media attention to the present day. With recognition as an official monument in the 1920s, the Giant had attracted the attention of both censors and advertisers by 1931. Threats from the former saw villagers keep watch on the Giant, but the earliest known defacement wasn't encountered until the Swinging Sixties.

A timeline for the modern Giant

Brian Edwards

- 1537 Event: Closure of Cerne Abbey; demolition begins
- 1617 Event: Giant not mentioned in survey of Cerne Abbas by John Norden and son
- 1694 Restoration: Churchwardens of St Mary's paid 3s to repair the Giant, 4–5 November (first record of the Giant)
- 1735 Culture: A local curate repeats folktale to the villagers of pinning down the sleeping giant that had feasted on (their) sheep. The apparent link with the popular *Gulliver's Travels* (1726) was overlooked until the following century (Britton & Brayley 1803, 483)
- 1742 Publication: Rev Francis Wise highlights the Cerne Giant was 'not to be seen at any great distance' (Wise 1742, 48)
- 1754 Event: On visiting, Richard Pococke notes that 'The lord of the manor gives something once in seven or eight years to have the lines clear'd and kept open' (Cartwright 1888)
- 1763 Publication: Anonymous letter account of Cerne Giant, *Royal Magazine* (Anon 1763a). Mentions three figures between the Giant's legs, possibly a date (Morgan Evans 1999)
- 1763 Publication: Anonymous letter in *The St. James's Chronicle* (Anon 1763b)
- 1763 Event: Letter from Rev Hutchins to Stukeley, 22 October, details the Giant 'furrows are about a foot or more deep' by two wide, and that 'The people sometimes cleanse the furrows and fill them up with fresh chalk' (Lukis 1883)
- 1764 Publication: Description of a gigantic figure. First printed image of the Giant, in *Gentleman's Magazine* (Anon 1764)
- 1772 Event: Three scarcely legible letters or figures are recorded between the Giant's shins below the previously noted row of numbers (Hutchins 1774)
- 1774 Publication: Posthumous publication of Hutchins' *History of the County of Dorset* (2 vols). Second printed image of the Giant: no penis, two rows of numbers/letters and symbols between the shins

7. The Giant's story: the archaeological results considered

1835	Publication: Earliest identified example of 'Giant's Hill' superseding 'Trendle Hill'
1842	Publication: John Sydenham, *Baal Durotrigensis*. The frontispiece: 'Colossal Figure at Cerne, Dorset' is censored and the navel is the same diameter as the nipples
1864	Publication: Report that 'of late years the chalk has not been laid bare' (*Southern Times* 1864, 4)
1867	Publication: Reported 'his mightiness didn't appear to us in good form' (*Dorset County Express and Agricultural Gazette* 16 July, 2)
1868	Restoration: Lord Rivers gives orders to have 'The Giant cleaned and restored, so that it may again look it did many years ago' (*Western Gazette* 20 March, 8)
1880	General Pitt-Rivers inherits the Giant
1883	Publication: A letter authored by Viator (pseudonym), outlines concern about the condition and preservation of the Cerne Giant (*Southern Times and County Herald* 1883, 3)
1885	In a revision of the 1835 account, a local labourer says the villagers tied down a sleeping 'Danish giant who led an invasion of this coast' (Gomme 1885)
1886	Event: Manufacture and installation of iron railings by master blacksmith Charles Way of Cerne Abbas, creates the coffin-shaped enclosure around the Giant
1888	Restoration: Restoration of the Giant completed in October.
1888	Culture: No association between the Giant and fertility is mentioned in Dorset folklore 'jottings' by Henry Moule, but J.J. Foster (London) promotes the idea that the Giant had a 'remarkable Phallic superstition' with 'counterparts' in 'Brittany and all over India (Foster 1888)
1893	Culture: Fertility features for the first time in the form of a 'superstition' posing as a perfect cure for barrenness in women if they sit on the Giant (Udal 1893)
1905	Publication: Media report that the Cerne Giant is under threat of vanishing 22 August
1905	Restoration: Guerilla gardening tidy up of the Giant in October
1907	Restoration: Some ladies cleaned the Giant (Vale & Vale 2000, 13–14)
1908	Restoration: A committee raise funds to restore the Giant under direction of Jonathan Hardy. Probable loss of the Giant's navel
1920	Event: Cerne Giant National Trust acquisition from Pitt Rivers Estate, 23 June
1920	Restoration: Giant cleaned
1922	Publication: The long awaited publication of Udal's *Dorsetshire Folk-Lore* injects belief into the superstitious link between the Giant and fertility (Udal 1922, 57)
1924	Restoration: Giant cleaned at a cost of £5
1924	Event: Cerne Giant listed as a Scheduled Monument, 15 October
1924	Event: Sir Henry Hoare endowment of £150 to maintain the Giant

1931	Culture: Shell illustration of *The Giant of Cerne Abbas* within his coffin enclosure by Frank Dobson
1932	Event: Following threats, villagers maintain a watch over the Giant
1940	Event: Giant's trenches are backfilled with bracken following decision in June to camouflage certain hill figures (e.g., Cerne Giant, Uffington and Hackpen)
1945	Restoration: Giant's trenches cleaned and the outline restored (led by Stuart Piggott and Eardley Knollys)
1945	Publication: Court action threatened after publication of aerial photograph in *The National Trust. A Record of Fifty Years' Achievement* (Milne 1945)
1949	Publication: It is remarked that 'the figure thus appears from a distance in greyish brown upon the green hillside' in Marples, *White Horses and other Hill Figures* (1949, 160)
1956	Restoration: Contractor restoration of the Giant
1968	Prank: Green and white hair, a bra and a vagina 'painted' on to Giant with the slogan 'Barnard was here' [reference to Dr Christian Barnard]
1974	Restoration: Six days of cleaning the Giant by a team of four
1977	Event: Giant re-fenced in much larger area including the Trendle (Keithley *et al.* 1999)
1978	Culture: Launch of the Lay-by Party held in September, the longest running annual calendar fixture focused on the Giant in modern times
1979	Restoration: Contractor cleaning/restoration of the Giant
1983	Restoration: Heineken sponsors British Trust for Conservation volunteers to rechalk the arms and vertical lines of the Giant for ITV Telethon event, 2 August
1988	Prank: A condom made of plastic sheeting covers the Giant's penis for an April Fool
1988	Restoration: British Trust for Conservation volunteers add fresh chalk for Telethon 88, a 27-hour non-stop charity appeal, 30 April–8 May
1988	Event: Fence removed 'from immediately around the Giant so that he can clearly be seen' (National Trust press release 5 May)
1989	Publicity stunt: Durex balloon lands next to the Giant, 7.00pm, 3 July 1989 (NT files)
1993	Restoration/Authorised promotion: National Trust 'restores' the Giant's nose
1993	Authorised promotion: Linking hands with the Giant (West Dorset Infertility Support Group) to mark National Fertility Week May (NT files: 25 March, 2 April)
1993	Authorised promotion: Giant holding a broom and sheeting to create logo for 'Great Dorset Clean-Up' in October (NT files: 'Public Affairs with the Cerne Abbas Giant' undated doc).

7. The Giant's story: the archaeological results considered

1993   Restoration: Rewhitening by Dorset Countryside Volunteers at request of Cerne Abbas Parish Council

1995   Prank: Condom painted on penis with white emulsion paint (*Express & Star* 29 June; 'The Johnny White Giant', *The Daily Star* 30 June)

1995   Restoration: National Trust centenary rechalking and top dressing in April

1997   Culture: National Trust agree to Men Behaving Badly 'Sofa' filming (see Chapter 13)

1997   Event: Giantess created by plastic sheeting next to the Giant, Bournemouth University project 10–11 July (Chapter 10)

1998   Publicity stunt: jeans made from purple plastic mesh added to promote Big Smith Jeans on 14 May. National Trust demand £5,000 access fee

2001   Authorised promotion: Giant sports a red kipper tie to shins and with a white knot, for Beating Bowel Cancer charity, for Loud Tie Day 4 November 2001

2001   Restoration: Dorset Countryside Volunteers and Cerne Abbas First School assist

2002   Publicity stunt: Placard as if held in the Giant's left hand reading 'I've got the balls to boycott Esso' by Greenpeace campaigners (16 May)

2002   Publicity stunt: Condom of grey latex sheet added re: Family Planning Association promotion of Sexual Health Week (*Bristol Evening News*, 6 August; *Birmingham Post*, 6 August)

2003   Restoration: Partial rewhitening of the Giant by the National Trust

2006   Authorised promotion: Rainbow added for Breast Cancer Awareness by Cerne Abbas First School (*Dorset Echo* 19 October)

2007   Publicity stunt: Homer Simpson in underpants holding up a doughnut, next to Giant using water-based biodegradable paint (*Dorset Echo* 2007)

2007   Publicity stunt: Giant's penis painted purple, and slogan added at the edge of the field to the Giant's right: 'Read Family Court Hell', 27 August

2008   Restoration: National Trust rechalking and refurbishment, with Dorset Countryside Volunteers, Association Orchis (Normandy), and Cerne Abbas First School

2009   Authorised promotion: Giant's nose covered in red plastic sheeting a month ahead of Red Nose Day

2010   Authorised promotion: British Legion poppy on left hand, 29 October

2012   Authorised promotion: Club converted into the Olympic torch, with the flame represented by pupils from local schools dressed in red, yellow, and orange

2013   Authorised promotion: Grass handlebar moustache added by British Seed Houses as Movember promotion for men's health issues in November

2015	Culture: A convincing off-site April Fool's Day prank sees a badger digitally added to an aerial photograph by Dorset for Badger and Bovine Welfare (*Dorset Echo* 1 April)	
2017	Publicity stunt: 'THERESA' election slogan added to shaft of Giant's penis, 'Standing up for Britain' (*Metro* 6 June)	
2017	Publicity stunt: Tennis racket added to the Giant's club and a tennis ball above his left hand with 'PP' encircled level with his right foot. Paddy Power publicity (*Dorset Echo* 3 July)	
2019	Publicity stunt: Flower added to top of penis, and note left in Cerne Abbas Village Stores cites International Women's Day, 8 March	
2019	Restoration: Rechalking and refurbishment by the National Trust for centenary of ownership of the Giant the following year	
2019	Authorised promotion: Heart above Giant's left hand comprised over 150 people for the launch of the Landscapes for Life Festival – 60th birthday of Dorset AONB (now Dorset National Landscape) and 70th anniversary AONBs and National Parks	
2020	Event: National Trust research excavation and OSL dating. 16–20 March	
2020	Prank: Coronavirus mask added to Giant (April)	
2020	Publicity stunt: Mankini styled on a disposable pandemic mask with slogans 'Wear Mask' and 'Save Live' either side and level with his testicles (October)	
2023	Union flag in left hand for coronation (*Dorset Echo* 7 May)	

Acknowledgements

We would like to thank Alfie Lumb, Catherine Bowles, Eric Howard, and Steve Fudge for assistance with the Giant's timeline.

Part 2
THE GIANT IN CONTEXT

Sheep grazing the Giant in July 2004 © Peter Stanier 2004

CHAPTER EIGHT

The Saxon Abbey of Cerne: an introduction to the abbey and recent archaeological research

Michael J. Allen and others

The Benedictine Abbey of Cerne, founded by Athemar or Ailmar, AD 987, being richly endowed with lands, became the seat of learning, and, no doubt, of much dissipation also: hence it is quite within the bounds of probability that, as 'Satan always finds some work for idle hands to do' the monks of that establishment (without having to impute to them any such dark inspiration), in conjunction perhaps with some of the townsfolk, may have occupied some of their vacant hours in portraying the lineaments of this legendary personage (Smart 1872, 69).

The Benedictine abbey was probably founded in the reign of King Edgar (858–975), but greatly augmented by the nobleman Æthelmær (or Ailmer), son of Ealdorman Æthelweard (Ælward). A document dated to 987 is one of the few written records from Anglo Saxon Cerne. There is evidence that Æthelmær the stout ('the Fat') was of royal descent (e.g., see Yorke, Chapter 5), and he and his father as 'members of the royal circle by both birth and public office … were naturally influenced by the new enthusiasm for the church and Christian leaning' (Yorke 1998a, 15). This was a time of monastic building all over the country (Farmer 1988), and although we introduced the abbey as a part of the archaeological background (Chapter 1), the OSL dates now bring this aspect of the Giant's life and landscape much more into the foreground.

There is much written about the abbey in staple reference works such as Victoria County History (Page 1908), Royal Commission volumes on Dorset, and papers in the *Proceedings of the Dorset Natural History and Archaeological Society*, and not least *The Cerne Abbey Millennium Lectures* (Barker 1988c), and contributions in the Cerne Historical Society Magazine (e.g., Wilkin 2021; Copson 2022; Foulser 2022; 2023; Willmott 2022; 2023; see in particular Yorke, Chapter 5). The history of the abbey seems to have been uneventful according to Page (1908), but by the late thirteenth century was the third wealthiest in south-west England (Knowles 1966, 702–3). The abbots of Cerne are known from 987 when Ælfric was appointed probably chaplain to Æthelmær, to Thomas Corton from 1524 to the dissolution of the monasteries in 1539 when he surrendered his abbey (Smith 2001–2008). Whilst there are numerous historical records for the abbey, there is very little in the way of documentation for the Saxon period. There has been little work on the archaeology of the non-extant remains of the abbey until the recent and current *Work of Giants* project (https://workofgiants.org) undertaking research, survey and excavation in the abbey (2022) led by Hugh Willmott (University of Sheffield) and Helen Gittos (University of Oxford), see below.

Saxon abbey

The Saxon origins of the abbey probably lie in a royal estate centre with its own church, at which the monastery was founded in the tenth century. Its income then placed it amongst the medium sized monasteries in the country, but one of the most important in Dorset; in the tenth century there may have been two priests and initially five or six monks, rising significantly in the eleventh century. The abbey was, however, founded towards the end of a phase of renewal of monasteries initiated by Edgar (r. 959–975) and Dunstan, archbishop of Canterbury, who died the year after Cerne's first foundation. Their renewal promoted the Benedictine customs of monastic life but while these monks may have expressed an affiliation to the Rule of St Benedict, the degree to which they followed it strictly, routinely, cannot be known. But the scale of the monastery would have been at its dissolution one abbot and 16 monks (Bettey 1999, 76). It exceeded the combined income of Sherborne and Abbotsbury (Farmer 1988) and was endowed with estates spread right across Dorset with a very sizable income (Bettey 1988). In the eleventh century the abbey acquired the relics of St Eadwold (reputedly the brother of the East Anglian St Edmund King and Martyr) and he became the main medieval saint of the abbey. Virtually nothing survives above ground of the Saxon abbey buildings, there is a relatively good record (summarised in Barker 1988c), and other accounts.

The abbey is known in particular for two exceptionally important works of Saxon literature. Ælfric, the first Abbot, also known as Ælfric of Eynsham (c. 950–c. 1010) and his pulic (Ælfric Bata) are the authors of an important educational text. While at Cerne, Ælfric produced two parallel sequences of writings; one, two collections of homilies, aimed at the spiritual instruction of monks and clergy, the other, three teaching books on Latin, aimed at the instruction of those just entering on a career in the Church. The Colloquies are edited and added to by Ælfric's pupil, Ælfric Bata, to such an extent that some scholars now identify Bata as their author. These writings, *Colloquy on the Occupations* (aka Ælfric's Colloquy), are a dialogue between a schoolteacher and his pupils (written between 992 and 1002). They are said to be warm, lively and revealing (Farmer 1988) and were an aid for pupils (in both the monastery and village) to speak Latin (Harris 2003). The text survives as a number of manuscripts and the translated text was edited by Garmondsway (1991); it includes narratives with, and about, the daily lives of ploughmen, cooks and merchants as well as huntsman and knights, and includes comments from a fisherman, smith and carpenter (Harris 2003). This is the best, but not the only such Anglo-Saxon work in, and from, England (Farmer 1988, 7).

The other is one of the treasures of Cerne Abbey; a beautifully illustrated Saxon book of prayers, the *Book of Cerne*. It was owned by the Bishop of Lichfield from AD 818–830 and was later bound, possibly as late as the sixteenth century, with medieval manuscripts into one volume at Cerne Abbey. It was taken from the abbey library at the Dissolution in 1539 and then sold by the Bishop of Norwich and Ely to George I. The king presented the book to Cambridge University, where it is now stored (Brown 1996). It is a composite manuscript made up from fragments that originated from other books, and largely comprises a ninth-century prayer book written primarily in Latin, with some sections in Old English. It is thought to have been written in the West Midlands (Mercia), and although bound at Cerne Abbey and once held in their library, none of the Anglo-Saxon writings themselves are associated with Cerne.

The importance of the abbey is demonstrated in that the privilege of wrecking (for the many vessels wrecked on the coast) was granted by royal lease to the 'Monastery of Cerne' from the reign of Edward III (r. 1327–1377) to that of King Henry VIII (r. 1509–1547) after which that privilege passed to Sherborne Abbey (Damon 1884, 72). The abbey was surrendered to Henry VIII on 15 March 1539 (Bettey 1988); all of the remaining Dorset Benedictine monasteries were closed in a 12-day period between 11 and 23 March 1539 – the last of which was Shaftesbury. After the Dissolution its lands were dispersed and the abbey was almost totally and comprehensively demolished, save for the gatehouse. Demolition seems to have started very soon after the departure of the monks and the village lost its livelihood (Bettey 1988). The manor of Cerne Abbas passed through a series of private landlords, who systematically stripped the abbey of its materials, to such an extent that the Rev John Norden, nearly 80 years later, recorded the abbey to be 'wholly ruinated' (1617). Much of the masonry was used in the rebuilding of Cerne Abbas and recognisable fragments of the abbey can be seen in many buildings and houses around the village and its environs, including the Royal Oak, in Cerne Abbas, which claims to have been built in 1540. One might have expected more to have survived and to have been incorporated into a local manor house. Abbey churches often became parish churches – but Cerne already had a suitable one. The gatehouse or porch, however, was curiously left standing after the Dissolution but fell into decline to become part of the farm buildings. Only in the 1990s was the roof finally replaced and the building restored.

At the Dissolution the abbey library was dispersed. Among the few books that survived was the *Book of Cerne*; manuscripts from the Cerne Abbey library are now to be found in Oxford, Cambridge and London.

Location of the medieval abbey church (Figure 8.1)

The general location of the medieval abbey with the surviving elements of the south gatehouse, abbot's hall and guest house is at the end of Abbey Street encompassing Abbey Farm (gatehouse), Beauvoir Field and the scheduled earthworks to the east. This area is, presumably, also the location of its Saxon predecessor. However, the precise location and arrangement of the medieval abbey church

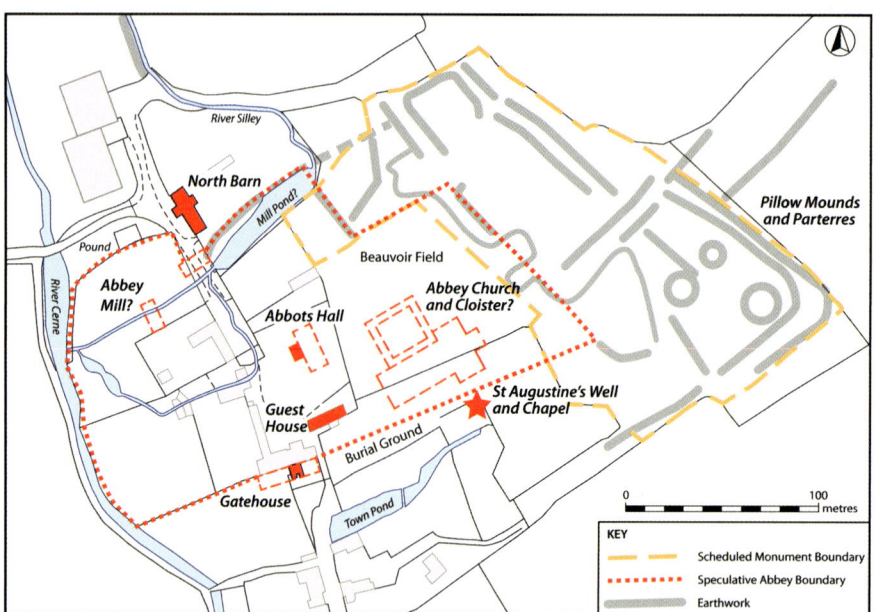

FIGURE 8.1. Plan of the abbey site showing extant earthworks, the Scheduled Monument area, and location of extant remnants of the abbey buildings (solid) and conjectural plan of other abbey buildings and abbey features (dashed), grey extant buildings. Image © P. Bellamy, Terrain Archaeology 2023

(and cloister etc.), has long been sought (Pope 1901), and even as recently as 2021 it was reported that 'We do not yet know where the Cerne Abbey Church sat' (Wilkin 2021, 5). The abbey was rapidly destroyed in the dissolution of the monasteries (Bettey 1988), and seems to have been razed to the ground, with few upstanding remains. Most of those few fragments of the abbey buildings that survive, with the exception of the abbot's hall porch (popularly incorrectly known as the abbot's gatehouse), became incorporated into later buildings. So total was the destruction of the abbey, that the location, let alone the full plan, of much of the complex has been, until now, speculative. The abbey has always been considered to be to the north of St Mary's Church and east of Abbey Street, largely occupying an area of open land encompassing Beauvoir Field, the current burial ground and St Augustine's well, and to a lesser extent the scheduled earthworks. In Figure 8.1 a number of the extant buildings are indicated, and the location of others postulated. The location of the cloisters has been taken from Willmott's GPR results (2023, figs 4 & 5: Fig. 8.3), which have been geolocated. As for the abbey church, from the geophysical results (see below) it appears most probably to lie south of the cloisters, rather than to the north as would conventionally be the case. Obviously the size and plan-form are pure speculation, but is consistent with other monastic churches. In its medieval phase it may, therefore, have taken a similar form of the plan of Sherborne Abbey but without the former parish church of All Hallows on its west end (Fig. 8.2). Note that in Figure 8.1 the barn at Silley Court is purposely labelled North Barn as this is what it had been known as for the past century or so, before the late twentieth century.

Nevertheless, as we have seen (above and in Chapter 1), the surviving elements of the medieval abbey complex are: the North Barn (now known as Silley Court), probably located outside the former north gatehouse and the

8. *The Saxon Abbey of Cerne: an introduction to the abbey and recent archaeological research* 187

FIGURE 8.2. Plan of Sherborne Abbey as an indication of the nature of that at Cerne Abbas. Image: Justin Russell 2024

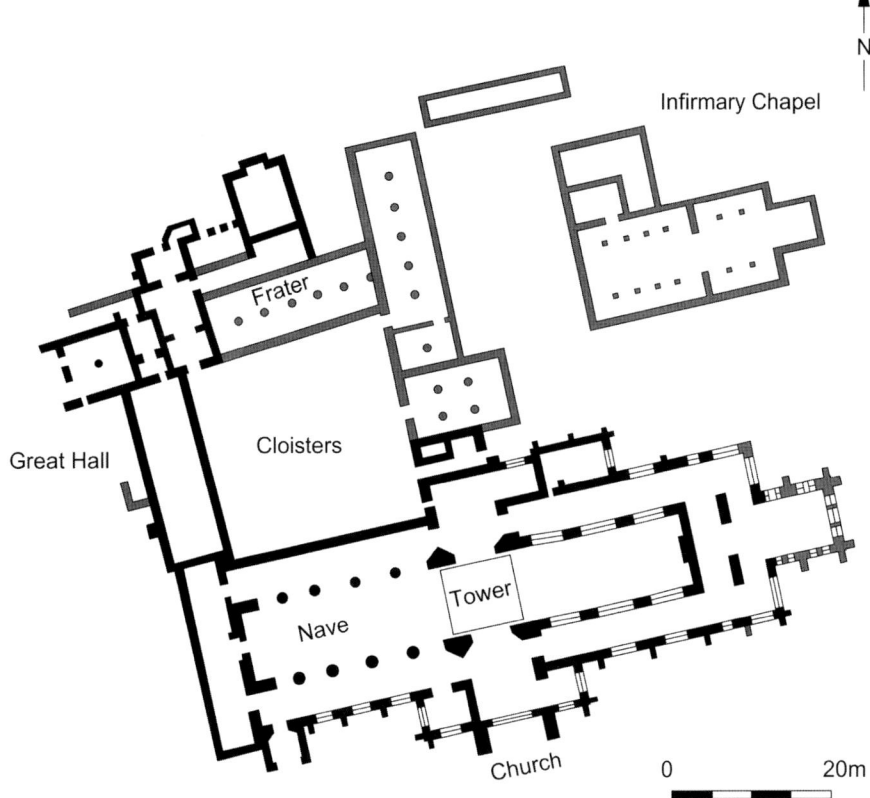

abbey precincts; the Guest House, probably the abbot's lodging; and the South Gatehouse, a portion of which is now incorporated within Abbey House; and the Abbot's Hall porch standing alone in the grounds of Abbey House, which according to some is one of the finest pieces of architecture in Cerne Abbas. In addition, there is the partial survival of a double row of houses of jettied construction that lined both sides of Abbey Street. These date to *c.* 1520 and were almost certainly built by the abbey, lining both sides of the approach to the South Gate. A large fifteenth-century Tithe Barn, now a private house, lies just to the south of the village. What we are missing are all the typical components of a Benedictine abbey such as the abbot's hall, the impressive abbey church and cloisters with reredorter and refectory, the north gatehouse, and an abbey mill and other possible sundry buildings and features, which might include a brewhouse and fish ponds.

In search of the abbey: new research in 2022 and 2023

Although there are a few finds from the abbey, such as masonry, carved statues, and floor tiles (Copson 2022), the general location of the site is clear, a significant portion of which falls outside the scheduled area. Copson reports that a concentration of fourteenth- to fifteenth-century decorated encaustic tiles (high status tiles decorated with an impressed design filled with white slip) have been found close to the north wall in the burial ground and may indicate the proximity of the abbey church (2022, 15), some bearing the royal arms of England from 1189–1360, which have been reconstructed and is now in the County Museum (Bishop 2021).

In grave cuttings in the same area, heavy spreads of demolition rubble up to 1 m deep has been noted containing mortared flints and fragments of Purbeck limestone and Ham Hill stone (Copson 2022). Another clue to the location of the abbey church is a funerary effigy of an abbot of Cerne of *c.* 1215, found just north of the north wall of the Burial Ground in 1815 in a ditch (Hutchins 1870), and now in the Cleveland Museum of Art in Ohio, USA.

In 2020 the Cerne Historical Society decided to make a renewed effort to discover where the buildings of Cerne had stood (Bishop 2021); 'It's difficult to believe that the Abbey Church, a building possibly on the scale of Sherborne Abbey, can have disappeared without leaving any visible evidence as to its exact location' (Clark 2020). The whole complex would have included the abbey church and cloisters, but also a chapter house, a refectory, a kitchen, a cellarium, a dormitory, a warming room, a reredorter (latrines), following a typical Benedictine pattern (see Bishop 2021). A full archaeological research programme, *In Search of the Abbey*, commenced in 2022 led by Hugh Willmott and a team of archaeologists from the University of Sheffield, in conjunction with the Cerne Historical Society. This comprised geophysical survey (GPR) in 2022 and excavation from 2023 onwards.

Key amongst the research questions were: ascertaining the precise location of the church, characterising the development of the cloistral ranges, identifying evidence for the pre-Conquest monastery, and understanding the impact of the Dissolution and later developments on the site (Willmott 2023), and importantly finding and characterising the Saxon monastery.

Previous geophysical surveys by Bournemouth University (gradiometer and resistivity) in 2011 (Bournemouth University 2014), and magnetometry by Leigh-Smith (2018; see Wilmott 2023) had limited success. However, in 2022 using ground penetrating radar (GPR) the survey produced clear outlines of the medieval monastic buildings for the first time (Fig. 8.3). They provide a definitive location for many of the key buildings (cloistral range) associated with the later medieval monastery, but only answer relatively superficial questions concerning the location and composition of the abbey complex. It cannot provide any chronology. Consequently, geophysical survey was followed in 2023 by targeted excavation to provide a chronological and social context. Although the project is in its infancy in 2023/4, the initial results have been stunning. The geolocated position of the cloisters and speculative location of the abbey church is given in Figure 8.1. Excavations of two trenches (Fig. 8.3) indicated the presence of monastic buildings: the cloistral walkway, chapter house, and a part of the abbey church. By the end of the first excavation season (2023), Hugh Willmott's research had not yet found traces of the Saxon, pre-Conquest monastery, although it is expected that future seasons of fieldwork will.

8. The Saxon Abbey of Cerne: an introduction to the abbey and recent archaeological research

FIGURE 8.3. Cerne Abbey: approximate location of the 2023 archaeological excavation trenches (A & B) in relation to Hugh Willmott's GPR survey: based on Willmott 2023, figs 5 & 6. Includes OS data © OS licence 100023974. Image: Justin Russell 2024

The abbey and the Giant

What is important to the Giant about the abbey is that if the general OSL dates are to believed, he was established on the hillside at about AD 940–980 and there are two sets of dates (from the right elbow and right foot) centred on *c.* AD 1250. Obviously the former seems to be coincident with the establishment and the foundation of the abbey in AD 987. It also seems coincidental that the later pair of dates from different elements of the Giant are so similar (AD 1250), but nothing obvious can be seen to be associated with the abbey, at least around this date. The potential relationship of the Giant and the abbey is explored in other chapters.

Acknowledgements

This chapter was originally agreed to be written by Hugh Willmott who was just starting his second year of the research fieldwork on the site, but unfortunately that commitment prevented him from contributing. Key members of the Cerne Historical Society were similarly committed. Instead, we thank, in particular, James Clark (University of Exeter), Barbara Yorke (University of Winchester), and Chris Copson (Cerne Historical Society). We also thank Brian Edwards, Sarah Fry, and Katherine Barker, for comment and information, and Peter Bellamy for producing Figure 8.1 especially for us.

CHAPTER NINE

The tenth-century Cerne Abbey: Benedictine ecclesiastical reform and land management

Katherine Barker

The dating of deposits in the lowest part of the gullies that form Giant 1 to the later tenth century (Toms & Wood etc., Chapter 3) immediately attract attention with reference to the founding of the Benedictine Abbey sited along the Cerne valley below this steep south-facing hillside; an Abbey community and estate that were legally 'registered' by written charter in AD 987 (Finberg 1964, 174, no. 613). Following the Benedictine reforms of the tenth century, monastic communities were to form an important part in what we now refer to as 'local government' to 'the general life of towns, villages and estates around them'. The role of the king here 'was extremely important; so also was that of laymen who helped to found monasteries or restore old ones' (Farmer 1988, 1). Suggested here is that the dating of these samples relates *not* to the 'making' of this overtly pornographic, 'non-Christian' hill figure, but to the management of this hillside, which was to follow on the establishment of the Benedictine Order. *The Cerne Abbey Millennium Lectures* (Barker 1988c), celebrating the thousandth anniversary of the Cerne Abbey charter took as their focus the monastic reform of the tenth century (e.g., Barker 1988b).

The Cerne Abbey charter of 987 was drawn up in the final decades of a troubled century; the pagan Viking threat from the north and the non-Christian Arab, Saracen threat from the south were seen as the 'harbingers of the beginning of the End' (Barker 2005a, 49). After a number of years of peace, Viking raiders had resumed their activities in the 980s, but a more substantial force had arrived in 991 (Keynes 2005, 58–9). Entries in the *Anglo-Saxon Chronicles* record that in 982 'three private crews landed in Dorset and ravaged Portland' and in 997 'they entered the estuary of the Tamar . . . there they burned and slew everything the met', and a year later in 998, 'turned eastward again into the mouth of the Frome and pushed up into Dorset in what every direction they pleased'. These were years that were also to experience both great famine and 'great pestilence among cattle' (Barker 2005b, 129–30).

Vikings had long been regarded as instruments of divine punishment for the sins of the English, and their activities made the king and his councillors more acutely aware then ever of the need to do what was pleasing to the sight of God (Keynes 2005, 59–60)

To the Christian world this was one of the signs of the 'last days' leading up to the Day of Judgment, the year 1000. The final decades of the tenth century were to witness the approach of the 'Second Coming'; Christian communities had, in short, to 'set their respective houses in order' (Reuter 2005).

It was King Edgar who was patron of the monastic reform of the tenth century. The pioneer of this monastic revival was the nobleman St Dunstan who, after a chequered career at the royal court, was installed as Abbot of Glastonbury in 940. Exiled to Ghent by King Eadwig he soon returned to become bishop of Worcester, London, and in 960, Archbishop of Canterbury. He was to be assisted by two other men of almost equal ability. The first was Æhelwold who became bishop of Winchester in 963 and both he and Dunstan were Latin scholars and able craftsmen. Oswald, the third leader of the revival, became Bishop of York in 972 (Farmer 1988, 2–3).

When the monastic life of all three founders was united through the customary called the *Regularis Concordia*, monks from Fleury and Ghent took part in the council at Winchester, which drew it up in about 970 (Farmer 1988, 3). This was an important document that safeguarded the reformed way of monastic life, relating not only to the Liturgy, but to daily study and work, and which included not only the preparation of parchment, book writing and book binding, but also teaching the young, and provision of hospitality and alms-giving – 'the last two were especially important and were often on a considerable scale' (Farmer 1988, 3). Farmer goes on to note that wealth was then mainly in the form of land and it was by way of foundation charters that founders made over estates, as did the king; monies that assisted in the life of those local communities and to ensure Divine blessing. St Benedict was the guide as to these far-reaching ecclesiastical reforms, which, as noted above related not just to the Liturgy, but to what we would now describe as wider 'estate management' and the legal chartering of ecclesiastical and monastic estates. This set the scene for the regulating of rents, often of market rights and, as also noted above, to teaching and care of the sick and destitute.

The Cerne charter of 987 records major endowments made by Æthelmær who belonged to one of the leading families of the tenth century, and both he and his father Æthelweard 'were naturally influenced by the new enthusiasm for the Church and Christian learning' and which promoted the drawing up of a legal charter … 'there was frequent discord among the nobility' (Yorke 1988, 15–6). Whilst Æthelmær may not have been the original founder of the Abbey, his rights to the tithes of Cerne and its surrounding area provides support for his inheritance of a former church property. It appears that the monastery at

Cerne was already in existence when Æthelmær produced his charter of 987, which records his reorganisation of the Cerne lands. Attention is drawn to the absence of a 'home estate' from this charter, which may imply that there was already a Christian community here and which he already held (Yorke 1988, 23–4). His death was to be followed by the reversion 'of the vills of *Cernel*' the estates to Cerne Abbey … 'the tithes of all the yearly produce in Cerne [and Cheselbourne] to go to the abbey, with tithes of honey, cheese and fat hogs from his other lands' (Finberg 1964, 174, no. 613; Barker 1988b).

The later tenth-century dating of the soil samples from the Giant also draws attention, not only to the dating of the abbey charter, but also to the literary legacy of Ælfric. Both Æthelweard (son of Alfred) and Æthelmær were patrons of the Winchester monk Ælfric who was sent to Cerne at the request of Æthelmær. Ælfric, one of the best scholars of the time, was transferred from Winchester to Cerne around 987. He was an accomplished and prolific writer, the greater part of his work dates from the years he spent at Cerne, 987 to 1002. He has been described as 'one of the great luminaries of Benedictine letters' (Swanton 1975, 106), and duly makes reference to the approaching 'apocalyptic year' of the millennium, in his declaration that 'Thousand is a perfect number' (Campion 2005, 33).

It was the 'Trial' conference of 1996 that prompted a careful reading of Ælfric's work (Barker 1988b; 1999; Chapter 10). We learn from Ælfric that tithes were to be paid to the priest and to be divided into three parts; one for repair, one for the poor, and one for God's servants' (Selbourne 1892, 260). He is quite specific about the administration of heaven; his pre-occupation with numbers and number patterns we find reflected in the Domesday hidage of the Cerne valley. Attracting attention is his reference to what we are given to read as 'pagan practices' relating to Blackmore, the open tract of common land just north of Cerne. He also makes reference to Biblical 'giants' but there is, in short, no allusion to anything we may read as relating to the Giant hill figure (Barker 1999).

The dating of these soil samples to the (later) tenth century thus coincides with both the forthcoming millennium and to the monastic reform of the church in the (re-) establishment of the Benedictine Order in complement of what, as suggested above, we would now describe as to estate management. This may suggest that the dating of these soil samples relates, not to the 'making' of this giant hill figure, but to the 'management' of this hillside as to further its grazing potential for sheep. Sheep were of major importance to the economy, 'they are our oldest domestic animal … their tenfold purpose – meat, fat, blood, wool, milk, skin, gut, horn, bone and manure – provided us with food, clothing, housing, heating and light … and parchment, which for centuries was the material upon which a permanent written record could be preserved' (Walling 2014, xix–xx) and which attracts attention to the literary legacy of Ælfric. We no longer use tallow for candles as we did during the eighteenth century when demand across Europe was such that the fat from a sheep's carcass was worth twice as much as the meat (Walling 2014, xx).

Introduced into Britain by the Romans, by the time of the Domesday Book of 1086, a century after the Cerne Charter, there were more sheep across England than all the cows, pigs and poultry put together. Many of the large-scale owners of sheep were monasteries, with monasteries acting as 'central depots' for fleeces from outlying farms and granges. One of the earliest records in 679 shows King Wihtred of Kent granting pastureland for 300 sheep in what is now Romney Marsh (Coulthard 2020, 178). Dorset has been a sheep county for many centuries and the Dorsets a hardy breed whose 'manure maintained the fertility of the light flinty Dorsetshire soils ... hillsides grazed by flocks' (Walling 2014, 178–9).

It's worth noting that the tradition that Hutchins (1774, 293) presents with reference to the Giant 'brings together' the Giant, Blackmore – and sheep. He notes that according to tradition there was a giant 'who resided here in former ages' and this local 'pest and terror ... made an excursion into Blackmore, regaled himself with several sheep, retired to this hill and lay down'. The locals then 'seized the opportunity, pinned him down and killed him and then traced out the dimensions of his body'.

It is reference to sheep grazing that attracts attention here relating to the width of the gullies that define the Giant hill figure and – not least – to their depth presented by recent work (Chapter 3; Fig. 3.14). Recorded as being at least 2 ft deep (60 cm) since the 1920s (Petrie 1926, 9) these gullies ran across hundreds of yards of this steep-sided hillside; an aerial photograph of the Giant taken in the 1920s presents a distinctly faint outline (Fig. 10.1). Also noted above (Chapters 1, 3 & 7) Hutchins (1774, 293) records that scouring took place 'about once in seven years by cleaning the furrows and filling with fresh chalk'. But this infill was not to bring the Giant outline to the surface until more recently. In short, that these deeper troughs have since been filled with a succession of rubble chalk and kibbled chalk leaving, for the most part a white chalk figure 'painted' on the hillside though he has not always looked so well defined (Chapter 7).

Not hitherto considered is the serious risk presented to grazing sheep injuring themselves in falling into these deep gullies – and being unable to climb out. This may go far to suggest that the digging out of this hill figure on a Benedictine estate in the tenth century was only shallow trenches, but by the post-medieval period were deeper slots and potentially a hazard to livestock.

Further to matters of dating, the recent work has demonstrated there are two giants: the tenth-century infill of the shallow Giant 1, which is in part sealed by later hillwash accumulating against the outline of Giant 2 in the mid-thirteenth century. That a shallow tenth-century cutting was re-dug creating a narrow trench against which later colluvium accumulated. This may relate to mid-thirteenth-century management of the hillside, and possibly to changes of land-use or farming practices. This presents a case for that presented by Hutchins, that this hill figure was cut by Denzil Holles, by 'lord Holles's servants during his residence here' in the 1640s, all but a century after the Dissolution when this estate was no longer held by the Church (Hutchins 1774, 293).

As noted above, the tenth-century dating of these deposits does not present evidence for the origins of this hill figure, but relates to the management of the hillside at that time – and again in the mid-thirteenth century. The dates all come from the colluvium (hillwash) spoil and soils eroding and coming down this steep-sided hillside coming to rest against the chalking and soils forming downslope below the chalking. The deposits dating to the 1250s also relate to that of colluvium coming downslope and which may – or does – relate to the later management of this hillside; these are deposits stratigraphically over the G1 soil-filled Giant that gave the tenth-century date.

CHAPTER TEN

The Cerne Giant: an antiquity on trial 1996: a summary

Katherine Barker

What prompted the Trial?

How old is the Cerne Giant? Living in Dorset from my schoolgirl days the Giant was understood as being prehistoric. In 1987 Bob Machin of the Bristol University Dept of Extra-mural Studies organised a conference in Cerne Abbas in celebration of the Cerne Abbey Benedictine millennium of AD 987–1987 (Barker 1988c). At that conference I first met Vivian Vale who, with his wife Patricia, on his retirement from Southampton University came to live in Cerne Abbas, and who drew my attention to the first known written reference to this hill figure in the Cerne Abbas churchwarden's accounts of 1694, which reads 'For repairing of ye Giant 3s' (see Morgan Evans 1999; Vale 1999; Edwards, Chapter 6). Without regular maintenance hill figures will grow over quite quickly and we are given to understand that if the Giant had already been in existence there is no hint as to who might have 'repaired' him over those earlier years, nor indeed at the time of the Trial (1996) as to his age. Its siting on the south-facing hillside above Cerne Abbas Church attracts attention here to the use of ecclesiastical funds to maintain this overtly 'non-Christian' hill figure.

Originally fenced in 1886 in his own coffin-shaped enclosure, over 100 years later, in 1977 the Giant was re-fenced and enclosed in a much larger area as to assist in its protection. On the hill slope above the Giant are Bronze Age barrows and a rectangular enclosure, which has never been excavated, suggested as a late prehistoric site or possibly a primitive Late Iron Age/Romano-British temple (see Chapter 1). There is, however, no evidence to associate this with the Giant. A photograph of the 1920s shows a much fainter Giant than we see now (Fig. 10.1), its shape defined by the grassed-over gullies.

It has recently been suggested that if the Giant does date from before the seventeenth century it may have periodically been 'invisible' and only in hot, dry weather would these gullies have presented this distinct outline (see Chapters 1, 3 6 & 7). Perhaps most attention is drawn to the location of this overtly pagan – pornographic – hill figure overlooking the former Benedictine

FIGURE 10.1. Detail of the Giant from an aerial photograph on a 1920s postcard

Abbey, its estate formally chartered in AD 987 (Yorke, Chapter 4). Since the eighteenth century the Giant has been identified with various mythological figures going back as far as the end of the sixth century when it is claimed that St Augustine met with a hostile reception in his founding an abbey to 'oust' the pagan god Helith. In 1938 Stuart Piggott made an association between the Giant and the Roman god Hercules noting the naked form and the club as common characteristics. A persuasive case was made by Joseph Bettey (1981; 1999) that this hill figure found its origins during the Civil War and is a 'lampoon' of Oliver Cromwell. The outrage of local Dorset communities at the wide-scale damage affected by both sides, King and Parliament, prompted the rising of the 'Clubmen' during 1645 in defence of their estates. At this time the Cerne estate was held by Denzil Holles who played a leading role in the political life of the nation since the 1620s and who was infuriated by the triumphant progress of Cromwell. Whilst Hutchins (1774) informs us that he learned from the steward of the Cerne manor that 'the figure was made by Lord Holles' servants', he goes on to note that some local people 'averred it as there beyond memory of man'. In short, the origins of this hill figure were already uncertain by the mid-eighteenth century.

The Trial

In 1996 I led a Liberal Adult Education programme for Bournemouth University – then still a 'new' University – and this prompted me to organise a day to find out what we know about this Dorset hill figure and what his age was, modelled on the lines of a public inquiry. In short we would test the assumption 'does something which *looks* prehistoric have to *be* prehistoric'? We would make adversaries face each other and place their case, in public for the public. The Trial was held in the old Cerne Abbas village hall on 23 March 1996, attracting a capacity audience (over 120 packed into the hall), and was filmed by BBC West (Beeston 1999).

Organising this along the lines of a public inquiry required a legal professional to conduct the proceedings, along with others familiar with the proceedings of a public inquiry. A leading Bournemouth barrister, Colin Patrick, was happy to take this on accompanied by David Morgan Evans, then General Secretary of the Society of Antiquaries of London, and Ivan Smith, Regional Land Agent with the National Trust. Leading academics led the three fields of inquiry; the first, that the Giant is of prehistoric/Romano-British origin, second that he is of medieval/post-medieval origin and third, the case for a 'living Giant' – that this hill figure is of significance irrespective of its age. Professors Timothy Darvill, Ronald Hutton and Barbara Bender acted as advocates for each case respectively and each introducing a succession of expert witnesses. The audience were invited to act as a jury, to cast their votes at the beginning of the day – and again at the end.

In support of an ancient Giant were Timothy Darvill, 'A prehistoric warrior-god?', Paul Newman 'In defence of antiquity', David Miles 'The Uffington White Horse and its antiquity' (see Chapter 17), Rodney Castleden 'Iconography and the identity of the Giant' (see Chapters 6 & 17) and Bill Putnam 'The Cerne Valley in prehistoric and Roman times' and concluding with John Gale's paper on 'The 1996 geophysical survey of the Giant' (see Chapter 1, previous work) and that, in short, there was 'no evidence … for additional features recorded by previous resistivity surveys … however, this absence of corroborative evidence does not disprove their existence and could simply reflect the poor ground conditions' (Gale 1999, 62). Attracting most attention here was that the Uffington White Horse on the Berkshire Downs 'is the only hill figure known to be ancient', is 'first recorded in the 12th-century Abingdon Cartulary' and is a figure that presents a highly stylised outline (see Chapter 7; Miles 1999, 39). Small-scale excavations indicated a construction date close to the first millennium BC and which had remained 'a highly symbolic focus for local communities for millennia' (Miles 1999, 41). In short, a hill figure *can* last for a long time.

This was followed by the case for a post-medieval giant led by Ronald Hutton, 'A seventeenth-century marvel?' with evidence presented by Vivian Vale, 'Churchwardens and the other God', and David Morgan 'Eighteenth-century descriptions of the Cerne Abbas Giant'. I spoke to 'Medieval Giants: Bible, history, heritage and landscape'. Hutton noted that it was Harvey Darton in 1935 who first noted problems relating to the dating of the Giant.

Back in 1981 Joseph Bettey had first systematically worked through the wealth of documentary evidence relating to Cerne Abbas, in particular bringing attention to the remarkable absence of any reference to the Giant until the end of the seventeenth century (Bettey 1981). He attended the Trial but did not present for this case, instead reference to his published work was widely referred to, and in part introduced by him during the audience questions. He did, however, summarise his work for the Trial publication as 'The Cerne Giant revisited'; this was so comprehensive that it negated Ronald Hutton reproducing much of his presentation on the day. All the witnesses called to the Trial contributed to the publication except Dr Keith Walker whose views on the relevant Elizabethan literature were included in Ronald Hutton's section.

Following the Dissolution of the abbey the Crown employed John Norden whom Bettey describes as 'a noted surveyor' to conduct a thorough survey of the Cerne Estates, listing all the lands, pastures and common grazing – including the 130 acres on Trendle Hill overlooking the abbey site, where he makes no mention of the Giant, 'notwithstanding the fact that in other surveys he was at pains to note archaeological features or historical curiosities'. Sited along the main road between Dorchester and Sherborne this hill figure would surely have hardly gone unnoticed. John Leland made reference to Cerne on his travels through Dorset but makes no mention of any Giant hill figure in his *Itinerary* of 1535–1543 (Leland 1770; Toulmin-Smith 1964).

In Part 3 'The Living Giant', Barbara Bender noted that we have learned that 'The Giant is over 2000 years old … Professor Darvill may be right … or he is 2 or 300 hundred years old and perhaps Professor Hutton is right'. She goes on to comment that 'we are obsessed with the origins of the Giant'; my case is that biography is as important as date of birth. The paper supported by nine witnesses including Tom Williamson, who spoke to 'Memory, tradition and the Giant'.

The day was ably chaired by Colin Patrick who presided over both the proceedings of the Trial and the question time afterwards (Darvill & Barker 1999). Ronald Hutton's case for a post-medieval origin of the Giant attracted great interest, and who 'put on a brainstorming performance' and certainly played an important part in increasing the seventeenth-century vote. One report said that Hutton seemed to have 'an adrenaline rush' and 'put on a terrific show'; at one point he 'gestured extravagantly' towards Tim Darvill inviting him to 'walk with me on the wild side!', despite the fact that apparently he didn't believe in the seventeenth-century Giant.

And what was the outcome of the Trial? At the beginning of the day, members of the audience were invited to cast their vote as to whether the Giant was prehistoric, medieval/post-medieval, or other. The vote cast was overwhelmingly in favour of the prehistoric. At the end of the day they were invited to vote again. And this time the Giant was still prehistoric, but by a much reduced majority (Barker & Darvill 1999). The papers were published by Oxbow Books in 1999 as Bournemouth University School of Conservation Sciences Occasional Paper 5 (Darvill *et al.* 1999), which also included an additional paper by William Keighley, Martin Papworth and David Thackray of the National Trust on 'Owning and managing a Giant'.

An exercise in 'experimental archaeology': the taping out of a Giantess hill figure

The Trial raised much interest in the Cerne Giant and posed questions not only of his date, but also relating to the setting-out of a hill figure and what this would involve (Barker 1997). The writer was given to understand that this would not have been an easy exercise and that it would take some time to complete. Thus prompted by both the Trial and these issues, the setting out of a Giantess, as an exercise in experimental archaeology, was undertaken the following year (July 1997) by 20 Bournemouth University students. Permission was granted by Lord Digby of the Minterne Estate on condition that 'there should be *no* pranks' and who, as a boy back in the 1920s, remembered running up and down the deep gullies that formed the outline of the Giant.

A carefully prepared scale drawing, an outline, of a Giantess was produced, based on that presented by the Giant; its basic measurements set out on a grid plan (Fig. 10.2). Assembled were a roll of rope and a huge roll of white plastic sheeting, the latter provided by a Bournemouth company who made supermarket bags. Also gathered were a collection of meat skewers. These heavy items were unloaded into the field above the Giant and moved to an area just above him. This south-facing hillside is *so* steep that those at the top of the hill figure cannot see those at the bottom. Essential here was the laying out of a rope-based grid. Following which lengths of plastic sheeting were unrolled and systematically 'skewered' into the hillside carefully following the grid plan (Fig. 10.3). Surprisingly (but satisfyingly) this took only a few hours, and then walking down to

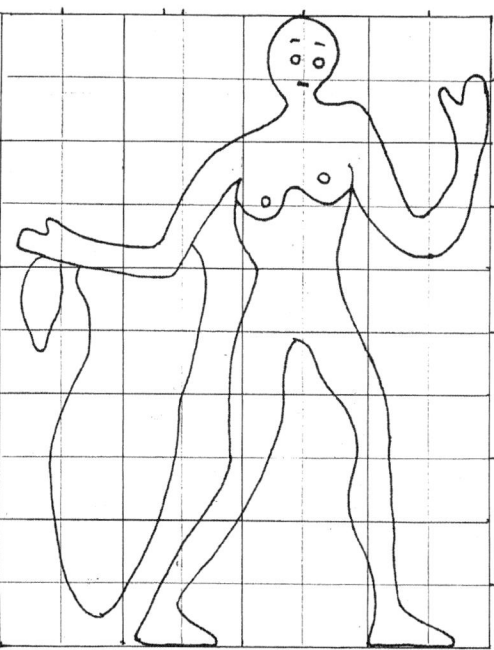

FIGURE 10.2. Giantess 'grid' © K. Barker 1997

FIGURE 10.3. Skewering the plastic sheeting to create the Giantess. Images © K. Barker 1997

the viewing point in the car park looked up and, for the first time, we saw the Giantess hill figure: a moment which will remain unforgettable. On completion and when most of us were well down the hill, one of the students admitted that he had climbed over the fence which protected the Giant and skewered in two short lengths of plastic sheeting on either side of its mouth ... the Giant was 'smiling'. Not only that, we noticed that also completed here were the Giantess' nether regions to complement the Giant.

10. The Cerne Giant: an antiquity on trial 1996: a summary

FIGURE 10.4. Aerial photograph of the Giant and Giantess. Image © Francesca Radcliffe 1997

FIGURE 10.5. Still from computer-generated animation by Evangelina da Sousa (Darvill *et al.* 1999, fig. 2). Image courtesy of Evangelina da Sousa and the School of Media Arts and Communication, now the Faculty of Media and Communication, Bournemouth University

The weather stayed fine and thus perfect for the aerial photographs taken by Francesca Radcliffe who flew over the site shortly after the Giantess was completed (Fig. 10.4). We had clearly shown that setting out a hill figure such as the Giant, is not a difficult exercise and does not take long. And a digging out of the trenches (depending on the number of people involved) could be completed within a single day. The Giantess remained in place for just one day and was then rolled up, but remains as a formal record in the county council archives, and published papers (Barker 1997; Barker & Darvill 1999, fig. 52); a reminder of which is given here which includes the Giant's smile.

A computer animation of the Giant

Back in 1996 computer-generated animation was a little-known technology. Evangelina da Sousa then of the Bournemouth School of Arts and Media Communication showed, though animation, the potential of presenting a moving Giant based on an aerial photograph. Those who attended the Trial were able, during the coffee breaks, to watch a moving Giant on screen, bending his knees up and down and waving his club (Fig. 10.5); an animation that was included in the BBC video of the Trial.

Legacy of the Trial

The Trial assisted in setting the scene, all but a quarter of a century later, for the latest scientific analysis and dating of soil samples from this hill figure. A unique, and distinctive, landscape feature the making of which will certainly have attracted the attention of the local residents at the time, but a 'word of mouth', an oral tradition, which was not to be handed on, nor indeed to be committed to writing.

CHAPTER ELEVEN

Why did we think the Giant was ancient?

Timothy Darvill

When the Cerne Giant was put on 'trial' back in April 1996 the overwhelming view of the audience at the start of proceedings, and still a majority view at the end of day, was that the familiar hill figure was ancient (see Barker, Chapter 10; Darvill *et al.* 1999): perhaps the image of a rampant warrior-god. As such it would have first been carved in later prehistoric times or, at the very latest, in Romano-British times with an emphasis on the work of the indigenous 'Britons' (Darvill 1999b). It was a strongly and widely held view at the time, supported at the trial by five expert contributions to the case. They built the argument for the Giant's ancient roots through the evidence of comparative iconography, parallels with other hill figures, local context, and insights from a then new geophysical survey of the hillside. What Team-Ancient were doing at the trial was simply developing an existing claim. So where did the idea of an ancient Giant come from? And why was it so pervasive?

Very often in archaeology advances come from making cross-comparisons and recognising patterns. And thus it was with the Giant. Back in the mid-eighteenth century the antiquary William Stukeley (1687–1765) was seemingly the first to make a comparison between the outline of the Giant and images of the Roman god Hercules, observations he reported to the Society of Antiquaries of London on 16 February 1764 (Stukeley 1764). What he saw was not of course what we see today. The Giant's features and form have changed over the last 300 years and here Leslie Grinsell's (1980a) plan-regression of the Giant back to the time of Stukeley is a useful reminder of how the modern familiar image came into being (see Fig. 1.7). In particular, the Giant's penis was rather shorter than now, and above it was a circular cut depicting his navel (Fig. 1.7; Grinsell 1980a, fig. 1a). Stukeley was an influential antiquarian in his day. He was instrumental in encouraging the more romantically inclined historians in the Royal Society to establish the Society of Antiquaries as a rival learned society in 1717 (Evans 1956, 49–53), and he was the secretary of the Society of Antiquaries from its inception in July 1717 through to 1726 when he moved to Grantham in Lincolnshire (Piggott 1985, 44). His observations on Stonehenge and Avebury are still important primary sources, and his interpretation of the Giant carried weight.

Where Stukeley pioneered, others later followed. But by the early twentieth century other approaches to knowledge-building were being applied and brought new insights. In the case of the Giant it was Sir Flinders Petrie (1853–1942) that added another layer of thinking and made the case for a much earlier, prehistoric origin. Best known for his work in Egypt and as a pioneer of systematic methodology, Petrie nonetheless spent a good deal of time working on sites in Britain, and in 1926 published *The Hill-figures of England* as Occasional Paper 7 in the developing series of topical monographs issued by the Royal Anthropological Institute (Petrie 1926). Although sounding comprehensive, Petrie's rather slim volume of just 16 pages lists 25 hill figures but only dealt with five in detail: the Long Man of Wilmington in Sussex; the Cerne Giant in Dorset; the Uffington White Horse in Berkshire; and the two crosses at Whiteleaf and Bledlow in Buckinghamshire. But it did include original detailed surveys of these five hill figures, and maps showing their wider landscape contexts. Mapping of the Cerne Giant (Petrie 1926, plate III) shows an enlarged erect penis with the navel fully incorporated into it. Grinsell (1980a, 30) argues that this fundamental change happened during the renovation and enclosure of the Giant in 1887 under the direction of General Pitt Rivers who had inherited the site as part of the Rivers estates in 1880 (Bowden 1991, 36). But it was not so much the form of the Giant that interested Petrie, rather it was the wider landscape context and the alignment and juxtaposition of various prehistoric tracks, field boundaries, and possible flint mines. These, Petrie concluded, showed that the Giant must 'be at least as old as the beginning of the Bronze Age' (Petrie 1926, 11).

From the start Petrie's argument was flimsy and even before the ink on the printed copies of his book was properly dry criticism for some of the views expressed started to appear. First off the blocks was O.G.S Crawford (1886–1957) who used editor's privilege to publish a biting review of Petrie's book in the journal *Antiquity* as a paper entitled 'The Cerne Giant and other hill figures' (Crawford 1929). As a highly proficient landscape archaeologist, Crawford was not impressed with Petrie's arguments and picks them apart in minute detail over several pages, but nonetheless concludes that:

> We have devoted much space to a single line of reasoning because much is made to depend upon it. If correct it could prove the Cerne Giant to be Neolithic or of the Bronze Age. Sir Flinders Petrie's other arguments are less cogent and seem to be inconclusive; but his conclusions with regard to the age may well be near to the truth, however fantastic his reasons (Crawford 1929, 279).

Left hanging in this way an ancient origin for the Giant found favour with some authorities, was rejected by others, and simply ignored by many. Just before the Second World War (1939–45) a young and energetic Stuart Piggott (1910–1996) took up the case of the Giant and published a short note reinforcing Stukeley's suggestion and reminding readers that local people referred to the Giant as 'Helis' or 'Helith', which therefore might link it with the Roman god

Hercules (Piggott 1932). Expanding his case he published a longer and more detailed paper entitled 'The Hercules Myth – beginnings and ends' six years later (Piggott 1938). Here he discusses the origins of the Hercules cult and its representation in Britain during Roman times before moving on to ask: 'What of the Giant of Cerne? Can he be a representation of what Stukeley called "our high admiral Hercules"?' (Piggott 1938, 326). Pointing out the obvious that the Giant was outlined with crude naturalism and is markedly phallic, Piggott used stylistic and art-historical grounds to conclude that 'I feel it almost inevitable that the Giant of Cerne must be Romano-British, and that it may possibly date from the years immediately following [AD] 191' (1938, 327).

With both prehistoric and Romano-British origins firmly on the table the Giant was rather neglected in archaeological narratives published in the early post-war period, perhaps because it didn't really fit with the themes that either prehistorians and Romanists were thinking about at the time. Certainly it very rarely appears in textbooks of the period, and Leslie Grinsell noted in forlorn tones that it was omitted from the well-respected *Ordnance Survey Map of Roman Britain* (Grinsell 1980a, 32). One exception was Anne Ross's meticulously researched *Pagan Celtic Britain* published in 1967 in which she effectively ignores the traditional division between prehistoric and Roman times and instead looks at religious sites, structures, and iconography on a broad cultural canvas spanning the later first millennium BC and early first millennium AD (Ross 1967). The Cerne Giant plays into her argument that classical and local deities were often interchangeable and indistinguishable, and she follows Stuart Piggott (to whom the book is dedicated) in linking the Giant with Hercules or a local variant of him (Ross 1967, 381).

Through the later twentieth century and beyond a niche literature developed that focused on hill figures and related themes. It flourished on the fringes of mainstream academic thinking, and well-worn arguments were variously rehearsed with approval, disdain, or, occasionally, augmentation according to the perspectives favoured by the authors. Morris Marples, for example, declared that the Giant is 'surely the most remarkable of the hill figures of this country' (1949, 159) before trotting through possible origins and concluding that Piggott was right in placing its origins in the second century AD during a revival of the Hercules Cult promoted by the Emperor Commodus (1949, 178). Nearly 40 years later, Paul Newman argues that rather than being an image of Hercules, the Giant is a depiction of Nodens who he sees as an equally important local deity with local roots 'an expression of the ancient phallic religion dating back thousands of years before the Romans' (Newman 1987, 101). Other deities have also been drawn in, Claude Sterckx, for example, suggesting it was a representation of the Celtic version of Jupiter and thus a product of the pre-Roman Iron Age (Sterckx 1975).

Nowhere in this literature is more attention given to the Giant than in Rodney Castleden's well-researched if slightly eclectic and fanciful book *The Cerne Giant* published in 1996 with a rather gushing Foreword by Rodney Legg

(Castleden 1996). Here the Giant was pushed firmly back into later prehistory as the protector-god of the pre-Roman Iron Age tribe living in the area who were known to the Romans as the Durotriges:

> The battle-ready figure of Helis, the tribe's head-hunting guardian god, war-leader, protector, provider, game-keeper, bestower of agricultural fertility, repeller of evil, bringer of good fortune. The Durotriges equipped him only with the most basic fighting gear, negligible even by prehistoric standards: a belt to hold a knife, a cudgel torn from the forest to make a rough and ready weapon, and a cloak wrapped round his free left arm in a dual gesture to ward off blows and to display the severed head of the enemy (Castleden 1996, 187).

Guidebooks are another important mirror of wider thinking. In this literature the Giant is well represented because since at least the late nineteenth century it has been highly visible, and since 1920 been owned and managed by the National Trust who list it in their annual handbook as a place to visit. Indeed, the steady trickle of visitors at all times of the day and night poses challenges to site management and necessitated the construction of viewpoints and parking areas. A guide to the site published by the National Trust in 2004 focuses on the controversies surrounding the date of the image, noting that 'the origin of the Giant is unknown and hotly debated' before briefly touching on a few possibilities both ancient and modern (Papworth & Keighley 2004, 2). Others are more assertive and more evocative.

The *Observer's Book of Ancient and Roman Britain*, part of a well-respected series of authoritatively written pocketbooks covering a wide range of field, tells readers visiting the site that it:

> has been described an Iron fertility figure, a Corina, a companion of Trojan prince who, expelled from Italy settled in Albion (Britain) and as a representative of the Roman emperor Commodus who assumed divinity and the title of Hercules Romanus. Whatever the explanation may be, many ancient legends and superstitions have grown up around the figure (Priestley 1976, 92).

Amongst the most fulsome coverage in a guidebook is that by Jacquetta Hawkes, impassioned archaeologist and member of the Bloomsbury set, whose description and analysis of 'the famous, the notorious, Cerne Abbas Giant' runs to nearly three pages in the first edition of her *Guide to the Prehistoric and Roman Monuments in England and Wales*, which also has an excellent aerial photograph of the Giant as its frontispiece (Hawkes 1951). Viewed with an artist's eye she notes how 'features, including the facial features (with the nose modelled in relief), nipples, ribs shown with a Rouault-like emphasis, and the erect phallus and testicles which are the source of so much interest and so little open comment' (1951, 125). On its origins she notes that 'the Giant with his knobbed club suggests Hercules; the undistinguished naturalism of the work would be appropriate to provincial Roman British Art' and goes on to follow the well-worn path that leads to Commodus's revival of the Hercules cult in the second century AD (1951, 126). But Hawkes was a romantic at heart, and a Romanticist in her theoretical leanings, so not surprisingly she saucily adds that:

> from the beginning the fertility element was strong, and the Roman god must have been partly identified, as so often happened, with some local cult or notion, let us say with a British Priapus, whose symbol was added unto Hercules, and whose power and significance were, inevitably, to prove the most enduring amongst a peasant population (1951, 127).

Returning to the question of why over the last 300 years the Giant has widely been accepted as ancient in origin we can see three main themes from the perspectives given here. First, is the reach and power of authoritative accounts. Stukeley, Petrie, Piggott, Hawkes and many others contributing to the story of the Giant were influential people and what they said and wrote coloured later thinking even when their views were disputed or overturned by subsequent findings. Second, is the sad recognition that if an argument is rehearsed often enough it becomes the 'established truth'. Perhaps not surprisingly, many accounts of the Giant reiterate a limited range of proposals that gain additional traction through repetition and an acceptance that since no-one really knows all options must be considered. Third, and perhaps the most interesting strand, builds on the observation that the image of the Giant is 'primitive', and connects with the deeply engrained notion that since people in prehistory were primitive that is where the Giant belongs. What this of course reveals is an uncritical acceptance of the kind of thinking about unilinear cultural evolution that emerged in the later nineteenth century in which society progressed from savagery, through barbarism, and on to civilisation (Morgan 1877). Such thinking is no longer acceptable nor is it appropriate or relevant. Culture change is nowadays seen as multi-linear and at times non-linear; societies across time and space interact with each other, and at any given time issues such as identity and power are contested and negotiated. Dating the Giant to the late first millennium AD doesn't so much take the Giant out of prehistory but rather brings prehistory into more modern times.

CHAPTER TWELVE

Giant assumptions: locating chalk figures within prehistory

Susan Greaney

The human understanding when it has once adopted an opinion … draws all things else to support and agree with it. And though there will be a greater number and weight of instances to be found on the other side, yet these it either neglects or despises, or else by some distinction sets aside or rejects (Francis Bacon 1620).

Armed with the knowledge that the Cerne Abbas Giant dates to the early medieval period, it is now worth looking back at the debates and discussion about the period of his creation, with the benefit of hindsight. An examination of the evidence presented to support the idea that the Giant was prehistoric or Romano-British, and why these arguments were wrong, has the potential to teach us how ideas emerge and dominate, sometimes to the detriment of our understanding of the past. It is well recognised that archaeological interpretations involve at least some element of subjective reasoning, and in the case of Cerne Abbas there was a scarcity of direct evidence. However, some giant assumptions have been made; we need to critically examine these so that we might better conduct archaeological research in the future.

Vaguely ancient

The anonymous letter accompanying the earliest known depiction of the Giant, published in *Gentleman's Magazine* in 1764, relates that the hill figure 'is supposed to be above a thousand years standing'. This conclusion was drawn due to a three-figure number, thought to be a pre-1000 date, which was visible, cut into the turf between the figure's legs. The letter goes on to say that 'some think it was cut by the Ancient Britons, and that they worshipped it; others believe it to be the work of the Papists, as here was formerly an abbey etc.' (Anon 1764, 336). This early account sets the tone for discourse about the Giant over the next 230 years; where a prehistoric origin is invoked, albeit with little direct supporting evidence.

The extremely ancient age of the Giant in local understanding is evident in William Holloway's 1808 poem 'Giant of Trendle Hill: A Legendary Tale', a creative origin myth for the figure. The poem tells the story of how a giant, who has long troubled the country by killing sheep, falls asleep on the side of the hill. To prevent the loss of more livestock, the local people come out and murder him, after which they cut the outline to mark where he lay. The poem firmly places the Giant not only in the pre-Christian era, but also in the distant pre-metal age, as the villagers attack with only wooden spears and darts.

Dominant voices

The first archaeological account of the Giant was written by Sir Flinders Petrie, the highly distinguished professor and meticulous surveyor, who published his account of the hill figure in 1926, at the height of his career. His account provided evidence that supported these local assumptions that the Giant was ancient. The field survey itself had taken place in 1919, when Petrie had taken 220 detailed measurements of the Giant with the help of his wife and son (Drower 1995, 340). His conclusion that the Giant must be 'at least as old as the beginning of the Bronze Age' (Petrie 1926, 11) was based on a number of factors. The first was the association with the earthwork enclosure, The Trendle, on the hillside above, which according to one dubious memory quoted by Colley March (1901, 106) was the traditional site of the village maypole. This maypole Petrie links to 'primitive pole workshop' being maintained at the enclosure, connected with the Giant. Another was the presence of a spring to the south, which he suggested to be sacred. A third reason was the presence of earthworks on the ridge above the Giant, a tumulus and a curving bank cutting off the ridge. And finally, on the western side of the river valley, Flinders Petrie noted several large banks and a roadway that he thought had been laid out with reference to the Giant, and as these were cut by deep pits, which he assumed were flint mines, that these banks and roadway dated within the period of flint mining, probably 'not after the early Bronze Age' (Petrie 1926, 10–11).

In the same short book, Petrie set out evidence that the Uffington White Horse, Long Man of Wilmington and two crosses in Buckinghamshire were also examples of prehistoric chalk figures. With this article, the date of the Giant was firmly established as broadly prehistoric in both archaeological circles (e.g., in the opinion of Ancient Monuments Inspector Charles Peers; Wilcox 1988, 524) and in popular understanding (e.g., Anon 1937). Three years later, O.G.S. Crawford, the much-respected Ordnance Survey archaeologist, dismissed Petrie's explanation relating to the banks and roadway, but agreed with him that the figure was prehistoric, highlighting the nearby scatter of long and round barrows. In particular, he thought the key was The Trendle, which he thought most likely Early Iron Age and therefore also the Giant (Crawford 1929).

Stuart Piggott was the next 'great man' of archaeology to publish his opinion on the hill figure. His first article (Piggott 1932) traces the possible name of the giant, 'Helith', via medieval legends of a 'wild Huntsman' to the figure of Hercules, thereby implying although not explicitly stating, a medieval date for the Giant. In a second cogitation on the subject, Piggott suggested that the

Giant's 'crude naturalism' might suggest a 'Late Celtic' style, placing him into the Romano-British period, and more specifically to the years immediately following AD 191 (Piggott 1938, 327) although why this particular date was chosen is unclear. Following Crawford, he suggested that The Trendle might be a Late Bronze Age or Early Iron Age enclosure, perhaps a primitive temple, and suggested that its local name, the Frying Pan, was derived from Beelzebub in local mummer's plays, a character who holds a club and a frying pan. He linked this figure to 'a true folk memory' of the Helith-Helis-Hercules character of the Giant that had been recorded locally. Piggott concluded that the Giant was 'the most amazing survival of primitive religion in Western Europe' (Wilcox 1988, 526). As Hutton (1999b, 115) has highlighted, both Petrie and Piggott are adherents to the romantic view that traditions of rural England might preserve millennia-old pre-Christian beliefs and rites.

After years of silence on the matter, Rodney Castleden's 1996 book on the Cerne Giant brought together much of the archaeological and historical evidence, together with results from new geophysical surveys which showed that the Giant had once held a cloak and possibly also a severed head, to firmly conclude that the hill figure was a depiction of an Iron Age warrior. In this conclusion, there is no doubt that Castleden was influenced by the earlier eminent archaeologists who had come to roughly the same conclusion. The book provided much of the basis for evidence presented in the 1996 trial to argue for a prehistoric or Romano-British date. Castleden's investigations of the iconography, which in the hands of others might have been clinchers for the identification of the figure as Hercules, are instead used as evidence for the Giant being a depiction of a broader pan-European Iron Age warrior god, drawing parallels with belted and aroused Celtic figurines and statues from Slovenia, Germany and France (Castleden 1999, 46–48). One might be cynical and suggest that within the broad remit of several thousand years of the prehistoric and Romano-British period and the entirety of Europe, it is no surprise that parallels might be found to the iconography and style of the Giant.

The key proponent of the 1996 trial, arguing that the hill figure was of prehistoric or Romano-British date, was not Castleden however, but Timothy Darvill (Darvill *et al.* 1999). Another eminent archaeologist, who had already penned books for the general public on prehistoric Britain, his evidence now forms an interesting case study in the art of corralling together evidence to fit an interpretation. As well as Castleden, Darvill invited a number of other key witnesses to support his cause: Chartrand set out how the Giant was positioned with careful knowledge of local topography to dominate the Cerne valley and was surrounded by prehistoric archaeology; Newman demonstrated that the violence, nakedness and sexual arousal depicted were clearly pre-Christian; Putnam demonstrated that the area was densely occupied in this Iron Age period; and Miles showcased dating evidence from Uffington White Horse, which demonstrated that hill figures could indeed be prehistoric. Darvill went further and cited the carved stone heads and supposed 'Celtic head cult' of the

12. Giant assumptions: locating chalk figures within prehistory 211

Iron Age (Ross 1967), an idea about which serious questions had been asked within five years of publication (Billingsley 2016, 80). Darvill concluded that the Giant dates to between 1000 BC and AD 250, probably to before the Roman period (Darvill 1999b, 29).

Four 'great men' of archaeology, Petrie, Crawford, Piggott and Darvill, each highly respected, articulate, and confident speakers and writers, have dominated the archaeological debate over the age of the Cerne Abbas Giant. Castleden, as both a historian and archaeologist, has a slightly lower profile and presents a slightly different case, as his book in considerable depth examines (and dismisses) alternative dates for the Giant and collates a wealth of historical evidence, as well as archaeological (Castleden 1996). Despite the seemingly objective investigations of these authors, there is of course no direct evidence at all that the Giant is prehistoric or Romano-British. Part of their willingness to ascribe an early date to the figure appears to be a preference for imagining a remarkable survival of beliefs and practices over millennia, rather than considering a more prosaic historical date. The prevailing view of other eminent archaeologists must have also swayed their thinking. There are also a number of underlying assumptions that may well have influenced, albeit subconsciously, the convictions of these archaeologists that the Giant dated to the distant past.

Underlying influences

Megalithic remains and earthworks, as well as natural rock formations and hills, were ascribed to the work of giants in medieval and early modern times (Fox 2000, 238–42). The Old English poem *The Ruin*, written in the eighth or ninth century, refers to the ruins of a Roman bathhouse as constructed by giants. There are numerous examples of generic attribution: for example, the Giant's Ring applied to the standing stones of Stonehenge, Wiltshire and the earthwork henge at Ballynahatty, Co. Antrim, or Barclodiad y Gawres ('the apronful of the giantess') on Anglesey. There are also specific named giants, such as Wade's Causeway, a Roman road in Yorkshire, or Hautville's Quoit, a standing stone in Somerset. The legends of Goram and Vincent explain the hillforts and gorges of Bristol, where there may have been another carved figure of a giant (Clark 2016). Could this underlying link between mythical giants and prehistoric monuments have unconsciously influenced the easy placement of the Cerne Abbas figure into prehistory?

The naïve style of the Giant, his slightly surprised face, as well as his naked and erect state, may have led archaeologists to recall Christian beliefs about the original state of humans prior to the fall of Eden, as well as general fertility rites and rituals, firmly located in the vague but pre-Christian distant past. Perhaps carved chalk phalli of the Neolithic period, or phallic emblems found on a wide range of Roman objects, made it seem more likely that the figure might date from one of these periods, rather than the Christian medieval world.

The Giant was seen by archaeologists as strange, primitive and 'other', making his natural home prehistory. The medieval period, by contrast, appears to have been viewed as well understood, familiar and ordinary, no place for such an outlandish figure. It 'seems impossible that nude figures should have been cut in the medieval age' (Petrie 1926, 15); clearly Petrie was unfamiliar with the lewd marginalia of late medieval manuscripts (e.g., Mattelaer 2010).

The fact that the Giant wields a gnarled wooden club is another feature that immediately raises a subconscious idea of prehistory, despite Darvill's contention that it is actually a finely carved weapon (Darvill 1999b, 11). Entering 'prehistoric club' into an internet search engine will bring back a variety of blow-up plastic clubs, perfect for your prehistoric fancy dress party, many of them looking distinctly like the one brandished by the Cerne Abbas Giant. The club is commonly seen in cartoon and other depictions of prehistory, such as in *The Flintstones*. Actual evidence for such wooden clubs is of course, extremely rare. Even considering the fact that such organic objects rarely survive, there seems to be a real absence of evidence for such weapons in the archaeological record (Stoczkowski 2002, 79). Instead, this is an accessory borrowed from a mythological character from the Middle Ages – the Wild Man. This hairy, naked man, a popular figure at carnivals and pageants of the fourteenth and fifteenth centuries, lived outside of civilisation where he fought off wild animals with his wooden club. Ironically, given the immediate association in modern minds between prehistoric people and wooden clubs, it is possible that this character can be traced back to Greek mythology and to Hercules, who is constantly fighting wild animals armed with a club, according to mythical narratives and iconography (Stoczkowski 2002, 79–82).

Finally, there is the well-known practice, albeit varied, complex, and not quite so common as generally supposed, of the Christianisation of certain pre-existing 'pagan' monuments. Monuments or enclosures that remained significant to local communities as places of burial, of centres of power or because of their association with symbolism or stories, were sometimes chosen as the location for early Christian foundations (Semple 2013). Much-cited examples include the church at Rudston in East Yorkshire built adjacent to a tall standing stone, and the church at Knowlton in Dorset constructed within a henge. It could easily be thought that the site of the important Benedictine monastery at Cerne Abbas, founded in 987 but with earlier monastic foundations in the village (Castleden 1996, 84), might have been a reaction to the presence of a significant pre-existing pagan monument or religious 'cult' site.

Persistence of an idea

The first part of this chapter set out how highly influential and respected 'great men' of archaeology have argued for the Giant to be of prehistoric or Romano-British date, each of whom collated together evidence to support their arguments. None of the individual strands of evidence are particularly convincing, but cumulatively, and perhaps supported by other subconscious assumptions outlined above, these views have been generally accepted without question. When archaeology is written with conviction and confidence about a monument with little hard evidence it is difficult to compose counterarguments or to contend the prevailing view.

Even though most chalk hill figures have been shown to date from the seventeenth or eighteenth centuries, archaeological discoveries, such as the extent of prehistoric settlement activity in the Cerne Abbas landscape, or the dating of the Uffington White Horse, helped to maintain hopes through the late twentieth century that the Giant would turn out, after all, to be prehistoric. A somewhat wishful assumption was maintained by archaeologists that the Giant was a miraculous ancient survival and dated to 'their' period of interest and expertise. Here we can see the effect of confirmation bias, 'the seeking or interpreting of evidence in ways that are partial to existing beliefs, expectations or a hypothesis in hand' (Nickerson 1998, 123). Despite being a widely evidenced and well-established psychological concept, we like to think of ourselves as primarily scientific and objective archaeologists, aware of our own biases. The Cerne Abbas Giant story shows that this is far from the case. Within confirmation bias are pertinent strands: primacy effect, that information acquired early in the process is likely to carry more weight than that acquired later; and belief persistence, that once a belief is formed, it can be resistant to change, even in the face of compelling evidence to the contrary (Nickerson 1998, 187). These biases can clearly be pervasive when the archaeological evidence is sparse. But archaeological opinions are not formed by people working alone; archaeologists are influenced by those who come before them and by those who dominate discourse. All of us work within our own distinct social and political context. Rather than perceiving these discourses in terms of paradigms (Kuhn 1962), it is perhaps more informative to view them as assemblages of actors: archaeologists, institutions, and publications, who form social and political networks (Lucas 2017, 267). Within these assemblages there are 'sticky' ideas that are retained in the face of contradictory evidence.

Such assemblages of scholarly discussion and ideas can be bounded and separated from other disciplines and ways of forming knowledge. Despite the Trendle maypole association being ruled out by Darton in 1935 (Darton 1935, 320–30), in 1954 the archaeologist Jacquetta Hawkes would write about the Giant as a remarkable survival of pagan beliefs (Hawkes 1954, 143–4) and archaeologists continued to cite the erroneous 'fact' as late as the 1990s (Hutton 1999b, 116). A respect for other disciplines, and an acknowledgement that historians, particularly those who work in local history and folklore, have valuable contributions to make to archaeological debates, would do us well. On the one hand, archaeologists have been largely ignoring history, and on the other, forgetting that archaeology is at least partly a science, a discipline where hypotheses should be clearly set out and tested, and the evidence both for and against an argument fully explored.

Doing better archaeology

As long ago as 1938, V. Gordon Childe concluded that explanations tend to be shaped by the assumptions that archaeologists make, rather than the data they gather (Childe 1938). If we acknowledge that this is the case, how do we move forward to ensure that archaeology is not doomed to repeat the mistakes of the past? We need to be better archaeologists.

The link between archaeology and the law is pertinent here, as the Giant is one of the few archaeological sites ever to have been put on 'trial'. Roger Thomas, in an exploration of the similarities between archaeological reasoning and the practice of law, sets out the two ways in which both lawyers and archaeologists collect data and interpret evidence (Thomas 2014). The first is a problem-orientated approach, in which a particular hypothesis is pursued, giving a clear focus but perhaps leading you astray if the hypothesis is wrong, or resulting in confirmation bias. It is this method that the dominant voices set out in the earlier part of this chapter followed. The second is an inductive or empiricist approach, where evidence is collected in a more neutral way, which may mean not really knowing what evidence to collect and what to ignore (Thomas 2014, 257). Any good archaeologist will employ a mix of both strategies and conduct rigorous scrutiny of his or her own case before presenting it to others (Thomas 2014, 269), something that Castleden did in his book. It could be argued that the 'great men' failed to do this, in the case of the Cerne Abbas Giant. At the 1996 trial, presenting all the available evidence led to a partial shift of the views of the jury (the audience) away from a prehistoric or Roman date and towards a seventeenth century or later date (Barker & Darvill 1999, 162). We now know of course, that both arguments are wrong, but a more inductive approach clearly led some to question the prevailing view.

Why then, were those prominent, eminent and self-assured archaeologists of the twentieth century, so wrong about the Cerne Abbas Giant? They did not act, in Thomas's words, as good archaeologists, but favoured a problem-oriented approach, being led astray by a wrong hypothesis and by their own confirmation biases, and the influence of those who came before. Unfortunately, those archaeologists who weigh up evidence from both sides of an argument, and present counterarguments that may undermine their own conclusions, do not sound as authoritative or confident, as those who present a clear single line of argument. It is these cautious and restrained voices, often under-represented within academic discourses, that we need to amplify within archaeology. We need to value different, diverse and lower-profile approaches that are currently excluded from our assemblages of discourse, so that there are a variety of voices and opinions in the room (Pope 2011; Hamilton 2014). As a discipline we need to question those in senior positions, to be humbler about what archaeology can achieve, to acknowledge the work of other closely aligned disciplines, to be more uncertain about our ideas and more scientific in our approaches.

CHAPTER THIRTEEN

Images of the Giant

Sarah Fry

The discussion in this chapter is based on a number of online and social media sources that are not explicitly cited in the text, but are listed below to help the reader follow the author's arguments. The key references are listed at the end of the bibliography.

Since Barbara Bender's 'A living Giant' (1999) was written 25 years ago, digital means of communication have transformed how people create, distribute, and consume information. The Giant lives in the age of Man, but in technological terms he is an inhabitant of the digital age. From traditional paper-based forms of advertising, news, media and journalism, early online forums and websites developed in the 1990s, and have rapidly evolved into the social media, digital news, and streaming platforms that we have today, with increasingly fast internet, and used on smartphones and other electronic devices. This widespread use of digital media connects people, creates virtual communities, and influences opinions on a massive scale. It offers archaeologists and those managing the archaeology, such as the National Trust, new and effective ways to engage with the public, share discoveries, disseminate archaeological knowledge, and promote awareness and appreciation of our shared heritage, making archaeology increasingly accessible and inclusive to a wider, more diverse, audience.

Against this backdrop the living Giant's mysterious aura now captivates online audiences with a fully global reach. He is no longer simply a figure in a working landscape in the Dorset countryside and part of the identity of the village he towers over, but he has gained a powerful and meaningful social media influence, incorporated into people's lives worldwide due to his historical significance and his unique appearance. In the last 25 years the iconic symbol of the Giant has been used to advertise and market a variety of businesses and commercial products, including adding images next to him, unofficially physically changing his appearance on a temporary basis, or digitally altering his image in different ways (see Table 13.1 and Fig. 13.1). Worldwide audiences have been reached, provoking a mixture of positive messages and concerns. When a biodegradable paint image of Homer Simpson appeared next to the Giant enclosure in 2007, stood in his Y-fronts holding up a doughnut to promote

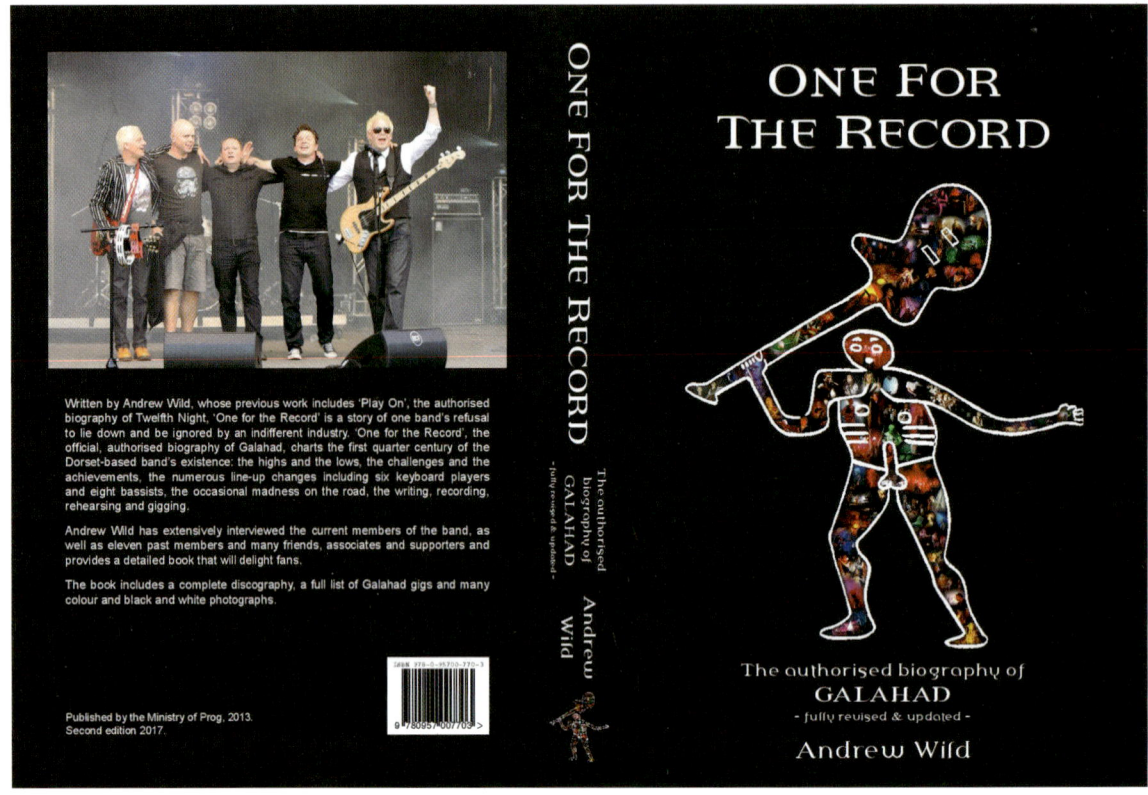

FIGURE 13.1. Digital cover image of *One for the Record: The authorised biography of Galahad*, the Dorset progressive rock band. © Andrew Wild 2011

The Simpsons Movie released in that year, it may have delighted fans of the American animated sitcom *The Simpsons*, especially knowing a Springfield Road can be found at the base of the Giant. But it also caused upset to the pagan community who vowed to use 'rain magic' to wash away the imposter.

In an increasingly accepting and inclusive world of the human body, identity, and relationships between people, promoted in all their forms, some advertising still removes the Giant's iconic manhood, or strategically covers it over with other objects in images (see Table 13.1). In 2023 the Oxford Cheese Company emasculated him to avoid causing 'woke' offence; whilst in 2021 after a paraglider was drawn and perfectly positioned over the Giant's erection on printed tea towels, it was seen as a nod to traditional British postcard humour by the artist, but the design was 'blasted' by *The Sun* newspaper as an attempt to censor the Giant. A couple of years earlier, in 2019, the Giant's 35 ft appendage received an anonymous floral covering, but not to censor it. This makeover was to spark the gender debate in celebration of International Women's Day, with its focus on a world free of bias, stereotypes, and discrimination. By making the Giant less about the traditionally masculine, the anonymous letter delivered to the Giant's local shop about this social activism called it an 'invitation for unity' to drive gender parity, and to create 'peaceful relationships within the sexes by finally creating equality'.

13. Images of the Giant   217

Images and comments associated with the Giant	Date	Key source(s)
Beating Bowel Cancer – Loud Tie Day	2001	Churchill 2007
Family Planning Association – Sexual Health Week	2002	Churchill 2007
Homer Simpson	2007	*Dorset Echo* 2007; *The Guardian* 2007; *The Telegraph* 2007
Olympic Flame	2012	Meech 2012
Movember	2013	*BBC News* 2013
International Women's Day – Flower	2019	*BBC News* 2019; Menendez 2019
Paraglider – Tea Towel	2019	Town Towels 2021
Rechalking the Giant	2019	Hartley-Parkinson 2019
Stephen Fry and National Trust	2019	Stephen Fry 2019; National Trust 2019a
Lockdown – Face Mask Covid	2020	Richard Osgood 2020a; 2020b
Have I Got News for You	2020	*Have I Got News for You* 2020
Dorset Council – Covid Face Mask	2020	Dorset Council 2020
Borat Movie	2020	Duell 2020
Oxford Cheese Company	2023	Maslin 2023; West 2023
Disappearing Giant	2023	Lumb 2023

TABLE 13.1. Selected images of the Giant referenced in the chapter

FIGURE 13.2. Giant sporting moustache to support Movember. Image: Germinal GB Ltd.

The image of the Giant as curated by the National Trust is one of social responsibility, providing a voice for change through collaborations to raise awareness of good causes and specific health concerns, using a variety of temporary manipulations and modifications to various parts of the scheduled ancient monument (Table 13.1). In May 2012, the year when the UK hosted the Summer Olympics in London, local Dorset children dressed in yellow, red, and orange to embody the spirit of the games, that of mutual understanding, friendship, solidarity, and joy, when they joined forces to recreate a living Olympic torch

flame on the Giant's club. Whilst in November 2013 the National Trust gave its permission for the Giant to sport facial hair in the form of a distinctive grass moustache to support Movember, the charity raising money each year to support and create an everlasting impact on men's health research, including prostate and testicular cancer, and to transform the way health services reach and support men (Fig. 13.2). These collaborations compliment earlier key health messages given by the Giant (Table 13.1): from 2001, when a tie adorned the Giant to publicise the Beating Bowel Cancer charity's Loud Tie Day; and from 2002 when a 21 ft plastic condom was placed on the Giant for the Family Planning Association to mark Sexual Health Week.

Through the COVID-19 pandemic the Giant played an important lockdown role. At a time of worldwide crisis, the lone figure of the Giant on the hill was shown setting an example to all by practicing his social distancing with a face mask makeover in April 2020 (Table 13.1). As a country under strict restrictions on social interactions, the Giant 'put a smile on some of the older people's faces' who were shielding and self-isolating, and attracted media attention. It was picked up as a story by the national press, who quickly tried to link the Giant's makeover to a hand drawing days before, of the Giant in a white mask, by the senior Ministry of Defence archaeologist Richard Osgood published on his social media accounts, who said it was simply a 'bizarre coincidence' (Fig. 13.3). And when the *Borat Subsequent Moviefilm* was promoted in October 2020, with a huge mankini style facemask draped over the Giant's manhood and the slogan, 'Wear Mask', 'Save Live' on either side, social media played its part, and Richard was jokingly asked on Facebook if this giant adornment had been his idea too. When Dorset Council needed to remind people in July 2020 it was soon to be compulsory to wear a face covering in shops and supermarkets, it took to social media and the Giant became the focus of its campaign. Sporting a blue digital face mask on its Facebook page, the council highlighted that 'Dorset's friendliest giant may not bother with pants, but at least he wears a mask' (Fig. 13.4).

Photos, both real and digitally enhanced, drawn images, and information shared on platforms like Instagram and X, formerly known as Twitter, continue to increase the Giant's visibility and spark debates from multiple voices about its origin and purpose. Increasingly acute observations about its very obviously erect appendage are now made. When the new dating for the Giant was revealed, *Have I Got News For You* (2020) commented that 'historians admit there's been a massive cock-up' alongside an unadulterated image of the Giant. And in 2019 the cleaning and rechalking of the Giant by the National Trust and its volunteers delivered an important message of the Giant's conservation and protection, but it provoked the comedian Stephen Fry to comment on X in response to the Metro newspaper publishing an image of the Giant being rechalked, titled 'Giant's erection to be polished by hand for two weeks'. It pictured a group of men 'concentrating on the shaft and balls', and generated 52,900 likes on X. This allowed the National Trust to show its good-humoured approach to the monument, it responded to Stephen Fry and highlighted the

FIGURE 13.3. A giant white mask. Image © Richard Osgood 2020

National Trust's 'top priority has always been taking care of our members' (Fig. 13.5).

With his enigmatic appearance the Giant has something unique to bring to the table. And with this niche his modern story contributes to his cultural significance, with social media and both the local and national press alert to stories about the Giant and the adornments that might be made to him, or the speed with which the grass may grow and cause the famous figure to disappear leaving tourists disappointed. What started out recorded in seventeenth-century writing, and later literature and art, and providing a focus for fertility rituals,

FIGURE 13.4. Giant faces up to the situation with his Covid mask: Dorset Council's reminder that wearing a face covering in shops and supermarkets was compulsory during the COVID-19 pandemic. Image: Dorset County Council

FIGURE 13.5. National Trust volunteers polishing up the Giant's parts. © National Trust 2019

celebrations and local festivals and events, is now a dynamic and complex culturally relevant phenomenon shaping public discourse and providing a tool for connectivity and public/social activism. His guardian, the National Trust, has curated the Giant's self-image and engaged with his audience, delivering valuable socially relevant messages, and creating new conversations with humour, whilst articulating the importance of conservation and respect for the monument.

Since the Giant was constructed a thousand years ago, he has existed bound up in folklore and legends as an intriguing and enduring figure captivating those that see him. Now in the digital age, the Giant's active social media presence gives him a voice to be heard in new ways, with new possibilities for his self-expression and shaping the interpretation of the past. The Giant is an enduring image that lives in popular culture as community focused and socially responsible. He also has the unique ability to stand proudly on the hillside overlooking the valley below, and not to get an automatic ban on social media for his happy, naked, un-censored, self-expression, now for all to see across the globe.

Acknowledgements

With very grateful thanks to Alison Dalby, National Trust, for all her sterling work on sourcing 'images of the Giant' that make this chapter. We also thank *Galahad*, and in particular Stu Nicholson (vocalist and manager) for permission to reproduce the Galahad image (Fig. 13.1).

CHAPTER FOURTEEN

A research agenda for the Giant

Michael J. Allen

Despite previous research on and at the Giant, and the production detailed research designs such as that produced by Oxford Archaeology in 1998, and for this project (Allen 2019), no research agenda has ever been produced for the Cerne Giant (as far as we know). This is not unusual for chalk figures; few, if any, have a research design. Nor does there seem to be any nationally for this class of monument (but see Allen & Scutt, Chapter 22). It is clear that the small keyhole excavations that we have done so far have addressed all of our original research aims, dating and palaeo-environmental history (Allen 2019). However, are they really enough? There is also the obvious need for more work on the landscape context and, now dated, some investigation of local contemporary sites (i.e., the abbey and others) is required. This research agenda starts from the very monument-specific before addressing some issues relating to the wider physical, cultural, social, chronological and archaeological context.

The Giant's outline

The new photogrammetry survey confirmed only a small, hardly detectable, variation in the Giant since Petrie's 1926 plan. Excavations demonstrated the enlargement of the width of the outline (from 40 to 60 cm) and the downslope drift of, in particular, both feet and the right elbow by as much as *c.* 30 cm (Figs 3.14 & 3.16). There, are, however, some key elements of the Giant that this research never intended to address, but have in some cases been investigated, albeit cursorily. A full detailed topographical and geophysical survey of the whole figure is clearly required, with attention paid to the following areas.

Belt

The possibility of a belt, potentially pre-dating the phallus, was suggested by Martin Papworth (2021) who noted a line between the chalked top of the legs and across the phallus, which would suggest that the phallus and testes were additions and that the Giant was not a single-phase build. The presence of this belt therefore had clear implications of whether the phallus and testes were part of the original design. The legitimate suggestion by Martin Papworth of a continuous belt behind or across the phallus was heightened by a tentative line in National Trust rendered data of the aerial photogrammetric survey. An animal track can be seen on aerial photographs from the 1970–80s and 1990 onwards running diagonally up slope to the Giant's midriff, across the chalked line at the top of the legs, crossing the penis, and exiting the figure running diagonally up the scarp slope towards the top of the hill. This could be seen during fieldwork in 2020 (March), 2023 (September and November) and January 2024, but does this animal track run along a former chalked line? However, others experienced in aerial photography interpretation, such as Bob Bewley, convincingly discounted this. Careful examination (geophysical survey and further augering) and site visits all show that this is natural terracette and sheep track running across the hill side, coinciding with, and joining the lines at the top of the legs (Fig. 3.18). In January 2024 augering showed no former chalk line (belt) existed.

Cloak and head

The presence and nature of the cloak/lion skin and severed head have long been a matter a debate and the 1979 resistivity survey by Tony Clark and colleagues (see David *et al.*, Chapter 3) clearly confirms a presence but not its character. The aerial photogrammetry did not reveal a cloak or lion skin draped over the outstretched left arm. If they exist they are clearly described on the ground in a completely different form to the rest of the outline. What form did they take? If not a deeply scoured chalk-filled outline like the rest of the Giant, then was this either more lightly cut, more ephemeral lines, or was a part of it upstanding chalk turf and soil, as the severed head seems to be? If so, then this questions why this difference exists. Was it aesthetic, constructional or temporal? At the end of the day despite the detailed survey by Petrie which did not record it in 1926, fieldwork endeavours of Castleden, geophysical survey by Gale, our photogrammetric survey (2020) and the GPR survey, we cannot even be sure of its existence, let alone its form or date. It could still be upcast from the original or subsequent cutting and scouring and soil, mud and chalk places in the 'open' space in the figure subsequently eroding down slope giving impression of folds in a cloak. Despite the tentative nature of the surveys, most of which have shown a strong confirmation bias, perhaps the most compelling argument for something over/beneath the left arm is, as Tony Clark suggested, that the figure looks unbalanced without something there, and its presence gives meaning to the position of the left arm, and improves the compositional balance of the figure (1983, 30). It could always still be an aesthetic pose, or a more purposeful gesture point towards Cerne Abbas hidden away just around the corner.

Clearly further work is needed to ascertain its presence and form. A good research design for which was commissioned by, and presented to the National Trust in 1998, and still stands as a sound option.

Letters, figures and symbols between the legs

Much has been said about the letters, numbers and symbols beneath the legs first reported by Hutchins in 1774. However, apart from Hutchins' record, they did not occur on plan before or since, nor have they been recognised in geophysical surveys nor the recent photogrammetry survey. The only slight evidence is the ephemeral indications from Rodney Castleden's micro-topographical survey, which could be said to suffer, in part, from confirmation bias. Again, despite a detailed walkover survey for this project, the photogrammetric survey and rendering by the National Trust of that survey, we still are unsure of their existence, let alone what they were. They still could be an intentional or unintentional addition by Hutchins to his illustration, but it is surprising that is their only origin and the only place they have been recognised.

Further research might entail more detailed examination of any surviving literature, documents, and letters associated with the research by, and writing of, Hutchins' book, or indeed its publication. Depending on their nature (possibly just soil cut figures that did not penetrate the chalk), then even excavation may not reveal them, but destroy any evidence of the existence. On balance any field to test for the existence, in this case, is probably better restricted to non- or minimally invasive (e.g., chemical sampling) of the area.

Reinstating lost parts

There have been calls and talk of reinstating the navel. Reinstating the navel is a very short-sighted aim and fraught with other implications. If, however, one is going to reinstate the navel and shorten his manhood by about 2.4 m, then why not his chin, which clearly did make the face a complete circle as seen in the sketch of 1763 and plans of 1764/94, and seen in all depictions until Petrie's 1926 image. Or indeed what about the cloak/lion skin and severed head some commonly discussed. Are they all contemporaneous anyway? On what grounds should one be chosen over another? Let alone arguments that he is a living monument (cf. Bender 1999) and these changes are a part of his life-path. Careful curation and maintenance is probably a better solution; even if this does freeze and fossilise the current outline with no obvious mechanism for significant change, evolution or development other than temporary additions or companions (see Barker, Chapter 10; and Fry, Chapter 13).

The Giant on the move: changes in the outline as a consequence of the scouring and rechalking: the excavated evidence

Obviously the figure has been treated with great respect, especially since being registered and made a Scheduled Ancient Monument within the National Trust's ownership on 15 October 1924. The scheduling aims to protect and preserve the hill figure, and in rare cases its overzealous description or application can result in unplanned and unwanted changes in the site. At Uffington the scheduled description prevents the removal of turf as this would change the outline of the (pre-)historic figure. This was taken too literally by those maintaining the figure, and grass and vegetation growing over the figure as new turf was not removed resulted in the horse becoming a much slimmer figure (now rectified, see Chapter 16, Miles & Palmer). Similar problems have not beset the Giant, however, despite the care and supervision in place during the volunteer-manned scoring and rechalking, our excavations clearly show the enlargement of the outline in the most recent chalking (2008, but largely 2019) from a consistent 50 cm to 65–68 cm (up to a 36% increase) in both feet and both elbows. This may, however inadvertently, be beneficial because as soil accumulated on upslope and downslope sides, the chalk lines, at the elbows, feet and base of the club in particular, have risen and no longer are at 45° on the line of the hill slope, but are almost horizontal. The consequence of that is that they are more difficult to see from afar, but the increased width goes some way to mitigating this.

More significant is downslope drift in the lines we have recorded at both elbows and both feet as the Giant slowly sliding down the hill. This is so small that it is imperceptible from the comparison of the photogrammetry survey with that of 1926. Between 1868 and 1995 the figure was very stable with very little change in the width of the trench or location of the lines; in general, changes seem only to be 1 to 3 cm (0.01–0.03% change in relation to the whole figure). Nevertheless the elbows have shifted as much as 12 cm and the feet as much as 28 cm downhill as result of the 1995 and 2008 rechalkings. The right foot has also drifted by about 28 cm, but that may have originated in 1956 during the rechalking by Beard & Co. This downslope drift is imperceptible in terms of the whole figure but is measurable and is nearly 0.3% in 25 years compared with as little as 0.01% in the previous 125 years.

Research agenda

1. The Giant outline in context

    1.1. Detailed full topographical and earthwork survey of the outline complete with accompanying micro-earthworks of the Giant

    1.2. Full geophysical surveys of the Giant (last undertaken in 1979);

    1.3. An accompanying earthwork survey and targeted excavation of the Trendle to define date and function and its (chronological) relationship to the Giant

    1.4. Establish the existence/presence and form of the cloak
        1.4.1. and its nature, outline and extent

    1.5. Establish the existence of the symbols and letters between the legs, via non-invasive research (aerial photographs) and fieldwork (e.g., chemical survey, GPR)

1.6. Define the nature and depth of the chalked trenches on vertical elements of the body, i.e., the Giant's sides/arms, and his legs
1.7. Club and left hand – examine the club to confirm the upper end was original and that this was not originally a book (the Charter, e.g., AD 987)
1.8. Establish the form and nature of the head, especially the chin and eyebrows (the later are a more typically seventeenth-century feature)
1.9. Outstretched left hand – confirm the absence or presence of anything held in the hand;
1.10. Proper CAD and digital interrogation of the topographical survey commissioned by the National Trust, especially of the following areas:
Head and cloak
Figures and symbols between the legs
Putative belt across the phallus
1.11. Limited excavation of
   a) the head
   b) the cloak
   c) phallus/testes to meet objective above
1.12. Post-geophysical survey turf strip and record (but no excavation) of the numbers and symbols to attempt to define their existence and nature.

2. Dating and chronology

Having established the existence of a first and second Giant, date these two elements more closely.

2.1. Date the pre-chalk-filled Giant trenches, and obtain an absolute date on the earliest chalk filling.
2.2. Open the debate of the age and reason for the Giant.

3. Identity

Despite recent research and comment (e.g., Edwards 2020; Morcom & Gittos 2024; Chapter 6; Yorke, Chapter 4) the identity of the Giant is still ambiguous, even if some of the previously proffered identifies can now clearly be discounted.

4. Historic context: relationship with the abbey

This research never set out to place the figure into its contemporaneous context. Nevertheless, following the Saxon date this book goes some way to doing that, but there is clearly much more that could be done, some of which may be alliterated by the recent research and excavation of the abbey.

4.1. Examine the historic/religious relationship of the tenth-century Giant with the (founding of) the Benedictine Abbey of Cerne, and any links between the abbey and the Giant in view of the coincidence for the creation and foundation dates.

5. Land-use and environment

As mentioned in Chapter 1, there may still be some footslope colluvium that could be explored; exploration in 2002 (see previous work) was limited to less than a day of probabilistic augering, and no fieldwork was done for this project on the wider geoarchaeological context.

4.1. Obtain long palaeo-environmental land-use record from colluvium at footslope locations in the vicinity of the Giant (not just immediately bellow the figure), and from both colluvial hillslope reservoirs and valley locations and from overbank alluvium in floodplain and watercourse margins. More detailed examination of the hillside for post-glacial colluvium at breaks of slope in the slope, at the foot of the slope and near footslope locations, and valley floor for alluvial/colluvial deposits.

4.2. Obtain a longer land-use and landscape history in which to place the Giant.

5. Historical absence

5.1. Re-examination of the lack of records pre-1694.

6. Social history

6.1. Thorough research into the social history of the figures going back to the late seventeenth or early eighteenth century as a minimum.

Future management issues

Scouring, rechalking and maintaining the figure is draining financially and in terms on manpower. The National Trust have reverted from large-scale commercial/semi-industrial scouring programmes of 1956 and 1979 to smaller scale volunteer-based activities (1995, 2008, 2019, etc.). It is, however, clear that regardless of how these are conducted, they need to be undertaken under a stricter archaeological supervision than previous scourings have been. Excavation clearly demonstrated the scouring and removal of the chalk emptied previous historic chalking events (the whole purpose of scouring), for which there was no archaeological record. It also removed some *in-situ* deposits and possibly compromising the preservation of key stratigraphical relationships.

More concerted consideration of a) the operations and its preservation of (scheduled) *in-situ* archaeological deposits and b) appropriate full recording of both those activities and the deposits removed (i.e., extent, depth, nature at locations all around the figure). Future scouring events might consider:

- Brief; and archaeological recording brief
- Rigorous archaeological supervision

228 *Michael J. Allen*

- Fuller record of the activities
- Full archaeological monitoring and recording
- Reporting internally, to the county archaeologist/archaeology department, and Historic England

If the cloak (and severed head) are considered *bone fide* archaeological features, then attempt to date them to define it they are a part of the original design.

Part 3

GIANT CONSIDERATIONS: WIDER REFLECTIONS ON THE RESULTS

This part of the book's overall discussion is divided into several sections; the first, 'Context and contrasts', discusses archaeological fieldwork on the chalk figures of the Long Man of Wilmington (2002 and 2004) and Uffington White Horse (1990 and 1994). We follow this with a series of essays and reflections placing the Cerne Giant into a much wider context in terms of himself, the landscape, and his date in Saxon/early medieval archaeology. The last set of reflections casts their net wider, examining chalk figures as an archaeological phenomenon. This section concludes with a research agenda that would be applicable to other chalk figures, especially those with a poorer historical context (i.e., pre-eighteenth century).

Galahad *In a Moment of Complete Madness* album cover © Galahad/Stuart Nicholson

CHAPTER FIFTEEN

The Long Man of Wilmington: a progress report on a giant conundrum

Martin Bell and Chris Butler

The Long Man is an icon of the South Downs (Fig. 15.1), looking out from the magnificent escarpment just east of the Cuckmere valley over the Wealden landscape to its north. It has provided the subject for artists such as Eric Ravilious in a painting of 1939 (Fig. 7.4) and has fascinated and puzzled generations in the ambiguous simplicity of its outline, which provides many slight clues but no positive evidence as to its date and meaning. For some it has conjured up images of a distant pagan past, speculation as to its date ranging widely. Some have suggested a Neolithic date; there is a possible long barrow on Windover Hill above and another on Wilmington Hill to the east. Some long barrows were called 'giants graves' (Curwen 1928). Curwen (1954, 306) was subsequently much more cautious as to date, noting 'it is not safe to assume the giant is very ancient'. A speculative case for a Neolithic date was made by Castleden (1983), and Flinders Petrie (1926) argued it was a deity of the Bronze Age or earlier. Others suggested a pre-Roman 'Celtic' association (Newman 2000), or pointed to similar figures on Roman coins holding two standards. More persuasive among parallels was a figure on an Anglo-Saxon brooch of the seventh century AD from Finglesham, Kent showing Odin, flanked by two spears (Hawkes 1965). Still others have suggested a medieval date, because it overlooks the Priory at Wilmington. Dating based on the figure itself is limited by its simplicity and by evidence that the outline has been significantly changed by multiple, sometimes not very accurate, restorations (Holden 1971; Castleden 2002). A preoccupation in the literature with changes to the outline probably reflects a hope that discovery of changes will provide hidden clues to date and meaning.

Comparisons with other extant giant hill figures only add to the possibilities and speculations. The White Horse at Uffington is recorded in an Anglo-Saxon charter, and has been dated to the Late Bronze Age or Early Iron Age (Chapter 16; Miles *et al*. 2003). The Cerne Abbas Giant, Dorset with its strongly phallic form and giant club has long been assumed to have early origins, and new evidence for its date is outlined in Chapter 3.

FIGURE 15.1. The Long Man in its South Downs escarpment setting. Traces of lynchets are seen along the slope above and below the figure of the Long Man. Fainter terracettes are seen running along the contours across the figure

Much of what has been written about the Long Man in the literature is highly speculative and includes a multiplicity of theories for which there is slim evidence. These diverse interpretations contribute to the wide public fascination with the site and its importance as part of the historic environment. There is obvious value in putting all ideas on the table for debate as demonstrated by

FIGURE 15.2. Excavating Trench 2 in 2004 with a circle of pagan worshippers below the hill figure on a flat platform created by post-medieval quarry spoil

the published debate on Cerne Abbas (Darvill *et al.* 1999). The significance that the Long Man has for some was very graphically shown on 26 September 2004, five days after the autumn equinox, and while we were carrying out the excavations. A large party of worshippers, some dressed in cloaks, wound up the path from Wilmington towards the Long Man assembling in a circle on the level platform, formed by spoil in front of a nineteenth-century chalk pit below the hill figure (Fig. 15.2). Thus was played out a fascinating juxtaposition between religious observance at the foot of the Long Man and our attempts, through excavation, to find out more about the figure itself.

The Wilmington giant

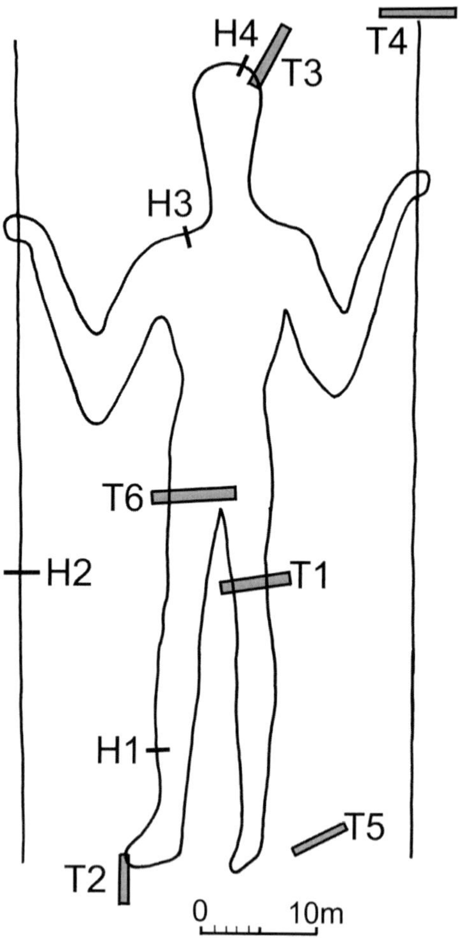

The Wilmington giant is the second largest ancient human figure in the world at 69 m tall. It is a simple outline of a human body with raised arms holding two staffs. Picked out in bricks in a restoration of 1873, these were replaced by concrete blocks in 1969. When this was done one of the writers (MB) assisted Eric Holden (1971) in small-scale excavations (Fig. 15.3). These showed that the outline was relatively superficial with only small marking out trenches visible in one, or perhaps two, of the four trenches and these only just penetrated the chalk. The first depiction is in 1710 by John Rowley (Farrant 1993, fig. 2), which shows the figure as rather more rotund, and in 1781 it is illustrated by W. Burrell with a scythe in its left hand and a rake in its right hand (Farrant 1993, fig. 3).

FIGURE 15.3. Location of section trenches across the Long Man by Holden (H1–4) and 2004 (T1–6)

The giant is in ancient chalk grassland, a Site of Special Scientific Interest on account of its natural historical significance. The Long Man is also a Scheduled Ancient Monument (Table 22.1). Cultivation on the escarpment and down top was limited, thus preserving archaeological sites of many periods; long barrows, round barrows, Celtic field systems, ancient trackways etc. (Curwen 1928). The area has recently been resurveyed by English Heritage as part of its investigation of South Downs National Park (Carpenter *et al.* 2013). That study and others have suggested that previously identified Neolithic flint mines (Curwen 1928) on the hill above the Long Man are actually post-medieval chalk quarrying.

Excavations at the foot of the Long Man in 2002

Puzzles concerning the date of the Long Man of Wilmington were identified as a theme for a programme in the series *Landscape Mysteries*, which was a joint production between BBC 2 and the Open University presented by the geologist Professor Aubrey Manning. The Long Man was seen as a suitable

15. The Long Man of Wilmington: a progress report on a giant conundrum

vehicle for the investigation of the history of the chalklands. The writers were delighted to be asked by the programme makers to carry out a short excavation at Wilmington to investigate the date and landscape context of the Long Man. The *Landscape Mysteries* programme was first broadcast on BBC 2 in October 2003, and is available in a box set of the series published by Demanddvd.

Our investigation took place between 1 and 4 November 2002 and involved the excavation of a trench at the base of the slope below the Long Man (Fig. 15.4). The purpose of this excavation was to apply methods that had been used elsewhere on the South Downs by Bell (1983) to look at colluvial slope sediments as evidence of the land-use history. A trench 15 m by 1.5 m was excavated downslope using a mechanical excavator under careful archaeological supervision, then a 0.5 m wide strip alongside the trench was excavated by hand and each of the artefacts, no matter how modern, was three dimensionally recorded. The trench revealed a long history of sediments and human activity (Fig. 15.5). At the base were chalk meltwater muds (Fig. 15.5, context 12) from a time of rapid physical weathering of the escarpment at the end of the last ice age. Cut into this were bowl-shaped hollows (Fig. 15.5, contexts 17–18), which contained some worked flints. Mollusc analysis by Alison Bell showed that these contained mainly woodland species, suggesting that they represent tree throw pits from a time when the chalk escarpment was wooded. Above this was a truncated buried soil (Fig. 15.5, context 13), investigated in micromorphological thin-section by Jodi Davidson. The soil contained 13 worked flints of Neolithic character and a rim sherd of a Neolithic bag-shaped bowl. This was overlain by a colluvial deposit derived from cultivation at the base of the slope. The layer contained worked flints of later Neolithic/Bronze Age character together with one Beaker sherd and three sherds of Late Bronze Age to Early Iron Age pottery. The old land surface and overlying colluvium were characterised by molluscs of open landscape indicating by the Bronze Age it was cultivated and grazed land. An optically stimulated luminescence date on the buried soil by Dr E. Rhodes indicated a date of 1650±940 BC. Luminescence dating is based on the principal that exposure to light, or heat, releases trapped electrons in crystals (e.g., quartz) and the quantity trapped since last exposure to heat (thermoluminescence) or light (optically stimulated luminescence) provides a measure of the age of sediments (Grün 2001): for a summary of luminescence dating see Chapter 1, and Chapter 2, Toms and Wood. Despite the standard deviation of almost a millennium, a Bronze Age date is strengthened by the discovery of pottery and flints of Neolithic and Bronze Age date in the buried soil and overlying colluvium. It seems there was cultivation, and perhaps a settlement nearby at the foot of the escarpment. Prehistoric activity seems, from about the middle of the first millennium BC, to have been followed by an extended period when nothing much happened; no artefacts between the later Iron Age and early medieval period were found.

FIGURE 15.4. The 2002 trench at the base of the slope below the Long Man. The old land surface is visible below a layer of chalk lumps

The section at the up-slope end of the trench produced a particularly significant much later stratigraphy (Figs 15.5 & 15.6). Here chalk was overlain by subsoil and then an earthworm-sorted stone-free humic soil (Fig. 15.6, context 4). Above this was a band of chalk pieces (Fig. 15.6, context 15), and then other layers of soil with traces of chalk bands. The chalk bands

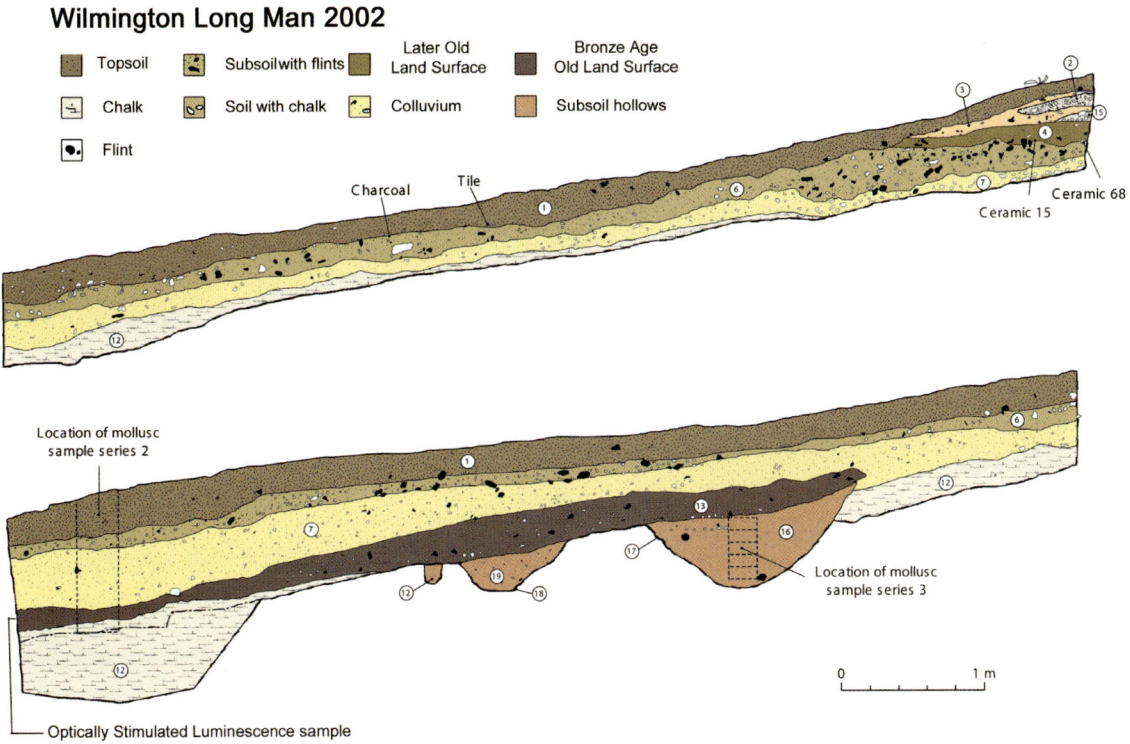

FIGURE 15.5. Section of the 2002 trench at the base of the slope below the Long Man. The upper section (right hand end) shows the buried soil (4) and overlying post-medieval layers; the lower section shows the fossil tree holes (17 and 18), Bronze Age buried soil (13) and overlying colluvium (7). Graphic: Shaun Buckley

may provide evidence of instability on the slope, perhaps associated with the making, or cleaning, of the Long Man. An alternative explanation for the chalk lumps recently proposed, is that they derive from the extensive post-medieval chalk/flint quarrying that is evident on the crest of the escarpment (Carpenter *et al.* 2013). However, none of the chalk pits lie immediately upslope of the Long Man excavation and they are separated from it by an incised hollow way (Curwen 1954, plate XXXII). There were two pieces of medieval pottery in the buried soil (Fig. 15.6, context 4) perhaps deriving from cultivation. Just upslope from the trench there is a negative lynchet prominent in photographs of the Long Man to its east (left on Fig. 15.1), perhaps representing the upslope end of arable land cultivated from the village/priory. The buried soil also contained a few fragments of fired clay and brick, discussed below, which were also present in the subsequent layers. Land molluscs indicate short open grassland conditions; they include two species, *Candidula intersecta* and *Cernuella virgata*, which are believed to have been introduced to this area in the medieval or post-medieval periods (Kerney 1999). An optically stimulated luminescence determination from the buried soil was AD 1420±620. The standard deviation is very large and would allow a date anywhere between the eighth century AD and the present day, but, together with the other evidence it makes a prehistoric or Romano-British date less likely.

Wilmington Long Man - South Section

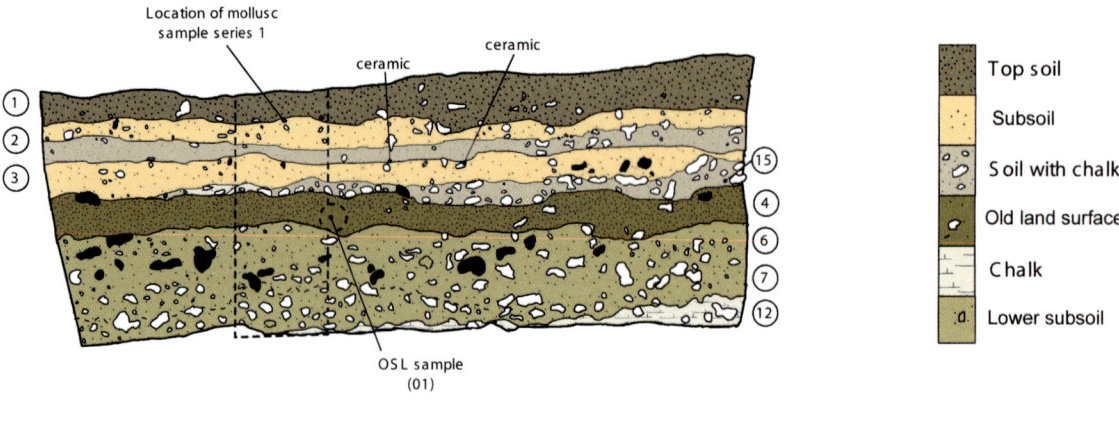

FIGURE 15.6. Section drawing of the upslope end of the 2002 trench showing the buried soil (4) overlying chalky layer (15) and position of the OSL sample. Graphic: Shaun Buckley

Excavations on the Long Man in 2004

A further four days' work was done at Wilmington on 25–26 September and 2–3 October 2004. This involved hand excavation of six sections across the outline of the Long Man to establish what traces remained of its original outline. All finds were three-dimensionally recorded and the sections were drawn. The trenches were located as shown on Figure 15.3. Most of the sections produced no clear evidence of a cut or chalk-filled feature indicating an original outline, although there were slight possible hints of shallow features in Trench 3 (Bell & Butler 2014, fig. 3.6), and Holden (1971) had found traces of a shallow gully in Trench H2 in 1969. Both possible features only just penetrated the chalk and may be the result of an experimental partial restoration by Wilmington vicar W.A. St John Dearsley in 1890, when missing bricks were replaced by a trench filled with rammed chalk (Farrant 1993, 134). The other trenches produced no evidence of cut features (Figs 15.7a & b). No evidence was found in any trench of features filled with clean chalk such as were employed to make the Uffington White Horse (Chapter 16; Miles *et al.* 2003) or the Cerne Abbas Giant (Chapter 3). The 10 trenches cut by Holden and the writers show the figure was never deeply cut into the chalk as often assumed, nor marked by chalk-filled trenches.

FIGURE 15.7. a) The upslope face of Trench 2 showing the soil profile and the absence of a cut feature, and b) Trench 6 (2004) across the Long Man's right hip showing the absence of any cut feature penetrating chalk

Ceramic building material

Ninety-six pieces of brick weighing 2,533 g were found during the 2002 excavation, with a further 330 pieces weighing 1,062 g from the 2004 excavations. Five complete bricks came from the 2004 excavation. The bricks are of four types:

Type A: A hard-fired dark red-brown sandy fabric with rare small white angular quartz inclusions, and occasional vesicles, presumably burnt-out organic material. A typical nineteenth- to early twentieth-century brick fabric. Some pieces have a remnant of the cement on the surface, others had white paint on the surface. Four complete Type A bricks were recovered in the 2004 excavations in Trench 3. They measure 220 mm × 102 mm × 64 mm with a shallow frog in one face. They appear to be hand moulded and are of poor quality (M. Beswick pers. com.) These bricks had been laid on their edges with cement adhering the flat/frogged faces of the bricks together, the exposed upper edge and one end (header) of the bricks had then been coated with a white paint. At least two coatings of paint could be identified on these bricks. There were many local brick making sites in operation during the later nineteenth century (Beswick 2001). A comparison with bricks of that date incorporated into local houses has found that there are many similar fabrics, but no exact match. Furthermore, this fabric is not the same as that found in the bricks from the brick-built kiln in the quarry to the west of the Long Man. It is possible that these brick fragments do come from a local industry and were incorporated into the outline of the Long Man during the various repairs that were carried out in the 1890s and later.

Type B: An orange grog-tempered, poorly fired, sandy fabric with a slight soapy feel. The grog is generally small and rounded and is present in various colours and sizes. This fabric also has frequent small rounded black ferruginous inclusions, together with very rare small angular quartz inclusions. This fabric corresponds to the 'fired clay' found in the 1969 excavation, and at that time identified as Roman tile (Holden 1971, 41). Fragments of this fabric were dated by the OSL technique during the latest project and were found to be of sixteenth-century date (see below).

Type C: A hard-fired, reduced, buff coloured sandy fabric with rare flecks of hard-fired red grog and occasional vesicles representing burnt out organic material. Almost certainly the 'yellow' brick used to line the figure in the 1874 restoration. One piece has two layers of cement, each coated with white paint, suggesting at least two episodes of repainting. Another example has a layer of khaki green paint applied over a single layer of white paint and corresponds with the known over painting of the Long Man during the Second World War to camouflage it from enemy aircraft. A complete 'yellow' brick was recovered from Trench 3 during the 2004 excavations. It measures 228 mm × 112mm × 68 mm. Commonly called a Gault brick, it has a double frog, which suggests that it may be machine pressed (M. Beswick, pers. comm.). Other bricks of this type were noted still *in situ* on the inner part of the left elbow of the Long Man. These 'yellow' bricks may have been produced near Pevensey and Westham Station, from where they could easily have been brought by rail. The brickworks was operating at the time of the 1874 restoration (Beswick 2001, 174).

Type D: Two pieces of very hard-fired grey-white sandy fabric, with occasional black ferruginous inclusions and very rare small angular quartz inclusions were found in the 2002 excavation. One piece also has a single layer of white paint applied to its flat surface. It is likely that these pieces of brick derive from the 1891–92 repair work, when white bricks were used.

A comparison was made with bricks incorporated into the structure of Wilmington Priory. The bricks in the walls of the Western Tower are grog-tempered but are hard-fired with larger grog pieces together with large black ferruginous and chalk inclusions. The Western Tower was dated to the sixteenth century (Godfrey 1928), but Abel Smith (2015) suggested a fifteenth-century date. These bricks are clearly not the same as Type B. However, the brickwork incorporated into the kitchen fireplace and in the chimney and fireplace in the room above the kitchen, is of an identical fabric to Type B. The bricks are handmade and vary in size between 180 mm to 240 mm long, 100 mm to 115 mm wide, and 45 mm to 50 mm deep. These rooms are mentioned in a will of 1606 (Godfrey 1928, 25) with the fireplaces dating to the sixteenth century. Finally in the Well House Court, which Godfrey identified as the 'newly built' kitchen of 1541, there are bricks of both the above grog-tempered fabrics incorporated into the fireplace. It is, therefore, clear that the Type B brick fragments can be linked to Wilmington Priory.

Discussion

The excavation evidence for the relatively superficial character of the figure is significant because the slope on which it is cut, although grass covered, is manifestly unstable. The escarpment is covered by terracettes, micro-terraces on a decimetre scale (Figs 15.1 & 15.4), which are the result of the small-scale downslope movement on shallow soils under conditions of animal grazing. Downslope movement is also evident from the concrete blocks that were emplaced in 1969; in places the lines of blocks had by 2004 buckled and contorted as a result of downslope movement (Bell & Butler 2014, fig. 3.7). We conclude that on this slope a relatively superficially cut figure will not have survived for millennia, thus earlier dates, especially prehistoric are unlikely. Additionally, there is a lack of activity between the mid-first millennium BC and the medieval period and subsequent evidence for slope instability represented by the chalk pieces above the buried soil containing medieval pottery and brick fragments is most economically interpreted as indicating a post-medieval date for the figure.

As it turns out the main clues have been there all along but have been overlooked, perhaps because successive writers had an unconscious prejudice towards earlier dates. Before the restoration in brick in 1873 the Long Man was marked by a faint depression in the downland turf and was visible in certain lights and after snow. However, a guidebook of 1791 in describing the Long Man says 'it is formed by a pavement of bricks underneath the turf which gives it thus differences in colour, in time of snow it is still more visible' (Shaw 1791, 376; Farrant 1993, 130). The notion of a brick pavement has been dismissed by most previous writers. Significantly the first depiction of the Long Man by Rowley in 1710, on an estate map at Chatsworth, shows the outline as a broken line (Farrant 1993, fig. 2), which could be consistent with either a shadowy outline or with an outline formed of bricks, on, or just below, the surface. Further evidence for bricks on the site comes from an account of the restoration in brick of 1873 in the *Eastbourne Gazette* on 29 April 1874, which notes the discovery of fragments of Roman bricks during the restoration (Holden 1971, 41). Holden also found brick during his excavations and at the time specialists suggested they were Roman.

During our excavations in 2002 and 2004 evidence of four distinct types of brick were found (types A to D) as outlined above. Four samples of brick of Type B were subjected to luminescence dating by Dr E. Rhodes at the Research Laboratory for Archaeology and the History of Art at Oxford University. Three were from the buried land surface, context 4 at the south end of the trench and one was from context 2. All produced very similar internally consistent dates and the mean age for the four samples is AD 1545±30. Bricks of Type B were used in Wilmington Priory dated to the sixteenth century including the 'newly built' kitchen of 1541, which has bricks of both grog-tempered Types B (and C) incorporated in the fireplace. Thus, the Type B bricks can be linked to Wilmington Priory, some in structures dated 1541 very close to the luminescence date noted above. That does not mean of course that the bricks were incorporated in the Long Man at that date and they could have become available as a result of subsequent rebuilding at the Priory.

Given the superficial nature of the Long Man, the absence of clear-cut features in most of the excavations, the evidence of several phases of brickwork and the statement of Shaw (1791), it seems possible that the Long Man was marked in brick from its inception and that the balance of probability points to a post-medieval date. There are various possibilities for its historical context and historical research by John Farrant (1993; 1995) advanced understanding but could not associate the figure with one individual, or historical context.

It is possible that the figure relates in some way to the religious turmoil between the Protestant Reformation (from 1529) and the restoration of Charles II in 1660. Hutton (2004) has suggested that it could represent the work of zealous protestants and depict the door to salvation. It might be associated in some way with the 17 protestant Lewes martyrs (AD 1555–7), burned at the stake during the Catholic restoration of Queen Mary (Castleden 2021).

Of particular significance is Hutton's (2004; Chapter 18) demonstration that in the late medieval and early post-medieval period there was a preoccupation with giants that was manifest in giant hill figures and could hint at an origin in secular antiquarianism rather than religious fervour. Figures of Gog-Magog are recorded at Plymouth Hoe in 1486 with another added in the Tudor period, and near Cambridge at the Gog-Magog Hills in 1640; a giant is also recorded at Shotover Hill above Oxford by the 1640s. When Bess of Hardwick (AD 1518–1608) rebuilt Hardwick Old Hall, Derbyshire, there were plasterwork figures of giants probably Gog and Magog above the fireplace (Worsley 1998). In Chapter 6 Brian Edwards argues, in the context of Cerne Abbas, that depictions of giants were (or became) associated with the Glorious Revolution of 1689 and the protestant William III was frequently depicted as Hercules. Professor Hutton (2004) suggests we should consider the Wilmington figure in the context of a Tudor and Stuart world in which secular landowners bought former religious houses and put their stamp on the land, perhaps rather literally in this case. Hutton highlights the need for an investigation of the owners of Windover Hill between the reign of Henry VIII and Anne to try to establish a context for the making of the figure.

Intervisibility

As part of the project an intervisibility survey of the Long Man and its landscape was undertaken. This was carried out in January and February 2005, and then repeated in May 2006. It is quite clear that the Long Man is meant to be seen from its immediate front. The best views are from the area of Wilmington Priory through to Wilmington Green, although the view deteriorates any closer to the figure itself. Good views can also be obtained from Arlington, especially from around the Church and Arlington Deserted Medieval Village. It can be seen from Raylands Farm, but not from the nearby deserted moated site. Further north a long-distance view can be achieved from the high ground at Upper Dicker, but there is no view from Michelham Priory, even from the gatehouse or upper storey windows.

To the west the Long Man can be clearly seen at Milton Street, from along Common Lane, from the area of Berwick Station, and from Mays Farm. Further north-west at Chalvington the view is obscured by rising ground, whilst at Selmeston it is obscured, although at Green House and along the footpath to Berwick Station it can be seen. Along the line of the A27 from the west the figure cannot be seen until the junction with Common Lane, and then only an oblique view. South of the A27 the Long Man is not visible at all from Alciston, Berwick, Berwick Court, the Rookery and Milton Court Farm.

On the east side the Long Man cannot be seen at all on the old coach road from Folkington, until you pass The Holt where it appears suddenly. Further north there are only oblique views from the higher ground north of Folkington, around Wootton Manor, although there are good views from the minor roads south of Abbot's Wood. From the south, the Long Man cannot be viewed from any direction. There is line of sight from places further afield, such as Laughton Place and the Dicker, but not without the aid of a telescope. Looking north from the Long Man, especially on a clear day, there are good views out as far as the Heathfield ridgeline in the Weald, some 20 km distance. This survey has confirmed the long-held view that the best place to see the Long Man is from immediately below the figure, at Wilmington itself. This would suggest that if one of the reasons for its original construction was to 'be seen', then the place from which it was meant to be seen was Wilmington, and perhaps more specifically Wilmington Priory.

Ownership

Given the emerging evidence for an early post-medieval date for the Long Man and the intervisibility study which puts Wilmington Priory centre stage, it is necessary to evaluate its ownership history. The architecture of Wilmington Priory has been analysed by Godfrey (1928) and its history by Budgen (1928), both aspects being more recently reviewed by Abel Smith (2015), following the lease and restoration of the priory by the Landmark Trust. It was an alien priory suppressed in AD 1414 and subsequently held by the Bishop of Chichester. In the fifteenth century a Great Chamber had been built on the first floor at the west edge of the priory buildings with a ground floor entrance facing south, flanked by two, still extant, corner staircase towers. In AD 1565 the priory was granted to Sir Richard Sackville and there is only patchy information on subsequent Sackville tenants which nonetheless includes characters of interest. Wills of 1602 and 1606 show these included members of the prominent Sussex Colepeper family (Attree & Booker 1904).

In 1618 the tenant was Sir Henry Compton (1584–1649) Member of Parliament for East Grinstead (Davidson 2010) who rented the Priory on favourable terms from Sackville family relatives. Abel Smith (2015) suggested on, stylistic comparisons with windows elsewhere of appropriate date, that it was he who inserted the 'magnificent transomed and mullioned window' on the first floor at the south end of the Great Chamber. Significantly, perhaps, this has window seats on either side (Godfrey 1928, 23) and looks directly at the Downs escarpment and the figure of the Long Man, the line of sight being 73 degrees to the window (Abel Smith 2015, 14). Perhaps, therefore, the Long Man was created as a landscape folly to be viewed from this window. If so the likely span of Henry Compton's focus on the estate is between his occupancy in 1618 and 1631 when he created a new mansion, now ruined, at the moated manor he had held since 1619 at Brambletye, near Forest Row, Sussex (Hannah & Peckham 1928). Compton's religion is also of interest; his two wives were both recusants, i.e., refused to attend Church of England services, and Compton himself was subject to Parliamentary investigation as a Catholic. Catholics, however, regarded him as schismatic, implying that, though not by conviction, he attended Church of England services (Davidson 2010). If Henry Compton was the author of the Long Man, it may have been a landscape folly but with a hidden religious meaning and origin that became conveniently lost in the mists of time and encroaching downland turf when his focus moved to Brambletye.

In 1700 the estate including Wilmington passed from Lord Northampton to his younger son Spencer Compton (1674–1743), nephew of Sir Henry. It was on a map made for Spencer Compton in 1710 and now at Chatsworth, that John Farrant (1995) discovered the earliest reference, a pictorial representation, of the Long Man. Spencer Compton was a significant, though not apparently especially able, politician. He also was a member of Parliament for East Grinstead, Speaker of the House of Commons and Prime Minister in 1742 (Hanham 2010). On his elevation in 1728 he adopted the title Baron Wilmington and in 1730 that of Earl Wilmington, reflecting a family association with the place which is perhaps puzzling given the comparatively modest Priory buildings. Compton's growing prosperity, partly through the lucrative office of Paymaster General, facilitated building of a mansion Compton Place in Eastbourne from 1726 (Berry 2021). As Earl Wilmington, Spencer Compton unwittingly projected this small Sussex village onto a world stage. His friend Thomas Penn seems to have been instrumental in 1739 in naming after him the settlement of Wilmington in the United States. Today, Wilmington is the largest city in Delaware and the family home of US President Joe Biden.

Spencer Compton was from a family of high church Tories and a member of the political Kit Cat Club whose political career was related to support for the monarch. He was clearly a man set on making his mark on the world who manifestly identified with Wilmington at a time when we have the first evidence for the Long Man, so he should perhaps be considered as a possible creator of the figure. He has not, however, been associated with significant developments at the priory unlike his uncle Sir Henry Compton whose involvement, religious background and dates are more in line with the archaeological evidence, which perhaps make him the most likely creator of the Long Man.

Conclusions

Work so far has clarified the origins of the Long Man. Although Neolithic and Bronze Age activity took place on the site that appears to be associated with agriculture and perhaps nearby settlement at the foot of the escarpment. Otherwise, there are no finds indicating there was anything that drew people to this spot in later prehistory, or the Roman period. We conclude that a figure, at the most superficially cut, is unlikely to have survived on an unstable slope for millennia. There were frequent post-medieval brick fragments and other artefacts from the excavations suggesting that in the last few centuries, but not before, something (i.e., the Long Man) drew people to the site. Although no piece of dating evidence on its own is absolutely convincing, if we put the various sources of evidence together, the artefacts, the bricks, the luminescence dating, the molluscs, the testimony of Shaw (1791) and the *Eastbourne Gazette* evidence for the presence of bricks before the 1873 restoration, all seem to point to a post-medieval date sometime between *c*. AD 1529 and the earliest depiction in 1710. The scientific dating evidence and possible religious and historical associations would permit an origin in the earlier part of this range. However, among the tenants and owners of the period Sir Henry Compton emerges as perhaps the most likely, though by no means certain, creator of the Long Man and if so the likely date is between AD 1618 and 1631.

Acknowledgements

This is an extended version of an interim report (Bell & Butler 2014) on our investigations first published in a book of limited circulation dedicated to Eastbourne Archaeologist the late Lawrence Stevens.

The 2002 excavations were carried out as part of the *Landscape Mysteries* television series, made for the BBC and Open University. The work was funded by TV6 and the Open University and we are grateful to Emma Cotton and Nick Metcalfe for arrangements and to the presenter of the series the late Professor Aubrey Manning for showing such an interest in our work and communicating that to the audience. Permission for the excavation was given by the Sussex Archaeological Society, English Heritage and English Nature and Mr Ray Ellis. We are grateful for the specialist input of Dr Ed Rhodes (OSL and TL dating); Jodi Davidson (sediments and Mollusca); and Alison Bell (Mollusca). For assistance in the field we acknowledge Dr Shaun Buckley, Dr Alex Brown, the Reading University MSc geoarchaeology classes of 2002 and 2004, and members of the Mid-Sussex Field Archaeological Team. Specialist advice is acknowledged from Luke Barber, Molly Beswick, Prof. Sue Hamilton, Pat Stevens, Dr Mike Allen, Mr Brian Edwards, Dr John Farrant, Prof. Ronald Hutton and Rodney Castleden. The Long Man is owned and cared for by the Sussex Archaeological Society.

CHAPTER SIXTEEN

I will survive: the continuing story of the Uffington White Horse

David Miles and Simon Palmer

The cutting of hill-figures in the turf is an obscure and little regarded art (Morris Marples 1949).

Call up 'geoglyphs' on Wikipedia and immediately there are images of Nazca Lines in Peru and the Uffington White Horse. All appear most clearly from the air.

The Nazca Lines, etched through the top layer of dust and pebbles, on flat surfaces, survive because conditions on the bone-dry, windless desert are as lunar as can be found on earth. Nazca figures on hill-slopes may have been almost extinguished by soil-creep. Gravity will have its way.

The English, with their fondness for Anglo-Saxon rather than Greek, usually refer to the Uffington White Horse (Fig. 16.1) and its companions as 'hill figures'. The topography of Southern England, with its succession of steep chalk scarps, provides a series of natural screens in the landscape on which figures can be displayed – startling white against a vivid green backdrop. And usually the highest and steepest slopes are selected, places where access is difficult for farmers but subject to the rigours of climate.

All over the world people are attracted to high places. These provide panoramic views of movements below – of neighbours, enemies, migrating animals, or changes in the sky. The 'cosmic' mountains bring us closer to the divine presence. High places can be seats of secular, but often, divine power. They are god-trodden. Some, like Mount Fuji, Mount Olympus, or Mount Ida, penetrate the heavens. Others, simply in terms of contours, are relatively modest – Jerusalem's Temple Mount – or even artificial mounds, from Babylonian zigurrats, Mayan pyramids to Silbury Hill (Wiltshire).

England's favourite American commentator, Bill Bryson, emphasised an important point: 'it seems to me what is truly remarkable about the White Horse is not that people some time in an immensely ancient past took the trouble to cut it into the hillside – though that is extraordinary enough, goodness knows – but that continuously for around twenty centuries others have made the effort to maintain it … The White Horse has been preserved simply

FIGURE 16.1. The White Horse above the Manger and Dragon Hill. © Oxford Archaeology Ltd. 2022

because people like it (I think that's splendid)' (Bryson 2000). It is a comforting thought, such continuity. Yet England has never been a fantasy of pubs, warm ale, cricket and village green preservation societies. As the historian Linda Colley points out 'But the past was another country. In the British case indeed the past was a great many countries' (Colley 2010). England has been created by immigration and emigration, land-grabs, plague, language shifts, religious upheaval and technological revolutions. A remote archipelago and the centre of a world empire.

As we started to write this article, horrific images appeared on our screens: of an attack by Hamas on southern Israel. Hamas named their brutal attack *Operation Al Aqsa Deluge* – a reference to the Al Aqsa Mosque compound, Islam's third holiest site and the Hebrew Mountain of the House of the Lord or Temple Mount. A geoglyph under the Dome of the rock marks the footprint of Mohammed. This has been one of the most contested fragments of the so-called Holy Land since Solomon's temple was destroyed in 582 BC.

The Temple Mount is not unique in provoking devotion as well as hate, hostility and rage. The Black Hills (Paha Sapa) of Dakota are the sacred, ancestral uplands of the Lakota. One of the authors (DM) was told by locals, 'we have been here since time immemorial'. Historical research, however, suggests the Lakota shifted their river-valley based way of life and took control of the Black

16. I will survive: the continuing story of the Uffington White Horse 249

Hills only between the later eighteenth century and the 1820s. By this time the Lakota were a horse people and their prey, the bison, was being driven into the cooler, wetter, lusher Black Hills by the onset of severe droughts on the prairies. Rock art indicates that people long before the Lakota were attached to this magnificent landscape. In the words of historian Pekka Hamalainen 'equine power ushered Lakotans into a new technological age'. So the Black Hills were vital economically to the Lakota as well as spiritually 'the heart of everything that is', albeit the focus of a relatively recent cosmology (Hamalainen 2019, 191, 382).

Today tourists flock to the Black Hills, often to see the land art memorialising the latest conquerors, the images of four white men: Presidents Washington, Jefferson, Theodore Roosevelt and Lincoln carved into Mount Rushmore between 1927 and 1941. Nevertheless the Lakota still actively pursue their claims. They continue to carve their own hero, Crazy Horse, into the rock. Like many sacred sites the Black Hills display a history of violence, co-existence and determination; contested places through the generations which nevertheless adapt to change.

So let us return to the apparently bucolic White Horse Hill; a place that also demands attention. Here the north-facing scarp of the Lambourne/Marlborough downs rises steeply to 261 m – the highest point in Oxfordshire. Not much by world standards. Yet it feels high, like surfing a rising wave close to the sky. Here we can literally stand on an ancient seabed thrust upwards by the tectonic plate of Africa.

The topography is complex, like a crumpled sheet. The scarp slope is carved by dry valleys or coombes. The most dominant is known as the Manger (Fig. 16.2), immediately below the hill figure of the White Horse. Its sides are rippled by the shutes where melting ice and mud slides carved into the slope, as permafrost melted at the end of the Ice Age. Above is the flattened pyramid of Dragon Hill, a natural chalk outcrop where St George is said to have slain the dragon and left a blood-soaked bare patch in the turf. St George, England's patron saint, who never set foot in England, was popularised in the fourteenth century, when his feast day was much celebrated. But no longer.

At the mouth of the Manger is a steep hollow, difficult to enter, where water flows cold and clear from below the chalk. This is the source of the River Ock (Celtic name for a fish), which flows through the Vale of the White Horse to join the Thames at the ancient site of Abingdon.

Flowing above the Manger is the sinuous figure of the White Horse, 120 m long, a continuous line from beaked head to tail, with segmented limbs. Up close the white lines seem to form abstract pathways to nowhere.

The highest part of White Horse Hill is dominated by the hillfort of Uffington Castle. To stand on the northern rampart is to experience one of the finest views in England – the country lies below us, the enclosed fields and thatched villages of the Vale, the ridge of Corallian limestone behind which runs the Thames. On a clear day the blue hills of the Cotswolds are visible to the north-west and the Chilterns to the east. The cooling towers of the Didcot

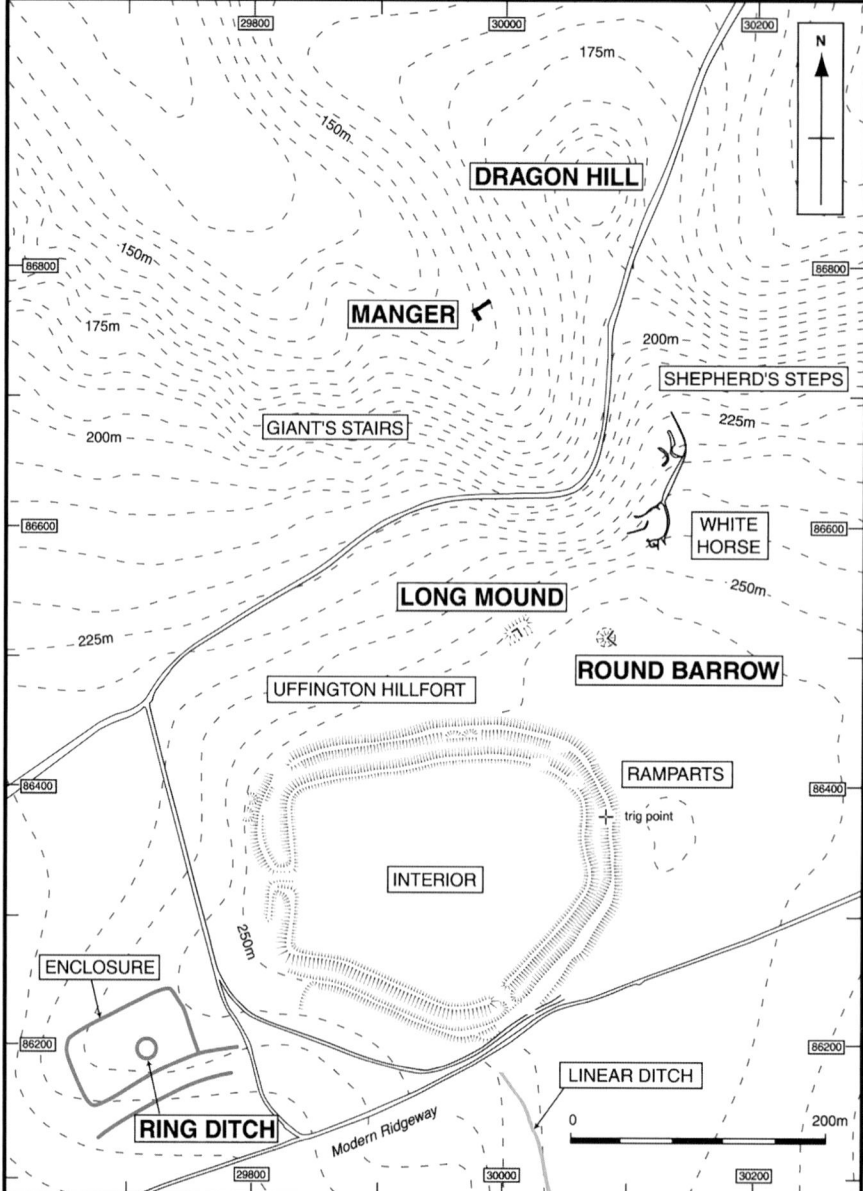

FIGURE 16.2. Location of the Monuments, © Oxford Archaeology Ltd. 2003

power-station with their dragon's breath are now gone – a twentieth-century memory of redundant energy. Instead wind turbines revolve on the edge of Swindon, where solar panels also gleam.

Elsewhere on the hill there are swollen mounds from the Bronze Age or even earlier. The Ridgeway, the long-distance path, which runs between the Wash and the south coast remains contentious. Was it a long-distance routeway dating back to the early Neolithic or even created by herds of migrating horses and reindeer 10,000 years ago? Detailed surveys, for example around Avebury, show the current line of the pathway running across Romano-British field boundaries and so, obviously, post-dating them (Fowler 2000). If the original Ridgeway was earlier then it was probably formed of a branching network of paths seeking out the drier, more navigable routes, rather than the single, hedged track we see today.

Along the Ridgeway there is a string of ancient sites and 2 km south-west of White Horse Hill is the great chambered tomb of Wayland's Smithy, monument to a slaughtered family, now guarded by its massive portals of sarsen stone – the hard sandstone blocks that erode from below the chalk and litter the floor of the dry valley by Ashdown House. Many were dragged off for use in megalithic monuments – the most permanent structures in the English landscape.

White Horse Hill now has every protective legal designation attached to it that the conservation bureaucracy can conceive (except UNESCO World Heritage Status). Yet these are not necessarily effective. The Ancient Monument Protection Act of 1882 was described as 'a molehill' emerging from a mountain of 10 years of debate (Miles 2019, 243). It certainly failed to protect thousands of archaeological sites, which fell like chaff before the increasingly powerful scythes of modern development. In contrast the relatively few sites in Guardianship (the highest form of protection) were wrapped in the rather sterile turf blankets or penned by the iron railings of the Ministry of Works. Thus protected from developers, archaeological researchers were also often kept at bay and well-known monuments remained ill understood.

In fact, while Uffington Castle and Wayland's Smithy were on the first 1882 list (or schedule) of 26 sites in England, the White Horse itself was not included. Why? Most likely because its origin remained contentious. In fact, legal protection only arrived in December 1929 – possibly because three years earlier Sir Flinders Petrie, the doyen of Egyptian archaeology, had surveyed the Horse and pronounced it to be Bronze Age in date. His argument was, frankly, unconvincing; but the great man bestowed respectability on the hill figure (Miles 2019, 244).

For many years there has been no question that the Uffington White Horse is of considerable antiquity. The surrounding land belonged to medieval institutions such as Abingdon Abbey and there are several written references to the Horse in monastic charters. In the twelfth century a medieval scribe included the Horse, along with Stonehenge and the Giant's Causeway, as one of the 'Wonders of Britain'. Such a wonder was seen as miraculous, not a work of man but of God.

As the study of antiquity developed, theories about the Horse abounded: it resembled the horse images on Iron Age coins or earlier Bronze Age animals; it was a memorial to the Battle of Ashdown, where, somewhere in the neighbourhood, King Æthelred and Prince Alfred defeated the Danes. For much of the eighteenth and nineteenth centuries this unlikely theory held sway, along with the rising reputation of Alfred the Great, who was credited with founding everything from the Royal Navy to the English University system. The reputation of the White Horse was also at its peak: a symbol of the foundation of Christian England.

The growing antiquarian belief in the pagan, Iron Age origin of the Horse seems to have undermined its pedigree as an English icon. However, a well-argued essay by Stuart Piggott in the influential journal *Antiquity* in 1931, supporting a Late Iron Age origin, was widely accepted (Piggott 1931). Rival theories persisted – that for example, the White Horse was the banner for those first English immigrants, Hengist and Horsa, and the image seen today was a shrunken version of the original (Woolner 1965; 1967).

While its origins remained contentious there is no doubting the fascination of the Horse's biography. Before Bill Bryson, Cambridge's distinguished Professor of Archaeology Grahame Clark emphasised 'This noble animal, it is well to remember, can only have survived though frequent scouring of the chalk, a very symbol of continuity between the prehistoric past and the present day' (Clark 1940). In spite of such support the White Horse, Britain's unique and largest prehistoric image, was often missing from textbooks. Its singularity created a problem. How to fit it into the story of the past.

In September 1980 a small group of us from Oxford Archaeology, assisted by local volunteers, began a detailed survey of the National Trust land at White Horse Hill. A section of farmland adjacent to the Horse had just been given to the Trust by the Hon. David Astor. We proposed to the Trust that a management plan was needed for the new holding. Only shortly before Peter Fowler and David Miles (representing the Council for British Archaeology's Countryside Committee) had warned the National Trust that archaeological sites were being destroyed on the Trust's properties because of the lack of archaeological survey and a reliable inventory of sites. To its credit the Trust responded promptly to our warning.

Now the opportunity arose to document the landscape, and at the same time improve the management of the land itself: stop ploughing in the area, reduce the intensity of grazing, remove the corrals of barbed wire around the Horse, remove the visible carparks, unnecessary signs and litter bins (they attract litter!). Not only would this protect archaeology but also encourage orchids, insects and skylarks back to the increasingly rare grassland of the Downs.

We did not put much emphasis in our project design on dating the Horse. In fact, in the early stages we were told by English Heritage not to think about digging into the hill figure. This would be pointless and damaging as the image was a sgraffito, made by simply cutting through the turf to reveal the underlying chalk. Clearly no one in 1980 gave much credit to Daniel Defoe who, in the 1720s, reported that a trench 'about two yards wide on the top' and 'about a yard deep' was dug in the slope of a Horse 'and filled almost up with chalk'. Admittedly Defoe was not the most reliable of reporters, but in this case he was exactly right. We first realised this by excavating Ministry of Works files in London and finding detailed drawings of a section cut into the 'beak' when the Horse was restored after the Second World War. This showed successive layers of chalk, altogether 1 m thick (Figs 16.3a & 16.3b).

For us this was a breakthrough – the White Horse was a deliberate construction with layers or strata, which would allow us to examine the shape of the figure through time and, theoretically, date it. However, finding dateable material in a figure designed to be kept clean remained a long shot. Until, over coffee in Oxford Research Laboratory for Archaeology and the History of Art, Martin Aitkin told us about a new dating technique that opened up possibilities: optically stimulated luminescence (OSL) (Griffiths & Stone 2022). It was 1994 when we finally excavated a narrow trench into the body of the Horse and extracted the OSL samples. The delay proved to be a godsend, as we were one of the first archaeological applications of the new technique.

Compared to radiocarbon dating, OSL estimates are relatively imprecise. However, Julie Rees-Jones and Mike Tite emphasised the relative reliability of the estimates for the White Horse itself: 'The luminescence dates from the White Horse suggest an approximate age of 1380 – 550 BC (at the 68% confidence level) for the construction of the first Horse. This range overlaps with the suggested date of 750 – 650 BC for the construction of the hillfort so that the White Horse and the hillfort could be contemporaneous' (Rees-Jones & Tite 2003).

Excavation revealed the complexity of the surrounding landscape – Early Bronze Age barrows, reused in the pagan Anglo-Saxon period. The Linear (Neolithic or Bronze Age) barrow was used as the focal point of a substantial late Romano-British burial ground where many bodies were decapitated post-mortem. So over some three millennia the Hill was an ancestral burial ground, the white chalk tombs carefully sited on false crests to be visible from below, the earliest of which pre-dated the Horse (Miles *et al.* 2003).

The hillfort itself, in contrast with others along the Ridgeway, was not intensively occupied and may have functioned principally as a ritual enclosure where scattered communities on the Downs and in the Vale came together. There is some evidence that this role continued through the Roman period.

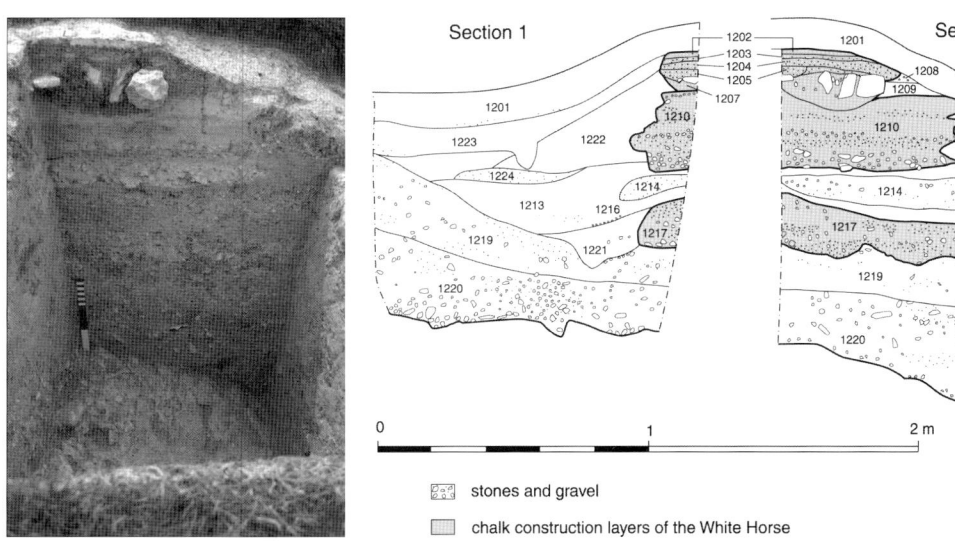

FIGURE 16.3. a) The post-war 'Beak' trench re-excavated, © Oxford Archaeology Ltd. 1990, and b) Section of the 'Beak' trench, © Oxford Archaeology Ltd. 2003

The study of British hill figures was, to a considerable extent, the preserve of an enthusiastic, often amateur, minority. Since the 1990s research on British hill figures has, fortunately, widened – as we can see in this volume.

The OSL dating of the White Horse settled the argument about a late prehistoric versus Anglo-Saxon origin and the excavation confirmed the original segmented shape of the image. This has led to more detailed discussion of its original purpose and context: Josh Pollard has considered the Sun Horse theory in more detail arguing that the figure is carefully sited in relation to the rising sun and its movement across the sky (Pollard 2017). The Trundholm sun chariot is the best know indicator of the importance of the sun in Bronze Age religion. The Uffington Horse, he argues, manifests similar beliefs in a daily drama of the rising and setting star.

Recently Ian Godfrey (forthcoming) has developed a more complex astronomical argument relating the White Horse to the annual solstices, proposing that it acted as a winter solstice sunrise observatory.

Miles has also argued that the White Horse should be seen in the context of the growing importance of the domestic horse in the economy, social status and warfare of the Late Bronze Age/Early Iron Age in Europe (Miles 2019, 160).

Not everyone is convinced. Believers in the British coin theory point out that the present OSL samples do not entirely exclude the possibility of a Late Iron Age origin. There is a 2.5% probability of the Horse being later than 210 BC. Certainly, in the mid-first century BC there was a period when abstract horses and sun symbols were important in coin iconography. And, for the first time, we have some idea of the political context of the White Horse area – on the boundary of three tribal groups (Catuvalauni, Dobunni and Atrebates), some, such as the royal house that ruled at nearby Calleva, showing a degree of loyalty to the encroaching Roman empire. Piggott argued that the coin images provided a model for the hill figure. Alternatively, the segmented beaked sun horse may already have been an ancient motif of British independence (Creighton 2000, 122–3).

As Professor Grahame Clark and Bill Bryson have pointed out: the survival of such a relatively fragile figure, requiring regular attention, is remarkable. Clearly the figure retained its significance to local communities through generations of change. There is substantial evidence for the continuity of worship at hilltop sites in Southern Britain. At Chanctonbury Ring, for example, a hillfort prominently sited on the South Downs and first constructed in the Late Bronze Age was also the home of two Romano-British temples. One, on the highest point, produced evidence of a boar/pig cult, another animal of great significance both to late prehistoric Britons and in the Roman empire (Rudling 2001).

No temple has been found at Uffington Castle, but there is evidence of cult activity and pagan burials in the fourth century, when pagan temples thrived in the Romano-British countryside. The horse also continued to play an important role in Roman religion – notably at the cult centre of Epona (the Celtic horse goddess) on the fortified hilltop of Alesia in Bourgogne. Nearer to hand, the remains of a Jupiter Column – topped by a mounted rider usually attacking a monster – have been found at Cirencester. Similar iconography is common on tombstones, reflecting the conquest of good over evil – an idea that persists into the medieval and modern world with St George. It is quite clear that icons of one religion are frequently translated into another and Gods become heroes.

Anglo-Saxon mythology permeates the White Horse landscape – Wayland the Smith at the nearby chambered tomb; Tell's Fort – the hillfort below White Horse Hill named for the Germanic archer god and Weland's brother (Miles 2019, 206–9). Various barrows are given Anglo-Saxon names or are adapted for Anglo-Saxon burial. By the eighteenth century we begin to have detailed reports of the great gatherings (the Pastimes) that accompanied Horse scourings (Fig. 16.4). These were the responsibility of local landowners, but attracted a wide range of people from the region and beyond to these 'rural Olympicks'. In spite of their scale they left virtually no archaeological trace.

By the later eighteenth century the White Horse was associated with King George III, who, in spite of his progressive mental illness and the loss of the American colonies, was a popular monarch. At his accession speech in 1760 he was the first Hanoverian to be able to claim to have been 'born and educated in this country'. The Uffington Horse, its Pastime and patriotism, seem to have inspired the creation of most of the other hill figures in Wiltshire, and beyond, around this time (Schwyzer 1999).

In the following decades England (or the United Kingdom since the 1802 Act of Union) saw considerable economic and social unrest both in the rapidly expanding industrial towns and in the countryside with unemployment, poverty

FIGURE 16.4. The scouring of the Horse in 1857 as depicted by Richard Doyle in Thomas Hughes' 'The Scouring of the White Horse'

and hunger flaring into violence and vandalism. Increasing class divisions seem to have alarmed landowners who, in 1778, had enclosed Uffington and adjacent parishes. Rural gatherings such as the White Horse Scouring were opportunities for antagonism and, in any case, were less fashionable as formal sports, such as horse racing, and urban entertainments flourished, made accessible by the new railways.

In an attempt to heal social rifts and hark back to the good old days, Thomas Hughes and Edwin Martin Atkins organised the Pastime of 18 September 1857 (Hughes 1859). This was successful but was, nevertheless, the last of its kind. In the later decades of the nineteenth and early twentieth century the Uffington White Horse was at its most vulnerable as the responsibility for its upkeep shifted from landowner to state. Yet the state had not quite got its act together (Miles 2019, 244).

Since the investigations on White Horse Hill (1989–1995) the hill figure has been in the care of the National Trust. There has been a cleaning exercise almost every year, with fresh chalk excavated from a nearby pit and hammered into the figure. Nevertheless there were rumours of problems. People who lived in the Vale claimed that the Horse appeared to be shrinking and was less visible than in the past. Those of us who had worked on the Horse rather agreed.

So in late 2022 Oxford Archaeology, with financial assistance from a lover of the Horse, carried out a drone survey of the figure and the hilltop. This data was then compared with all the previous surveys and aerial photographs. The results were striking. In places – such as the neck and the head – the lines of the Horse had shrunk by, on average, 40% in the past 40 years (Fig. 16.5). It was decided, with the co-operation of the National Trust, English Heritage and

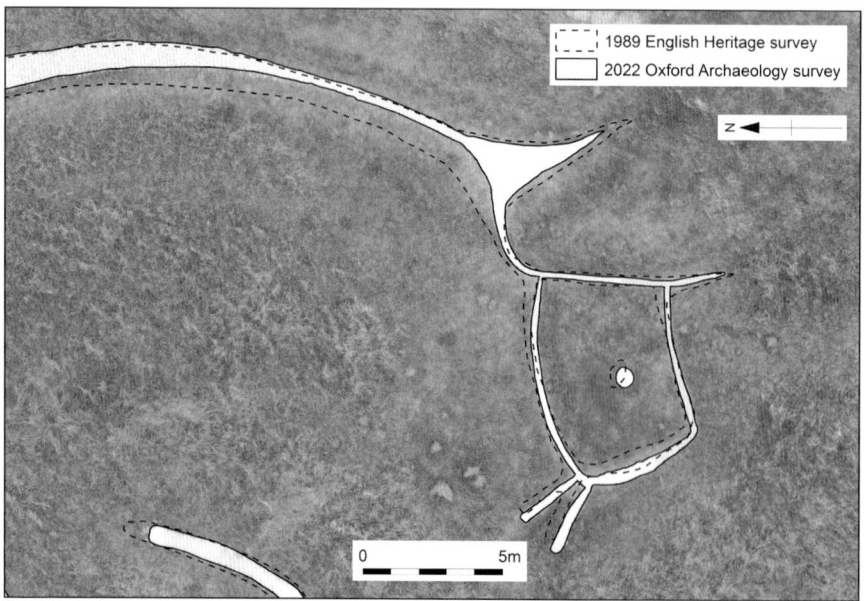

FIGURE 16.5. Comparison of the Horse in 1989 with the Horse in 2022, © Oxford Archaeology Ltd. 2023

Historic England, to excavate a number of small trenches across the figure in July 2023. The aim was to establish if the aerial mapping exercise was accurate. Were the lines of the Horse drastically shrinking in width? The results were absolutely clear. The survey was accurate. Beneath the turf the puddled chalk, forming wider lines, was apparent. The shrinkage was especially obvious around the head and neck. The legs had reduced in length, and the body, filled with new chalk, had risen to form a flatter plane, above the level of the turf.

At the time of writing this brief account, negotiations are taking place to restore and conserve the hill figure, at least to its condition of the 1980s. The neck and head should probably take priority, followed by the removal of chalk on the 'elevated body', and finally the lengthening of the legs and tail.

Aside from conservation, another issue is the dating of the Horse's origins. Our OSL sampling was some of the first. Yet OSL specialists regard the samples and dating as reliable, albeit covering a wide range. Since the 1990s OSL techniques and statistical analyses have improved. New samples could be taken with minimal damage to the hill figure by opening old trenches, potentially to achieve a more precise estimate of its date of origin.

Nevertheless the White Horse will continue to generate debate, fascination and loyalty – a living shape shifter, a Wonder of Britain, adopted and cared for by generations: the ultimate survivor, a true icon of the past and the future.

CHAPTER SEVENTEEN

Two chalk giants: Wilmington and Cerne revisited

Rodney Castleden

The giants compared

In Britain two hill figures represent the human form. There has been speculation that there may once have been more (e.g., Newman 1997a, 1; Castleden 2000, 102–6.). Claims have been made for giants, now lost, at Wandlebury, Plymouth, Penhill near West Witton and Shotover Hill near Oxford (Marples 1949; Lethbridge 1957; Taylor 1987; Newman 1997a; Castleden 2000, 98–101). However many there once were, only two survive, on hillsides at Wilmington in East Sussex and Cerne Abbas in Dorset.

The two figures are often described together, because they have points in common. Both are drawn in outline and are major landscape features. Both are frontal images with their feet turned to the right as if they are walking along the hillside, both bear emblems of some kind: the Long Man holds two tall staves, the Cerne Giant a club. Each has a medieval priory near its foot. Their size suggests that they are more than mere portraits of individuals. But they have many points of difference. The Long Man is very nearly symmetrical, the Cerne Giant is not. The Cerne Giant was drawn with a narrow trench packed with chalk rubble, while the Long Man appears from the start to have been made with bricks set in the turf. The Long Man measures 69 m from crown to heel, the Cerne Giant 55 m. The Cerne Giant offers significant detail – eyes, mouth, fingers, nipples, phallus, testicles, belt – the Long Man offers none. The Cerne Giant is chalk-cut and has been repeatedly maintained by the addition of chalk rubble. The Long Man has never been chalk-cut (Bell & Butler, Chapter 15).

The many differences render comparison pointless. They belong to different cultures, different times. These are important points to remember, as many writers have argued from the one to the other. For a long time it was assumed both were ancient. In the 1990s it became the fashion to argue that the Cerne Giant was modern, probably seventeenth century, and from this that the Wilmington Giant was also modern. But whatever can be proved about the one proves nothing about the other. They have to be treated separately.

The Long Man of Wilmington

Archaeological investigations on the Long Man 1969–2002

Forty years ago, I reviewed some of the theories about the origins of the Long Man and was persuaded by the landscape context that a Neolithic origin was possible (Castleden 1983). The main purpose behind the review was to provoke discussion, if possible encourage archaeological investigation; in this it failed. The evidence shows that Windover Hill was a focus for intermittent ritual and other activity over a long period, but it no longer seems likely that the figure was made at the beginning of that period, instead rather near the end. Intriguingly, though, in 2002 worked flints of Neolithic character and a fragment of a Neolithic bowl were found immediately below the Long Man (Bell & Butler 2014, 21; Chapter 15).

The Long Man was first investigated by archaeologists in 1969. A pioneering but inconclusive resistivity survey was carried out across the top of the figure by K. Gravett (Holden 1971, 50–5). At the same time, using a brief window of opportunity before the nineteenth-century outline bricks were replaced by concrete blocks, Eric Holden opened six small trenches to explore the nature of the outline (Fig. 15.3). Only two of the trenches suggested that the outline had ever been cut down to and slightly into the chalk rock. One was at the top of the head, the other two-thirds of the way down the eastern staff. If the rest of the outline was cut to the chalk but without eroding it, there was no archaeological evidence of it (Holden 1971, 38). These were the earliest indications that the Long Man might always have been a surface feature only – marked with bricks set in the turf. In the 1990s, sensing that it was worth attempting a new resistivity survey, I designed a meter that I hoped would produce a clearer result, and this was tested on the Cerne Giant (see below).

I found several small (1–4 cm) pieces of weathered orange brick very close to different parts of the Long Man outline. They appeared to be pieces of the 'Roman pavement' under the grass described in the eighteenth and early nineteenth centuries (e.g., Shaw 1791; Gough 1809). The fabric certainly looked like Roman brick. Two fragments of 'fired clay' found on the outline by Eric Holden in 1969 were at that time tentatively identified by Barry Cunliffe as Roman (Holden 1971, 41–2). Under a microscope, the fabric of my samples and Holden's looked very similar. In 1874, the local press commented, 'In the work which has lately been carried out, it was necessary to remove the turf in some places, and in so doing fragments of Roman brick were discovered' (*Eastbourne Gazette*, 29 April 1874).

In 2002 one of my samples was sent to Oxford for thermoluminescence dating. When the date (about AD 1600) came back from the Oxford lab, Ed Rhodes said he had shared the result with 'the others'. By chance my sample had arrived at the lab at the same time as samples from Martin Bell, who had dug a test pit below the Long Man. Ed Rhodes naturally assumed we were working together. I had not known about Martin's trench, which was opened and backfilled between my visits. Although John Manley, the Sussex Archaeological

Society CEO, had given both Martin and me permission to undertake work on the site, he had not mentioned to either of us that someone else was working nearby, but our results usefully converged on an important conclusion: suggesting that the Long Man had been made of orange brick in around 1550, and subsequently grassed over and disintegrated *in situ*. The 'Roman pavement' was, potentially, the weathering and disintegrating Tudor brick outline. My sample yielded a date 50 years later than Martin Bell's because it had been exposed at the surface for a time, while Martin's had remained buried; both would have been set in the turf in the middle of the sixteenth century.

The layer of orange brick fragments in Martin Bell's 2002 trench was associated with a near-contemporary layer of chalk rubble. Initially the rubble was assumed to represent material dug out to make a trench for the bricks, but that trench would have been only 5–8 cm deep, and it would not have yielded coarse chalk rubble, only topsoil and pea-sized chalk fragments. It is still unclear what the chalk rubble represents. Possibly it is the result of accidental spillage from the excavation of one of the chalk pits on the scarp crest, and therefore unrelated to the Long Man, but the chalk pits are not directly above Martin's excavation, and are in any case isolated from it by a hollow way above the Long Man, which would have acted as a sediment trap (Bell & Butler 2014, 23). The fragments could not have been spilt when Dearsley tried to 'chalk' the outline in 1889–90 as the layer of chalk rubble is covered by topsoil and is evidently roughly contemporaneous with the weathering of the Tudor bricks. The question remains open.

In the 1980s, the absence of evidence argument was used to promote a late origin for the Cerne Giant (Bettey 1981). Since then some historians and archaeologists have striven to prove that both giants are early modern (Hutton 1999b). But stylistic and constructional differences between the two figures suggest that they were made at different times for different purposes. An early modern date for one does not prove an early modern date for the other.

In the 1980s the earliest known drawings of the two giants coincidentally dated from the 1760s. This suggested to some that both figures were made not long before, though not later than 1742, when Francis Wise briefly mentioned the Cerne Giant while discussing the Uffington White Horse (Wise 1742). In the 1990s a drawing of the Long Man made in 1710 was discovered, putting the Long Man's latest-possible date back 50 years (Farrant 1993). Similarly, the hitherto earliest mention of the Cerne Giant was trumped when Vivian Vale discovered in a churchwardens' account a mention of the cost of repairs to the figure in 1694 (Vale 1999, 71–5). The dates for both giants have been gradually pushed back with the discovery of new documents. Now forensic dating techniques have pushed them back further.

2004 excavations on the Long Man

In 1996, I initially used the resistivity meter tested at Cerne Abbas to search for historic changes in the Long Man's shape (Castleden 1996), this time with probes set 50 cm apart (they were 25 cm apart at the Cerne Giant). The scale of the survey nevertheless necessitated switching to a commercial 4-probe meter, supplied by the County Archaeologist, Andrew Woodcock. The main variations to emerge were as follows: originally the left (west) foot turned outwards, the staves were 3 m taller, and the west staff had a diagonal line sloping down on the west side from the tip. This last feature is reminiscent of the scythe shown in the 1766 Burrell drawing, though the survey plot suggests a crook or flail more than a scythe (Fig. 17.1a).

A photo taken immediately after the 1873–74 bricking showed faint traces of earlier positions of the legs, which were more splayed, with feet turned outwards as in the Burrell drawing, and toes pointing downhill. A mistake was made in the 1870s bricking because the lower part of the figure had become very indistinct; indeed the increasing indistinctness was the reason for the bricking. Nineteenth-century drawings of the Long Man confirm that the feet were difficult to see. The engraving published in the Rev G.M. Cooper's 1851 article, for example, shows the Long Man without feet.

The results of the 1996 survey prompted Martin Bell and Chris Butler to excavate six small trenches on the Long Man in 2004, to ground-truth some of the changes suggested by the geophysics (Bell & Butler 2014, 24).

Butler's trench 4 proved that the left (western) staff did indeed once continue another 3 m up the hillside as my 1996 resistivity survey indicated but, frustratingly, excavation had to stop owing to heavy rain just as the uppermost 50 cm of the 'flail' emerged (though not mentioned in short summary by Bell & Butler 2014). Ann Downs saw the Long Man before the 1870s bricking and in the 1920s recalled its unbricked appearance: 'There was visible above the head of the Long Man a curved line running at right angles to the staff which he holds in his left hand. This was supposed to represent a scythe' (Shoosmith 1938). Holden rightly inferred that what Mrs Downs meant was 'above the level of the head', which in turn seems to imply that the western staff was originally taller, as indeed has now been proved (Holden 1971, 50).

The excavation on the site of the original position of the left foot (Butler's trench 5) was thought by Chris Butler to show no subsurface trace of an earlier outline, but a distinct lens of sand was visible in section buried within the soil, which might have been the remains of disintegrated Tudor brick (see Bell & Butler, Chapter 15). The main finding was that there is no surviving evidence underneath the modern concrete and brick outlines for anything existing on the site earlier than the mid-sixteenth-century orange brick outline. It is possible that the Tudor brick outline replaced an earlier brick outline, just as the 1960s concrete outline and 1890s brick outline replaced the 1870s outline, but it must be admitted that no evidence for it has been found. This leaves us with a provisional date for the creation of the Long Man in the 1550s.

On the head (Butler's trench 3) there was clear evidence of a semicircular trench cut to the solid chalk and filled with chalk rubble. This represents the attempt by the Rev Dearsley in 1889 to recreate a presumed original trench outline. Dearsley, who was Wilmington's vicar, assumed the Long Man had originally been marked, like the Cerne Giant, with an outline trench packed

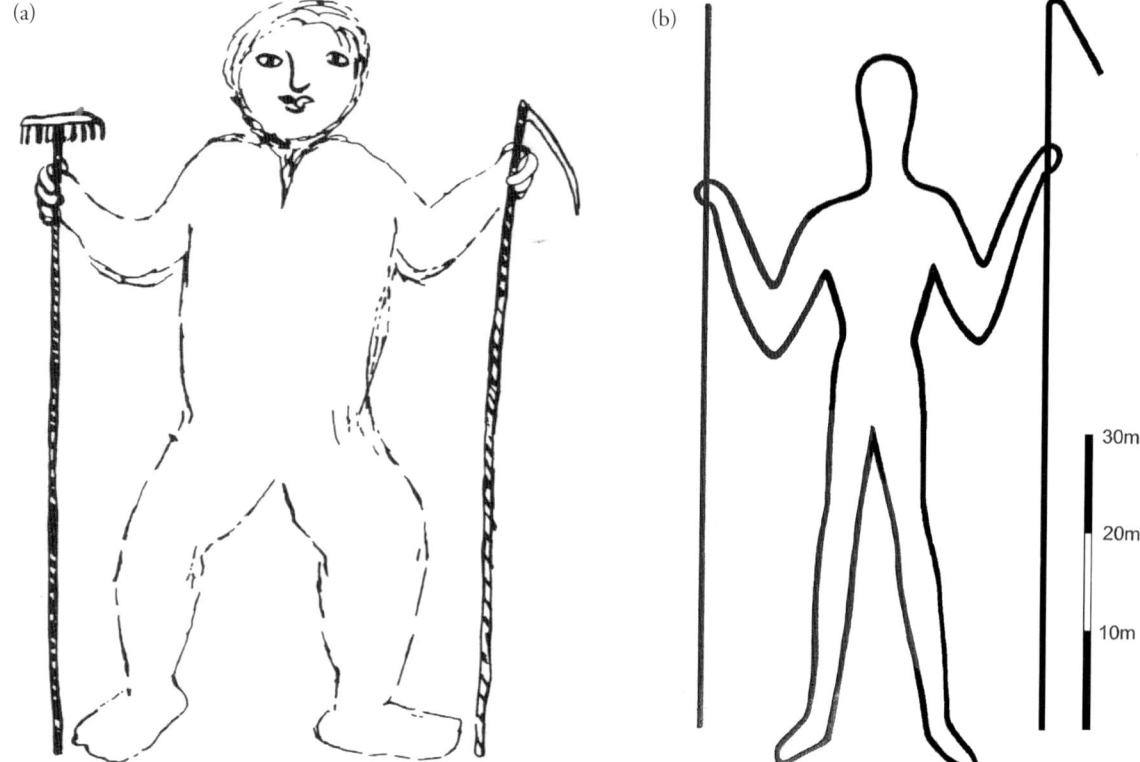

FIGURE 17.1. a) The 1776 Burrell sketch of the Long Man (Burrell Manuscripts, British Library Add. MS. 5697, f 342v.) and b) The Long Man before the 1873–74 bricking. The changes are inferred from resistivity survey and the 1874 Lavis photograph

with chalk rubble. Dearsley's restoration was halted by the Sussex Archaeological Society in 1891 when it was declared a failure; rain quickly eroded the rubble fill. In 1892–93, the outline was remade in red brick. According to documents held by the Sussex Archaeological Society, Dearsley had completed only the upper half of the figure and he described his trench as 'just 12 inches' (30 cm) wide, and 'in a V shape 12 inches deep', yet in 1969 Eric Holden discovered a rubble-filled trench *two-thirds* of the way down the right staff (Holden's trench 2), significantly lower than Dearsley's intervention. It was semicircular in section and 24 inches (60 cm) wide, not V-shaped and 12 inches wide. The implication is that it was not made by Dearsley (Dearsley 1890). This semicircular trench was not seen in the other sample pits, so interpretation is problematic (Fig. 17.1b).

Summary history of the Long Man

The Long Man has emerged as a succession of brick outlines set into the turf and made in the 1550s (orange brick), 1873–74 (yellow brick), 1891–92 (red brick painted white), 1896 (same) and 1969 (concrete blocks painted white). But what was there between the 1550s and 1873? In the eighteenth and early to mid-nineteenth centuries the Long Man was a faintly visible shallow depression. It may be that the Tudor bricking was preceded by a still earlier (medieval?) phase that has completely disintegrated, so the origin of the Long Man remains

17. Two chalk giants: Wilmington and Cerne revisited

FIGURE 17.2. a) The burning of Richard Woodman and nine other Protestant martyrs in Lewes in 1557, from Foxe's *Book of Martyrs* and b) a Protestant martyr in the flames from Foxe's *Book of Martyrs*. Note the position of the arms on the Long Man.

indefinite (see Bell & Butler, Chapter 15). But let us suppose for a moment that the Tudor giant was the original creation. If so, why was it made?

A new working hypothesis

If the Long Man is Tudor in origin it may have been made as a protest against the burning of Protestant martyrs in the 1550s. The most prominent Sussex martyr was Richard Woodman, an ironmaster from Warbleton. It may be significant that the Long Man looks out in the direction of Warbleton, 10 miles away in the Low Weald. Woodman's case was high-profile because unusually he was able to write an account of his interrogations and the account was circulated (Lower 1851). It began, 'Gentle reader, … First, you shall understand that since I was delivered out of the bishop of London's hands, which was in 1555, and the same day that Master Philpot was burned, which was the 18th of December, I lay in his [the bishop's] coal-house eight weeks lacking but one day; and before that I was a year and a half in the King's Bench [Prison in Southwark] for reprimanding a preacher in the pulpit, in the parish of Warbleton where I dwelt.' Many pages long, Woodman's account was later published in *Actes and Monuments* (Foxe 1563) (Fig. 17.2a).

Woodman was twice interrogated in London by Bishop Bonner, and sympathisers in Sussex may have assumed Woodman was racked, though in truth he was not. Bonner was a pioneer of the nice-and-nasty interrogation technique. His victims were imprisoned in his coal-house to consider their situation, brought out to dine with the bishop, then re-incarcerated in the coal-house (Fig. 17.2b). Possibly the Long Man's staves represent the rack, the 'flail' one of the rack's levers and the elongation of the figure the imagined stretching of Woodman's body (Fig. 17.3). Open protest against the persecution was dangerous; an anonymous, enigmatic gesture would have been safer. Those who made this iconic protest against religious persecution understood the need for secrecy,

FIGURE 17.3. The rack, as depicted in Foxe's *Book of Martyrs*

and may even have kept from their families what they had done. With the benefit of hindsight, we know that the death of Queen Mary in 1558 heralded over a century of Protestant supremacy, and that it would have been safe to divulge the secret on Elizabeth's succession, but the local people had no way of knowing this at the time. The assassination of Elizabeth and the unforeseen accession of another Catholic monarch might have meant a new wave of persecutions and retribution for the creators of the Long Man. The knowledge expired when the Long Man's creators died. This would explain why there is no surviving story to explain the Long Man. The absence of a local tradition could also explain why so many of us, myself included, assumed the Long Man was ancient.

The Cerne Giant

Archaeological investigations 1979–98

As at Wilmington, archaeology came late to Cerne Abbas (1979) and in the form of a resistivity survey, this time by Tony Clark and Alister Bartlett (Clark 1983; see David *et al*. Chapter 3). A large area of low readings below the left arm was immediately hailed as Hercules' lionskin, but the smoothness of its outline rather suggests a folded, draped, non-Herculean cloak.

My own resistivity surveys in 1989–90 and 1995 were carried out to explore the possibility that lost details of the image might be recovered (Castleden 1993; 1995a). It seemed likely that the original image included something hanging from the left arm, or perhaps from the left hand (Oxford Archaeological Unit 1998, fig. 7.2). The soil on that part of the hillside was uniformly shallow, so that closely set probes (0.25 m apart) and readings taken at close intervals (0.25 m) might produce a clear image of lost sections of outline trench dug into the chalk if any existed. The surveys recovered evidence of a double sinuous line

FIGURE 17.4. Changes to the phallus. As it stands today (left), and as it was before 1908 (right), based on the 1764 drawing and 1995 resistivity survey

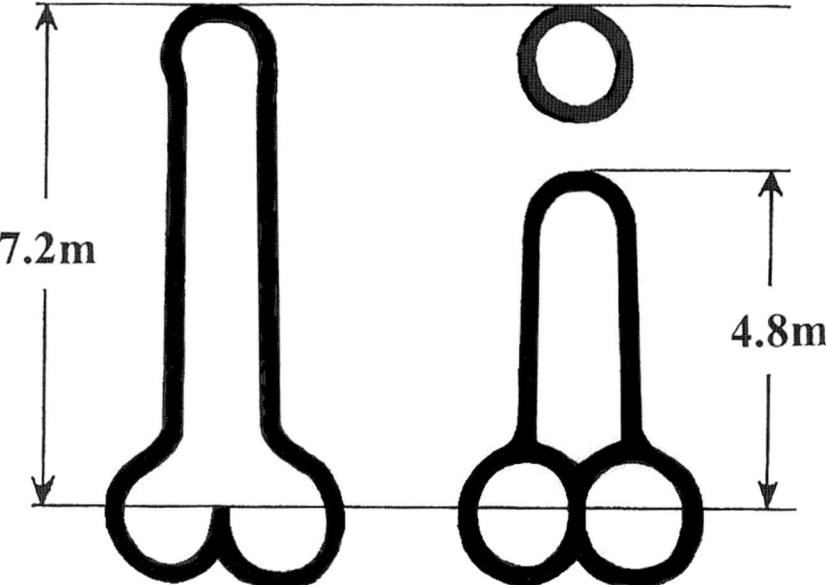

suggesting a cloak folded over the left arm. They also confirmed that the left hand had originally been a fist, as shown in the 1764 drawing in the *Gentleman's Magazine*. Below it was an enigmatic low built feature that may have represented a severed head (Castleden 1996, 162–170). The 1764 drawing and a 1902 picture postcard show that the top of the phallus was originally a separate ring representing the navel. The resistivity survey showed two lines of low resistivity 1.25 and 2.25 m down from the present tip of the phallus. The original phallus was therefore originally unexaggerated in proportion to the rest of the figure (Castleden 1996, 176–7 & 219; Fig. 17.4).

The iconography was consistent with an Iron Age tribal protector-god (Castleden 1999). A Gaulish bronze figurine shows a naked warrior with his cloak wrapped twice round his left arm, with the two ends hanging down, like the Giant's cloak. Another Gaulish warrior figurine, also naked, has his cloak wound several times round his arm to make a ball of padding. In both the wrapped arm is half-raised, as if parrying a blow. The cloak was in widespread use as a primitive shield. In Iron Age Europe this type of cloak-shield and a homemade weapon evidently formed the routine equipment of non-professional warriors (Fig. 17.5).

At the close of this research phase there was still no absolute date, so a further phase was proposed. At a site meeting in 1996 David Thackray, the National Trust's regional archaeologist, nominated me and Martin Papworth as joint co-ordinators of the next phase (Fig. 17.6). I contacted Mike Tite at the Archaeological Research Laboratory in Oxford, establishing that he was ready to undertake the OSL dating, while Martin was to prospect for funding.

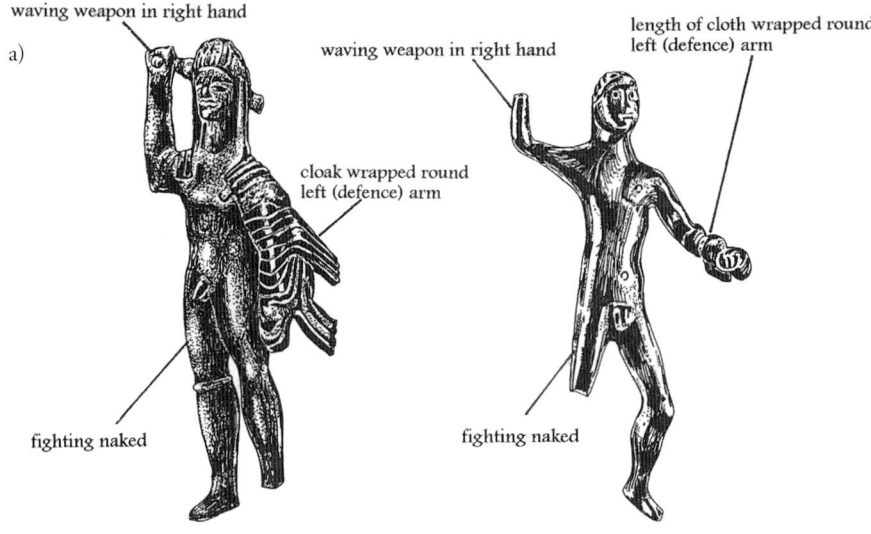

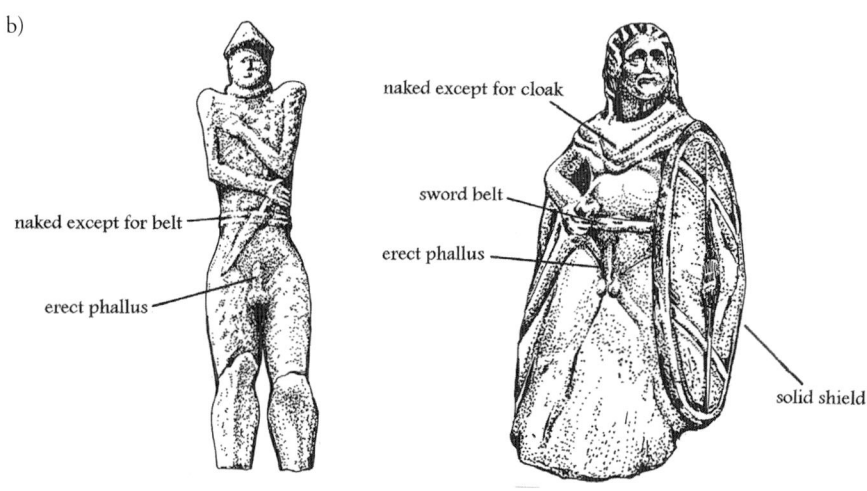

FIGURE 17.5. a) Two bronze figurines of Gaulish warriors, first century BC (from Bernard Pickard collection) and b) Two Iron Age warrior images, (left) Life-size sandstone statue from Hirschlanden, Germany, c. 500 BC, (right) Terracotta figurine of a Gaulish mercenary, third or second century BC

The National Trust commissioned and funded an OAU research design, which incorporated my proposal for OSL dating of the oldest surviving silts in the outline trench at the elbows and feet, which I identified as likely sediment traps (Castleden 1995b; Oxford Archaeological Unit 1998, 17, fig. 10).

Then the project stalled. One problem was cost; the funding needed to fulfil the OAU research design (£60,000 in 1998) was unachievable. A second, related, problem was uncertainty about the survival of ancient outline silts needed for dating. I remained optimistic that some silts might survive, but David Thackray, who had overseen a thorough scouring in 1979, was pessimistic; he thought it likely that the outline trench had been thoroughly cleared of ancient silts at that time.

FIGURE 17.6. The 1996 site meeting. From left to right, George Walkley (student), David Thackray (National Trust), Rodney Castleden, William Keighley (National Trust Warden, back to camera), Michael Clarke (current (2024) National Trust area warden), Lawrence Keen (Dorset County Archaeologist, hidden behind Martin), Martin Papworth (National Trust), Paul Gosling (English Heritage). Photo: Gerald Pitman

The 2021 OSL dating and its implications

In 2020, the National Trust decided to fund OSL dating, coincidentally focusing on the sampling sites proposed in 1995. It transpired that the 1979 scouring had cleared only the upper half of the accumulated silts, and the fills from several older recuttings had survived below. The OSL date announced by the National Trust in 2021, in the range AD 650–1310, eliminated the generally favoured seventeenth-century scenario (see Toms *et al.*, Chapter 3). It also discounted a Late Iron Age origin, though the iconographic evidence (Castleden 1999) still points to a culture rooted in the Late Iron Age.

The phallus presents a particular problem in the context of a Christian community. The proximity of the Giant to the site of Cerne Abbey understandably worries some commentators. If the abbey was founded in 978, it is unlikely that the unequivocally pagan image was created at that time, or later. Ælfric, chosen to be abbot of Cerne in 989, was a liberal teacher who believed that, to make choosing God a virtue, people should be free to choose between God and Devil. The Cerne Giant may have been available as a visible Tempter. Ælfric and his successors might have tolerated the Giant, but they would not have sanctioned its creation. The time window suggested by the OSL date might therefore be narrowed to AD 650–978.

Pagan images and temples still existed in sixth-century England. Gildas, writing in the 540s, mentions 'devilish monstrosities' of pagan cult images still to be seen in the countryside (Winterbottom 1978). Medieval writers such as Goscelin of Canterbury told a story about a confrontation between St Augustine and idol-worshippers at Cerne in around AD 600 (Gameson 1999). In 601 Augustine was sent advice by Pope Gregory, who had decided 'on the matter of the English that the temples of the idols in that nation ought not to be destroyed' (King 1952, *Ecclesiastical History of the English People*, Book I, chapter XXX). Gregory's advice was to adopt them because these were the places where followers of the old religion were accustomed to worship. The location chosen for Cerne Abbey, close to a pagan idol, is therefore significant.

Uncomfortable with the thought of the Giant so close to the abbey, Martin Papworth has resurrected an idea aired by Tim Darvill at the 1996 'Trial' conference (1999b), that the Giant was made by a pagan community (National Trust blog), then through neglect became overgrown and disappeared. Eventually the faint outline in the grass was noticed and recut (Hilts 2021, 15). A problem with this is that even a grassed-over hill figure remains visible and identifiable. The Long Man of Wilmington was marked out in brick in the 1550s and afterwards neglected. The 1710 Rowley drawing shows the Long Man with a dotted outline – it was still legible after 150 years of neglect. The figure remained grassed-over until the bricking of the 1870s. Even 300 years of neglect left an image at Wilmington that was intermittently visible under certain lighting conditions. A Green Giant at Cerne could not have escaped notice either; nor, significantly, could he have been recut unless his form was traceable.

Another speculative scenario to explain the survival of the Giant through the Christian era holds that the phallus was added late, superimposed across the belt in the seventeenth century (Brown 2021). This too is unconvincing. The way the genitals are drawn stylistically matches the rest of the image too perfectly for them to be a late addition. The tip of the phallus, lengthened by adding the original circular navel to make it into an exaggerated cartoonish feature, is a very late feature, probably as late as 1908 (Grinsell 1980a; Castleden 1996, 176 & 198). The fact that nipples, navel and testicles were all originally depicted in the same way, as 2-m-diameter rings, and the eyes were similar but smaller rings, strongly implies that they were all part of the same original design. The image must be treated as a single integrated concept. A second problem is that the aerial photogrammetry data does not really show the belt crossing the phallus; nor does the resistivity survey (nor the 2024 augering). If it had crossed, the belt would have been a straight line. Instead the lines on each side slope down to the phallus, suggesting they were drawn separately, one to each side of a phallus that had already been drawn. A belt on a naked figure may today look redundant but, as several contemporary figurines show, some Iron Age warriors fought naked except for a belt to hold a knife, presumably to despatch a wounded adversary or decapitate a dead one (Castleden 1996, 147–48; 1999, 46–47, 50–51; Fig. 17.5b).

From the evidence we have, the figure has suffered no major distortion since it was first made. The recent excavation and dating prove that a hill figure can remain in position on a steep chalk hill side for a thousand years without migrating downslope more than 10 or 20 cm (actually up to c. 35 cm from 2008/2019 rechalking). In conversation, Martin Bell easily persuaded me that an ancient origin was unlikely for the Long Man because of soil creep, the gradual movement of soil downslope (an idea repeated in Bell & Butler 2014, 25 & 28). It was a credible objection based on a sound knowledge of landscape processes, and there is plenty of evidence of soil creep at the Long Man, yet the observation does not apply to the Cerne Giant. The 2020 excavations at Cerne show that the four sampled points on the outline, the elbows and feet, have somehow remained in virtually the same places from the beginning. Equally surprisingly, the outline of the Cerne Giant has remained a constant width to within a few centimetres: a remarkable level of continuity.

A new working hypothesis

It is natural to assume that the likeliest date for the Cerne Giant's creation is in the middle of the OSL range, around 980, but, given the phallic, non-Christian nature of the image, the beginning of the range is likelier. A new working hypothesis consistent with the OSL dating and the iconography is needed. The image is a pre-Christian icon, possibly left over from the late pre-Saxon period. The Late Iron Age people of Dorset, the Durotriges, strongly resisted the Roman occupation and, although partial Romanisation followed, many Durotrigian beliefs will doubtless have persisted. Some Iron Age sacred sites continued as places of worship until near the end of the Roman occupation, even those that were close to Christianised sites (Salway 1981, 735). After the Roman withdrawal, the old refuge of Badbury Rings was refortified; other Dorset hillforts too were reoccupied between 400 and 600 in what looks like a revival of Iron Age tradition (Seaman 2023). It seems likely that Durotrigian beliefs were reasserted at this time.

By the mid-seventh century Saxons occupied East Dorset: West Dorset shortly afterwards. Britons of pre-Roman descent no doubt lived and worked resentfully under Saxon masters, maintaining customs and beliefs with a Late Iron Age ancestry. A pagan resurgence lasting a decade began with the domination of Wessex by Mercia when Cenwalh, king of the West Saxons, was driven out in 645. A hill figure created then might represent a revival of a Durotrigian tribal god. A protector-god in Late Iron Age tradition remains plausible in terms of iconography.

It has been suggested that successive writers have been unconsciously prejudiced, that they have *wanted* the giants to be ancient (see Darvill, Chapter 11; Greaney, Chapter 12), perhaps in the face of the evidence, (e.g., Bell & Butler 2014, 25–6) but the iconography of the Cerne Giant argues for antiquity. The erect phallus was a widespread 'good luck' symbol in Late Iron Age Europe. The phallic Giant is emphatically pre-Christian in spirit and, as argued in 1996, fits well the idea of a representation of a protecting guardian-god of the pre-Saxon

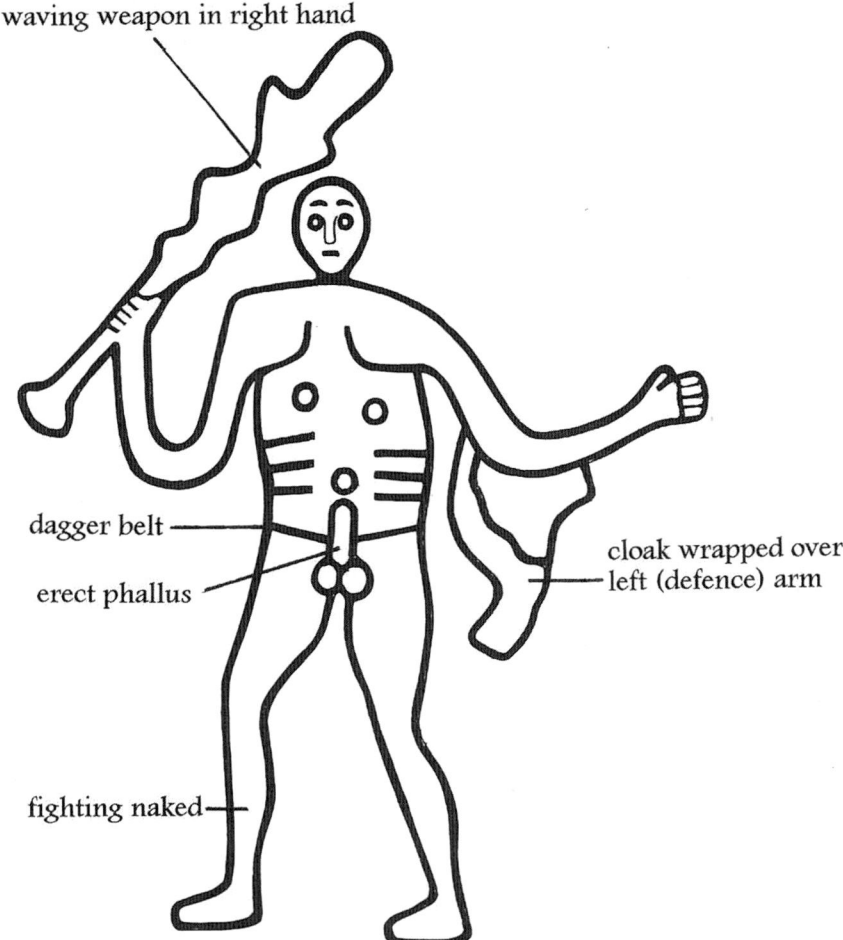

FIGURE 17.7. A reconstructed Cerne Giant. This is how he might have appeared in the seventh century, a portrayal of the tribal protector-god of the area round Cerne Abbas

people of this part of Dorset, ever-ready to fight for them and overcome their enemies (Castleden 1996, 49–5; Fig. 17.7).

In summary, I propose that the image of the Cerne Giant was added to the hillside in the seventh century AD by native Britons, descendants of the Durotriges, as part of their attempt to fend off the incursion of the Saxons. Christianity was another enemy, and the depiction of the ancient god may have formed part of the resistance to forced conversion. This was a time of double crisis for the native Britons in Dorset, and may have been the moment when the image of the ancient protecting deity went up as a call to arms, either as an icon of defiant resistance – or as part of a short-lived celebration of pagan resurgence.

Either way, the two giants now have something significant in common: new dates necessitating new explanations and new identities.

CHAPTER EIGHTEEN

Implications of the hill figure dates

Ronald Hutton

Of our more venerable carved chalk hill figures, the two apparently oldest of all, the Uffington White Horse and the Cerne Abbas Giant, have now been submitted to sampling and OSL analysis to the highest professional standard (although the process of sampling and analysis itself has changed significantly during the decades between the two projects). Anyone interested in these figures should be grateful to those involved in these projects. With the dates in the bag, the biggest single puzzle about both monuments should now have been settled, and we should be able to embark on the subsidiary work of explaining why the apparent objective evidence of a time frame for each fits the period concerned. In other words, everything should now be starting to make more sense. In actual fact, it does not, and the reverse situation now obtains: that in each case the dating yielded by OSL makes a particularly bad fit with the cultural context with which it is connected. As a result, the difficulties associated with understanding both figures are actually much greater than before, and the problems in the Cerne case in particular now both many and serious. With each image, the pattern is the same: that the OSL results made it older than the cultural context that, in our present state of knowledge, would seem to fit it best. The Uffington Horse bears some stylistic resemblance to figures on Late Iron Age metalwork, especially of the style known as La Tène, and especially to horses as represented on coins from the end of the pre-Roman period. The dating however suggests a Late Bronze Age or Early Iron Age provenance, and it is much harder to find any British images from that epoch that resemble the hill figure. In the case of the Cerne Giant, the problems that now confront us are, as said, much worse.

An early modern dating for the figure would have got us round all sorts of difficulties that an early medieval one now presents. The period between the fifteenth and seventeenth centuries was a notable one for the place of giants in English culture, manifested in their appearance in urban pageantry and (indeed) as carved hill figures. In the latter guise they are recorded at Oxford, Cambridge, Bristol and Plymouth, with two together at the last site. None have survived, and neither have any drawings or descriptions of them to compare with the Cerne specimen, though one of the pair at Plymouth was recorded as brandishing a club in the manner of the latter. The popularity of giants at this period was the consequence of a mixture of increasing knowledge of the Bible, with its references to them, and of the widespread impact of Geoffrey of Monmouth's *History of the Kings of Britain*, which portrayed them as the original inhabitants of Britain and with which the Cambridge and Plymouth figures were explicitly linked (Barker 1999, 94–106). By contrast, there was no such vogue for them in Anglo-Saxon and early Norman England, and of course no link with Geoffrey's book, which was not yet written.

The earlier dating also forces us up against a huge question that a later one would have solved: what place a major Benedictine abbey would have had for a huge, highly sexualised figure right next to it, and why that place would have justified the effort of maintaining it upon the abbey's land, decade after decade. Indeed, that dating seems to suggest that the giant was there from the first establishment of the abbey, and therefore was maintained all through its history, till its dissolution and beyond. There is no need to worry about why the monks would have fashioned a pagan image, as after all giants were perfectly Christian, being (as said) Biblical. Highly sexualised carvings are also fairly common in medieval churches, though more particularly from the eleventh and twelfth centuries onward, apparently to warn against the dangers of lust (Weir & Jerman 1986). The stumbling block at Cerne is the complete lack of any parallel example, from anywhere in the Christian world, of a medieval monastery fashioning and maintaining a figure of that kind, or anything like that kind, in the neighbouring landscape. What relevance it could have had to the religious life at Cerne, one of the most important, well regarded and well behaved of English monastic communities, is very hard to see, to account for such an apparently unique and anomalous initiative.

The OSL dating also poses, even more starkly than before, the crucial problem of why nobody mentions the giant before the late seventeenth century, despite the fact that Cerne was situated on a major regional highway and is exceptionally well served by local records from the thirteenth century onward, as a function of the importance of the abbey estates. Coyness and embarrassment cannot account for such a resounding silence, as those most affronted by the figure would have been those most expected to draw attention to it. The Tudor antiquary John Leland recorded various legends attached to the early history of the abbey, but none make reference to the giant. When Henry VIII's commissioners raked up complaints about the abbey in order to justify its reform, and then dissolution, they did not take advantage of the obvious ammunition that the maintenance of the hill figure would have represented. The numerous legal documents concerning the land at Cerne in the late Tudor and early Stuart periods mention much of the topography and its markers, but always seem to miss what we see as the most obvious and striking of all; as does the detailed survey of the neighbourhood made in 1617. All this evidence has been considered before most comprehensively in *The Cerne Giant: an antiquity on trial* (Barker 1999; Bettey 1999; Morgan Evans 1999; Vale 1999; i.e., Darvill *et al.* 1999, 1–124); but it is now all the more glaringly in need of explanation, as we now apparently have around 700 years of silence to explain instead of any lesser span.

As an identity for the giant, Hercules is still the most likely candidate, and the more so in that he could fit almost as readily into an early medieval context as into an early modern one. The identification is however not wholly secure, and there is no known representation of Hercules that matches the Cerne figure. Furthermore, some of the iconography of the Greek hero might have been appropriated for a different being. In an early medieval context, it may seem at first sight that the best candidate would be Helith, as a pagan god local to Dorset, mentioned in a medieval text and directly associated with Cerne. Unfortunately, on closer inspection it is precisely that context into which Helith does not fit. Nobody associated Helith with Cerne until Leland did so in the 1530s, followed by later Tudor and Stuart antiquarians such as William Camden. Leland took the god in turn from a medieval text, the *Memoriale* of Walter of Coventry, writing at the end of the thirteenth century (Walter of Coventry 1872, edited in the Rolls Series 1872, volume 1), who stated that Helith had been the pagan deity of Dorset at the time of its conversion to Christianity by St Augustine of Canterbury. Walter's passage on Augustine's mission to the county (Walter of Coventry 1872, 60) – a mission itself almost certainly fictional – was in turn an abridgement of one in the *De Gestis Pontificorum Anglorum* of William of Malmesbury (edited in the Rolls Series 1870, p. 184), a very good scholar writing in the early twelfth century: but there is nothing about Helith in William's work, and he represents an addition by Walter, from some unknown source and for some unknown reason. William drew in turn on a work from the 1090s, the *Vita Sancti Augustini* by Goscelin of Canterbury (sometimes called of Saint-Bertin), which describes how the saint had a vision of Jesus at the future site of Cerne, and this gave the place its name (edited in *Patrologiae Latina* volume 80, 1863, 44–5). He does not mention Helith either. Moreover, none of the writers who did, from Walter to the Elizabethan and Jacobean antiquarians, connected that god to a hill figure. Indeed, it would be hard to see why any medieval Christian would have wished to commemorate the downfall of a pagan deity by representing that deity in apparently triumphal pose. Helith therefore gets us no closer to finding a context for an Anglo-Saxon carved giant.

Could the apparent silence of the records concerning the figure be explained by a loss of it during some significant period of time, after which it was rediscovered and renewed? This would certainly make a fit with the new discovery that parts of the Giant were possibly redesigned at some point, the phallus cutting across an earlier belt, and so perhaps a later addition. It would also fit the molluscan evidence, of shells from a species of snail that reached England in the Middle Ages and favoured woodland habitats, referred to elsewhere in this volume. It is a sensible suggestion, but once again a proposal that seems to make sense of the new data raises new difficulties in turn. How would later generations know of the existence, and position, of a former carving on the hill? The best means would by written records, but that is precisely what we lack for the figure in all the quite abundant written evidence that we possess for medieval and early modern Cerne. Could oral tradition alone have preserved the necessary information for centuries? If it did, would woodland growth have spared enough of the original outline for it to be located and restored after such a lapse of time? The answers to both questions might be affirmative, but this still stretches credulity a bit.

There is one possibility that might still be considered when facing this crop of new problems, and one that I raise with great trepidation and reluctance: that we may not yet have got the OSL process itself quite right yet. This is not to fault its application at Cerne, or indeed in the other cases in which it has now been used. It is a question about the process itself. This is relatively novel, and it may be recalled how radiocarbon dating was applied for two decades before it became apparent that there was a fault in its methodology. This had produced dates that were generally too young, in the case of the Neolithic by about a thousand years. Once discovered, it was duly rectified by calibration, and has given us the dating mechanism that we have used ever since. As it stands, OSL has put both the Uffington Horse and the Cerne Giant into periods in which, on all present evidence, they make less sense than others, and by the opposite effect to early radiocarbon analysis: the results seem unexpectedly early. We really need some kind of test mechanism, as we obtained for radiocarbon decay, that will either knock these doubts out of court or provide a means of rectification.

Let it not be supposed that all these comments represent a howl of indignation or horror at the current outcome of the dating process. On the contrary, they are hugely exciting, because if they really are accurate then they make medieval and early modern England a much less familiar place than we thought that we knew by now. If the Cerne Giant really is an early medieval Christian monument, at least in origin, then we do not need to add this detail to what we thought we knew about the people of these periods and the ways in which they saw the world. We need to revise what we think about those things, quite drastically. That is an exhilarating and provocative prospect.

CHAPTER NINETEEN

Heroes, kings and giants at assembly places

Stuart Brookes

In light of the arguments put forward by Tom Morcom and Helen Gittos (2024), and other authors of this volume, it is a great shame that there isn't any mention of a muster or assembly at Cerne recorded in the sources. Nevertheless, there are intriguing clues regarding the character of the site that bear comparison with other assembly places of early medieval England. On the one hand, as explored by Barbara Yorke in her contribution, there is a real possibility of a link between the symbolism of the Cerne Giant and martial behaviour. Like Cerne, there are several places in the early English countryside where scholars have highlighted an intriguing connection between assembly places and mythical or historical heroism, as evidenced by place-names, monuments, and folklore. On the other, is the landscape of the Giant itself. Analysis of other assembly places has demonstrated topographical relationships between sites of military muster, the local character of the landscape, and the administrative geography documented in Domesday Book. It is argued that the Giant both correlates with the corpus of known sites, and that the new dating for the Giant fits within the chronological model of assembly place development as understood from elsewhere.

Establishing a corpus of early medieval mustering sites

Several military mustering sites are noted in the sources (Fig. 19.1). Bede mentions a mustering in 651 at *Wilfaræsdun*, 'ten miles north-west of Catterick', in preparation for a battle that never took place (*HE*, iii. 14). Evidently near Gilling (North Yorkshire), a convincing argument has been put forward identifying 'Wilfare's Hill' with Diddersley Hill (NZ 173078) (Breeze 2005), a prominent dome-shaped eminence beside the Roman road (Margary 1973, road 82) close to the presumed border of Bernicia and Deira. The *Anglo-Saxon Chronicle* mentions further musters at Aylesford, Kent in 1016 (ASC 1016), on the Pilgrim's Way at the lowest crossing point of the River Medway, and between Godwin, Swein and Harold in AD 1051 at a place in Gloucestershire referred to as either Beverston (ASC E 1048) or Langtree (ASC D 1052). In fact, the two places are only 3 km apart, and probably both refer to a location between them at the crossroads of a ridgeway and Roman road (Margary 1973, road 544). At this

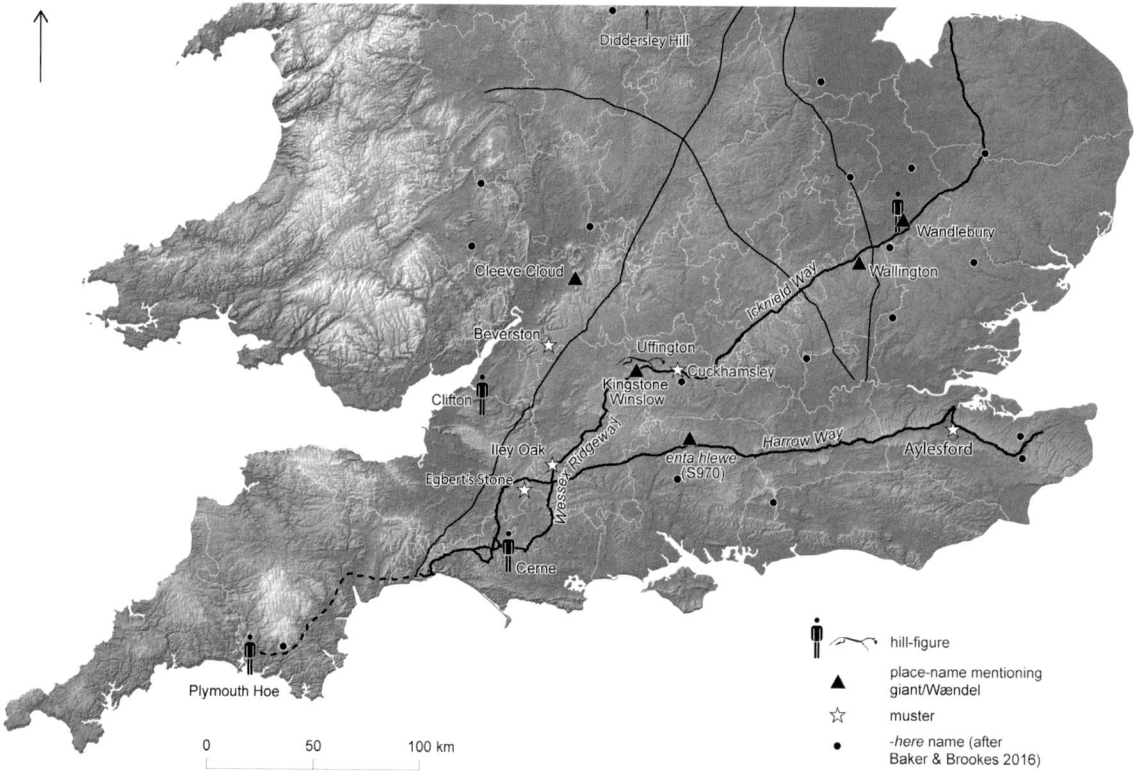

FIGURE 19.1. Map showing the boundaries of the Domesday shires, and the course of Icknield Way/ Wessex Ridgeway and Harrow Way, and their relationship to sites mentioned in the text. The course of the two ridgeways in Wiltshire and Dorset follows Hindle 1993, fig. 2. Data: SRTM https://dwtkns.com/srtm30m/ [accessed Dec 2023] (Brookes 2020)

point are located several prehistoric barrows, including at least one that was reused in early medieval times for secondary inhumation (Pantos 2002, 94–5). The meeting at *Swinbeorg*, mentioned in Alfred's will, may also be added to the list if, as suggested by Marren (2006, 13), it was the site of a muster prior to the battle of Ashdown in 871 (ASC A, 871). Both *Swinbeorg* (either *swīn* 'pig', or *swin* 'a creek, a channel' + 'hill, barrow': Toller 1898) and the battlefield are unlocated, but suggestions for the latter, including Lowbury Hill and Kingstanding Hill, both near Moulsford (Oxfordshire) (Marren 2006, 118–21; Peddie 1989, 83–5), would suggest a location somewhere on the Wessex Ridgeway near the Berkshire/Oxfordshire border. Two further musters are mentioned in the *Anglo-Saxon Chronicle* in the lead up to the Battle of Edington, 'to the east of Selwood' at *Ecgbrihtesstan* 'Ecgbert's Stone', and *Iglea* 'island of high ground' (ASC A, 878). The first of these places is unidentified, although John Baker and I have suggested it may be Moot Hill Piece in Gillingham hundred in Dorset (ST 761305) – a site we refer to as a 'hanging promontory' on account of its topographical character (Baker & Brookes 2013, 153–4; and see below). More convincingly, *Iglea* is probably Iley Oak in Sutton Veny, Wiltshire (Gover *et al.* 1970, 154–5). This location served as the meeting-place of the two hundreds of Warminster and Heytesbury in AD 1439, perhaps preserving a memory of assemblies at an early date. Although not precisely locatable, tradition places it on the boundary between Sutton Veny and Longbridge Deverill, near Lord

Heytesbury's Lodge at Southleigh Wood (*ibid.*, 155). Notably, this is the location of a Neolithic henge monument forming a *c.* 30 m enclosure within a 1 m high bank and 0.5 m deep ditch. Finally, the *Anglo-Saxon Chronicle* (ASC D, E) entry for AD 1006, describes how a Danish military party took up a position on *Cwicchelmes hlæwe* or 'Cwichelm's Barrow', now Cuckhamsley or Scutchamer Knob in Berkshire, seemingly in an attempt to provoke a West Saxon military response – an event that might denote the site also as one of English military mustering (Halsall 2003, 157; Williams 2006, 208–11; Semple 2013, 1–2, 159–92; Baker 2019, 37–9; Yorke, Chapter 4). There is little doubt that *Cwicchelmes hlæwe* refers to an impressive crescent-shaped mound on the Downs just south of the Icknield Way/Wessex Ridgeway, not far from its intersection with an 'army-path' (Old English (OE) *here-pæð*).

This relatively small corpus of mustering sites furnishes us with only few clues to the character of early medieval military assembly. To this list one could add the many hundred meeting places that can be derived from Domesday Book, on the presumption that these were also sites of local muster, as might be suggested by the 'Hundred Ordinance', but there are grounds for thinking these were in many cases new sites of the tenth century and later, and therefore perhaps not entirely equivalent to Cerne in date or character (Whitelock 1979, 393; Baker & Brookes 2015a; Brookes 2023). Also of potential relevance are sites denoted by place-name elements descriptive of military activity, in particular the OE elements *here* 'an army', *fyrd* 'army (service), military expedition' and their cognates (Toller 1898; Baker & Brookes 2015a). These have a more restricted distribution and, in several ways, bear some resemblance to the setting of Cerne.

The character of early medieval mustering sites

Of course, reading significance from a specific configuration of natural and geographical features is hugely difficult. Nevertheless, this short list of mustering sites introduces a few repeated patterns, both in the relationships between monuments in the landscape, and potentially also the cultural meanings ascribed to these places. Certainly, we can presume that there were specific locations where mustering was to take place, and that these locations were well enough known to troops uniting from over a large territorial area.

Most obviously, these places needed to fulfil a function as sites of mobilisation and civil defence. They needed therefore to be accessible to those assembling, well positioned for forward deployment, and well-enough known as distinctive places to groups travelling from a distance (Baker & Brookes 2015a). It is not surprising then to find a close correlation between mustering sites and long-distance routeways, with ridgeways especially prominent. Most contenders for the location of Ecgbert's Stone lie on the Harrow Way, while Iley Oak is on the Wessex Ridgeway (Peddie 1989, 128–34). The crossing point at Aylesford joins several drove-ways that became the Pilgrim's Way and are named already in the mid-tenth century (S1211, S1212; Witney 1976, 230). Both *Wilfaræsdun* and Langtree lie beside Roman roads. When considered in this way, the importance of the Icknield Way and its continuation as the Wessex Ridgeway is particularly striking, providing access to both Cuckhamsley and *?Swinbeorg*, as well as several other sites of note: the Blowing Stone near Kingston Lisle – traditionally thought to have been used by Alfred to summon his men before the Battle of Ashdown; Wayland's Smithy; Uffington White Horse; and north of the Thames, Harlawe, *here-hlāw* 'army mound' in Essex and Thetford, 'nation, people ford' in Norfolk (Marren 2006, 119; Baker & Brookes 2015a). While Cerne is not well connected by Roman roads, it does lie proximate to both the Wessex Ridgeway between Bookham and its intersection with the Harrow Way on Telegraph Hill, and the *here-pæð* from Dorchester to Sherborne mentioned in the late tenth-century bounds of S744. The latter has been suggested to be a branch of the Harrow Way that extends it to the south coast (Grundy 1937, 110; Hindle 1993, 19).

There is evidence that at least some of these musters were at places that were located to facilitate the rendezvous of forces coming from different territories. The *Anglo-Saxon Chronicle*'s account of the assembly at Ecgbert's Stone specifically refers to units organised by shire, and – although the site is not precisely located – it was evidently close to Penselwood, which forms the boundary between Dorset, Somerset and Wiltshire. The locations of Cuckhamsley, *Wilfaræsdun* and potentially *Swinbeorg* are also apposite, as is the tendency of *here-* places to often be located on boundaries where two, or even, three shires come together (Baker & Brookes 2015a). This impression that mustering sites related to the territorial arrangements of shires, and thereby the cadre of fighting units, is supported by other evidence. Although it is located on the territorial boundary of Berkshire, Cuckhamsley is identified in a charter of 990 × 992 as the meeting-place of the Berkshire *scirgemote* or shire-court (S 1454; Gomme 1880, 63–4; Stenton 1971, 381).

A similar liminal location cannot be recognised at Cerne, at least in relation to the shire boundaries as fossilised in Domesday Book. Located centrally within the shire, it bears a closer resemblance to Aylesford. The latter lies on the River Medway – the historic boundary of East and West Kent – and is just 4 km west of Penenden Heath, site both of the shire-moot named in Domesday Book and the great assembly of 1076 × 77 (Douglas & Greenaway 1981, 481–3). While there is no similar historical account identifying Dorset's shire-moot, Cerne's location close to the geographical centre of the shire and near both major north–south and east–west routes through the region would make it better placed than the shire centre of Dorchester for such a function. It may also be relevant that, like Kent, Dorset-shire originated piecemeal. Diagnostically Anglo-Saxon finds of the late sixth or seventh century are rare in Dorset, extending only as far west as Dorchester (Eagles 1994, 17–18; Hinton 1998b, 38–45) and this line may have been the limits of West Saxon control until its expansion in the later seventh century (Finberg 1953). For a time at least, this would place Cerne in something of a borderland, which, like Aylesford, retained some significance after the creation of the shire.

Cerne is also similar to this corpus in not giving its name to the local Domesday hundred: *Stane* OE 'Stone' (Thorn 1991). It is often the case that

Domesday hundreds preserve the names of their assembly places, so it is noteworthy that these names describe different locations, even when they are relatively close to attested mustering sites. In recognition of this fact, we have suggested that shire-musters, such as those described above, represented a higher-level of military mobilisation that may also predate the establishment of the hundredal system (Baker & Brookes 2015b; Brookes 2023). In this regard it may be significant that by the twelfth or early thirteenth century the three Domesday hundreds of Totcombe, Modbury and *Stane* had been grouped (or possibly regrouped) into a larger district under the jurisdiction of the Abbot and monks of Cerne (Thorn 1991, 39; Mills 2010, 121). While the meeting places of Totcombe (Tatcombe, south-east of Cerne) and Modbury (*(ge)mōt, beorg*, OE 'moot-mound' in Cattistock parish (near ST 607017)) are well known, different locations have been suggested for the stone from which *Stane* takes its name. As Morcom and Gittos (2024) suggest, the most likely candidate is Bellingstone, 2 km to the west of the Giant, which lay both on the boundary with Modbury, and was central to the larger unit of Tollerford, Modbury and *Stane* hundreds. In *Somerset & Dorset Notes & Queries* 1 (1888–9), 248, the Cross-in-hand stone (ST 632038) is proposed as an alternative.

Besides highlighting similar aspects of accessibility and liminality our discussion of these sites also drew attention to the physical character of these places, which we described as 'hanging promontories' on account of their topographical form, comprising slightly rounded eminences thrusting forward from higher ground to create natural platforms lying below the highest point of land (Baker & Brookes 2015b). Based simply on the shape of the land, Giant Hill and the Trendle earthwork are an archetypal 'hanging promontory'. As is common to other early medieval assembly places they may also be associated with distinctive features such as stones, trees and pre-existing monuments (Baker 2019). Should the Trendle enclosure indeed date to the Iron Age or Roman periods, this would also make Cerne comparable with other sites.

One further pragmatic aspect of mustering sites is that they need to be well known. Doubtlessly, when armies assembled the amount of time this took needed to be as short as feasibly possible: men needed to be provisioned, and if they had travelled from afar would have only been able to come with few rations as well as their weapons. According to the *Chronicle* account of 878, Alfred only needed to stay one day each at Ecgbert's Stone and Iley Oak, so this supposes that there was an efficient system of communication about when and where to assemble (Baker & Brooks 2016, 222).

There are hints that all the known musters were places that were singled out for specific attention in early medieval mental geography. In several cases scholars have highlighted an intriguing connection between the place-names by which they were known and historical persons. Personal names form the first elements of Aylesford, *Ecgbrihtesstan*, *Wilfaræsdun*, Cuckhamsley and perhaps Beverston, and some, if not all, of these names might have resonated with historical or mythical significance (Ekwall 1947, 139; Baker & Brooks 2016, 12).

Cuckhamsley may contain the name of the seventh-century leader of the West Saxon royal house: Cwichelm (e.g., Anderson 1939, 214; Halsall 2003, 156; Williams 2006, 207–11; Baker 2019, 37–9). Likewise, Ecgbert's Stone perhaps harkens specifically to a West Saxon king (r. 802–39), and grandfather of King Alfred, who ended Mercian supremacy in pitched battle (ASC E 825; Williams 2016, 48). By associating places with an ancestral figure, naming helped to situate myth in the landscape, perpetuating memory and history of the place, as well as attaching a superior quality to the location. Perhaps it was also a means to deliver more specific messages. Various authors have drawn attention to the role that certain personal names in places may have played in amplifying political culture: whether by asserting West Saxon authority over peripheral areas (e.g., Yorke 2001, 17–18; Reynolds & Langlands 2007, 33–4), or communicating allegiances and political unity (Mawer *et al.* 1927, xxii; Jones 1998; Baker 2019). Potentially, they were also mnemonic devices they drew on the power of past military victories as a way of tightening purpose and resolve.

Such a reading raises questions about both Aylesford and *Wilfaræsdun*. 'Wilfare' is an otherwise unknown person, though the element sits comfortably within the range of early recorded dithematic personal names (Insley 2003, 376–7). Aylesford could be derived from either the OE *egl* 'mote, beard, awn, ear (of barley); claw, talon', or the masculine personal name *Ægel*, which is not recorded independently, but appears also in place-names such Ailsworth (Northamptonshire), Aylesbury (Berkshire) and Aylestone (Leicestershire), and may derive from the Egil of Germanic legend (Wallenburg 1931, 286–8). In this case, the muster site would seem to be named from a head manor, but it may be significant that the *Anglo-Saxon Chronicle* records Aylesford also as the site of a battle in AD 455 where Hengest defeated Vortigern (ASC AE). The naming of this place as site of two English victories several centuries apart hints at some form of posthumous commemorative or literary memorialisation (or invention) rather than historical record – whether by using an established and significant place or by manipulating the symbolism of an older battle to contextualise contemporary actions (cf. Williams 2016, 48). In a similar way, Wilfare might literally derive from *Wil-* OE *(ge)will* n. 'will, wish, desire' + *faran* v. 'to set forth; to travel, journey; to go, proceed'; a personal name that could be taken to describe the outcome of the failed battle. In personifying the event, we may potentially see some attempt to explain the hill location and preserve its meaning.

We can't now know whether Ecgbert's Stone was thus named before the Battle of Edington or afterwards, but it seems likely that mustering in the shadow of ancestral heroes was intended to offer a psychological boost to an army preparing for war (Williams 2013, 28; 2016, 48). Likewise, Barbara Yorke's contribution has explored in some depth the potential meanings that might have been associated with the Giant, among which the power and courage of Hercules is easy to relate to the heroes associated with other mustering sites. Like Ecgbert or Cwichelm a warband assembling in sight of Hercules could draw on the power these figures engendered for the difficult tasks that awaited it, and in this regard the possible conflation between Hercules and other prominent West Saxon leaders would not be contradictory (cf. Yorke, Chapter 4).

A similar case for symbolic naming has been made of landscapes associated with major confrontations, hinting at some of the possible ways that places and events were named and remembered in connection to each other. Thomas Williams (2013; 2016, 45–9) has discussed the many monuments, names and folk-etymologies associated with the battles of Ashdown (871) and *Eðandun* (878), and a comparable landscape relating to an otherwise unrecorded battle has been postulated in Ewelme, South Oxfordshire (Mileson & Brookes 2021, 133–6). In all cases there is a sense of a wider symbolic landscape created by, and remembered through, ancient monuments, terrain, trees, stones and other features.

Giants, assembly and wayfinding

This observation alone raises intriguing questions about the reception of other giant-heroes in the English countryside. Sarah Semple (2013, 179–80) in her discussion of the evidence notes that such creatures are relatively rare, although some further examples could be added to her list. Perhaps the most striking comparison to be made with Cerne are a series of monuments not far from the Icknield Way in Cambridgeshire. Cut into the southern part of Wandlebury Camp, an Iron Age hillfort south-east of Cambridge in the Gog Magog Hills, are a series of hill figures seemingly depicting at least one giantess, ?horses and a possible chariot (Lethbridge 1957, 62; Fig. 19.2). While the outlines of the figures have been disputed, Richard Coates' (1978) discussion of the place-names in its vicinity – Gog Magog, Mag's Hill, and Wandlebury (*Wendlesbiri* tenth century, OE 'Wændel's stronghold') – make clear that there were folk myths mentioning a giant at the location going back at least to the tenth century. Significantly, the Wandlebury giants are only 4 km from Mutlow Hill: a hanging promontory atop the linear earthwork of Fleam Dyke, that we have previously discussed as

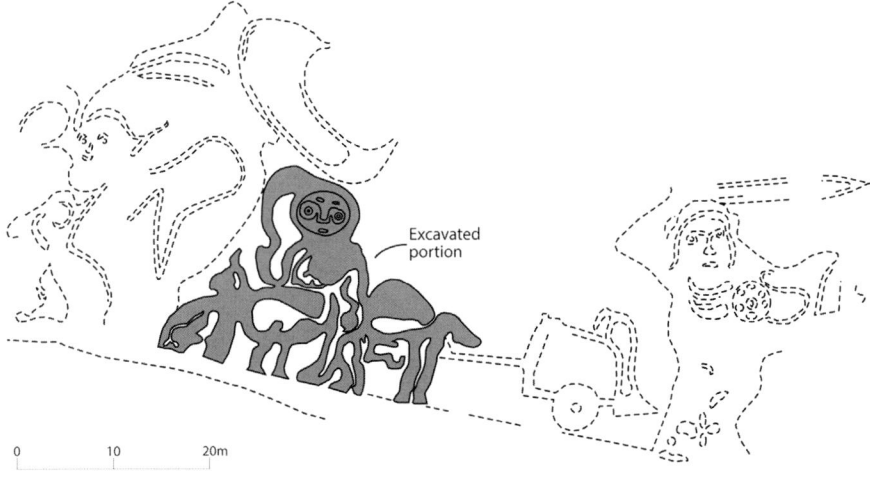

FIGURE 19.2. The figures of Wandlebury Camp identified by Lethbridge (1957, fig. 11a)

a probable supra-regional assembly site (Baker & Brookes 2015b, 14). Indeed, Mutlow Hill and Wandlebury are connected also in local legends. According to stories related by Lethbridge (1956, 193), giants were buried in Wandlebury, as was a golden chariot beside the road connecting the two places.

The example of Wandlebury Camp leads us into two further directions. Hill figure giants are also mentioned in other folk tales, intriguingly often at major geographical and territorial boundaries (Crawford 1929).[1] According to the fifteenth-century topographer William Worcestre a giant named Ghyston was once depicted on the sheer rocks of Clifton Gorge in Bristol (Coates 2011, 177). Also by the fifteenth century two chalk-cut figures on Plymouth Hoe apparently guarded the entrance to the port, in a reference perhaps to a battle recorded by Geoffrey of Monmouth between the Trojan Corineus and the giant Gogmagog (Clark 2016). City expenditures from AD 1494–5 for the recutting of the image attest to its early existence (*ibid.*). By contrast, less secure are the Long Man of Wilmington, East Sussex and the lost giant known as Bullingdon on Shotover Hill, Oxford, both of which appear to date no earlier than the seventeenth century (Bell & Hutton 2004; Bell & Butler 2014; Chapter 15).

The other fascinating observation to be drawn from the example of Wandlebury regards its possible allusion to giants known from Norse mythology (Mawer & Stenton 1926, 114–5). There, Vandill is the name given to a sea-king and a giant, and is also found as the second element of a giant named *Ǫrvandill* (*ibid.*). While Coates (1978, 77–8) is doubtful that *Wendel-* refers to a deity, noting its occurrence as a common personal name in Old German, Gelling (1993, 131, 136) is less circumspect, citing two further instances where the element may be suspected of being a mythological character. The first is Kingstone Winslow, on the Berkshire/Wiltshire border (*Kyngeston' Wendescleve* 1252–61, OE 'Wændel's cliff') referring to the escarpment rising to Wayland's Smithy beside the Ridgeway/Icknield Way, just west of Uffington. The second is Cleeve Cloud in Gloucestershire, a striking limestone eminence marking the western edge of the Cotswolds referred to as *Wendlesclif* in an authentic charter of 777 × 779 (S141; Gelling 1993, 131, 136). This same cliff gave its name to the bishop of Worcester's manor and the fourteenth-century hundred. To these two we may wish to add two further pre-Conquest examples: Wallington, Hertfordshire (*Wallington(e)* 1086; *Wandelington(a)* c. 1180 OE 'Wændel's farm') astride the Icknield Way, which also has Metley Hill (*Metelawe* fourteenth century OE *(ge)mōt-hlāw* 'meeting-place hill'; Coates 1978, 78) distinct from the hundredal assembly place at Odsey (Gover *et al.* 1938, 168), and Wensdon Hill, Bedfordshire (*Wendlesdun*, 969; S772; OE 'Wændel's hill'), also on Icknield Way.

When listed in this way, the association of many of the places described above with the long-distance ridgeways that cut across England is striking, as is the clustering of monuments and names at significant places along these routes. Beside the White Horse, within just 2 km of Uffington we find Wayland's Smithy, Wændel's cliff, the Blowingstone, Alfred's Castle and the probable assembly mound of Dragon Hill. Near Wandlebury Camp ('Wændel's stronghold') are the giant hill figures, Gog Magog Hills, Mag's Hill, Mutlow Hill, and Fleam Dyke. Lined between them along Icknield Way/Wessex Ridgeway are

Wallington, Wensdon Hill, Harlawe and *Cwicchelmes hlæwe* (Fig. 19.1). Situated on the western extension of this same route are Iley Oak and Cerne.

The identification of so many giants along a single long-distance routeway opens the possibility that they were connected by some kind of spatial orientation, potentially to aid wayfinding, or just part of a wider mental map. Potentially, such mnemonics helped to structure and store knowledge, to imagine fantastical worlds, to give directions, and prompt spatial behaviours. Importantly, such sites were also understood relationally, and as such the 'real' giant at Cerne may well have been referenced by other metaphorical giants preserved in place-names along the route. In this regard is an interesting interpretation made by Peter Kitson (*A Guide to Anglo-Saxon Charter Boundaries*, in prep.) of the etymology of 'Icknield' itself. He suggests a derivation from *Iceni* and *hilda* (OE 'slopes'), i.e., the road 'slopes/points' towards the Iceni, giving the sense of 'way leading towards the Iceni'. However, this would have very soon have become garbled in folk etymology of the Late Anglo-Saxon period to become Ickenhilda – a female name, so meaning a mythological giantess. Like the Harrow Way (*hearg* OE 'temple, site of worship') that passed by Stonehenge, these names and places hint at a conceptual navigation that stretched right across southern England (Reynolds & Langlands 2011, 417).

If these landscapes describe a world-view in which giants had meaning, we can only speculate what they represented to the people that encountered them. Semple (2013, 179–80) draws attention to the association in early poems of *enta* 'giants' with Roman relics. The correlation between giants in the landscape and long-distance routeways may be related to that. Another meaning may well be martial prowess, as the Giant's seeming depiction on Harold's flag on the Bayeux Tapestry might also suggest (Barraclough 1969, 120). Certainly the comparisons with mustering and battle sites documented in the chronicles of the ninth and tenth centuries suggest that the Giant attached a quality to the place by making connections with specific events that are now lost to us. Whatever these were, stories of mustering and battle were a celebration of community – a coming together in the face of adversity. If the Giant was indeed a renowned place of assembly, it was also a site of communal memory. The image should be seen, therefore, not only within the context of the local landscape, but also as part of a wider conceptualisation of early medieval political, military, and national identity.

Abbreviations

ASC: *Anglo-Saxon Chronicle*; Translation: Whitelock 1979

S: Anglo-Saxon charters are annotated with reference to the numbering outlined in Sawyer 1968

Note

1   Though not a hill figure the association off the south coast of St Michael's Mount, Cornwall with the giant defeated by Arthur in *Morte Arthure* might also be relevant.

CHAPTER TWENTY

Wiltshire's chalk equine hill figures: what's the problem?

Garry Gibbons

The hill-side figures of England … have never received any adequate attention … there is nowhere any accurate copy of these works. Their surroundings have never been examined, and scarcely anything has been written about them, except a variety of historical guesses on insufficient evidence (Petrie 1926, v).

Although referring to hill figures was thought to be rooted in antiquity, Petrie's appeal accurately describes Wiltshire's equine chalk horses today; they are under-researched, poorly understood and, as a result, they are under-valued across the heritage profession. This chapter presents the results of discrete, small-scale investigations conducted over the last 20 years that serve to underscore the degree to which our knowledge and understanding of Wiltshire's most visible landscape monuments are both partial and misleading. Two of the projects featured in this paper are the result of fieldwork and social history research which serve to demonstrate the value of joint investigations in testing taken-for-granted origin stories that have persistently attached themselves to Wiltshire's chalk horses for the past 150 years.

The results of field investigations have raised conservation issues relating to Wiltshire's chalk hillside monuments – which is particularly pertinent as none of Wiltshire's chalk horses are on the Heritage at Risk Register (see Chapter 22). In many cases Wiltshire's equine hill figures have been subjected to a dramatic loss of shape and character over the last century due, in part, to environmental and/or visitor degradation, and incremental changes accrued during routine maintenance.

Secondly this chapter briefly explores new pathways of linking individuals, groups and/or institutions with specific chalk horses. Recognising less powerful voices who are nevertheless strongly associated with the creation and maintenance of local chalk horses offers the possibility of a more nuanced and grounded narrative. Adopting a bottom-up approach raises the promise of enhancing – or challenging – the primacy of familiar origin stories that continues to shape a popular understanding of Wiltshire's equine hill figures.

Recent extensive, and welcome, remedial works at the Westbury horse have rightly stabilised the monument for the enjoyment of future generations but in all likelihood, the total cost of those works will overshadow any funding, from any source, directed to any other of Wiltshire's equine hill figures. This chapter is, in part, a call for parity of funding to better understand, value and protect Wiltshire's chalk equine hill figures.

Introducing Wiltshire's equine hill figures

A total of eight extant chalk horse hill figures can be viewed in north Wiltshire's landscape, whilst a further three, possibly five, are now lost to us (Fig. 20.1). Other equine hill figures do appear across England and Scotland, however, Wiltshire is unique in boasting a stable of chalk horses; a phenomenon that remains unexplained.

The Millennium Horse at Roundway, near Devizes (1999) and Pewsey's chalk horse (1937) are twentieth-century creations, the remaining surviving horses dating to the eighteenth or nineteenth centuries. Although precise dates when these early horses were cut are uncertain, some little progress has been made in recent years (as detailed below). All Wiltshire's extant equine hill figures are maintained by their local communities although Westbury, having recently been subject to extensive remedial works undertaken by English Heritage, stands alone having been completely concreted over in 1957.

Whilst a number of Wiltshire's chalk horses are situated close to Iron Age hillforts (Bratton, Cherhill, Devizes), others are located near other forms of ancient monuments – Alton Barnes (long barrow) and Marlborough (prehistoric mound). Still others are located alongside toll roads (Broad Town,

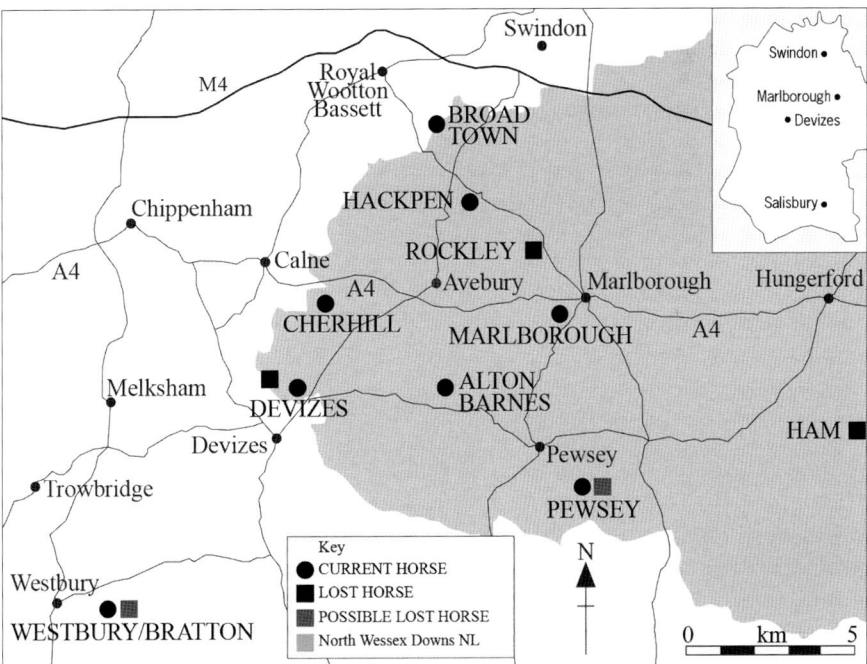

FIGURE 20.1. Wiltshire's extant and lost chalk horses. Image: Steve Cheshire

Hackpen, Rockley). The remaining pre-twentieth-century hill figure at Ham appears to have no obvious associations, nor do the twentieth-century creations (Millennium Horse, near Devizes and Pewsey). Little attention has been given to the siting of Wiltshire's chalk horses and their cultural associations; equally, there has been no analysis of their visual envelopes. All Wiltshire's chalk horses face sinister (left) save for the Millennium Horse, near Devizes which, although originally designed facing sinister, was cut facing dexter (right). There is no standard representation of a horse, some were created 'standing' showing all four legs, or at a trot. Tails are either cob or docked, suggesting different breeds and function of each horse. Steering away from simply linking representations of chalk horses with contemporary artists (i.e., Stubbs), horse iconography in the early modern period is an area of research that is underdeveloped in equine hill figure studies, and which holds the potential for better understanding multiple readings of these highly visible landscape monuments.

Three now lost horses are known to have existed. The Rockley White Horse was revealed following ploughing of pasture land in 1948, when it was fortuitously captured by aerial photography (Marples 1949, 219). A chalk horse just inside Wiltshire's boundary near the village of Ham was featured on an OS 6-inch County Series map (Crawford 1922, 73). Devizes' first chalk horse was thought to have been located below Oliver's Castle, near Devizes and is discussed at greater length below. Two other 'lost' chalk horses are less certain to have existed. Wise (1742, 48) reported that local people recalled the Bratton (now Westbury) horse having been cut within living memory, dating the existing horse to the late seventeenth/early eighteenth centuries, thereby challenging claims the current horse was created in 1778 (Edwards 2007, 17), which itself was a replacement of a much earlier version with Alfredian connections (Plenderleath 1874, 25). A horse cut in the late eighteenth century is said to have existed close by the current Pewsey horse, but other than a vague outline described by a local resident, as reproduced in Marples (1949, 89), little other firm evidence exists. Claims for an early Pewsey horse was likely due to it being confused with the nearby Alton Barnes horse, in the Vale of Pewsey.

Origin stories

Key to understanding these figures is the date of inscription and of revisions and scourings, reasons for their inscription, funding and the person/s responsible for their instigation and manufacture. The true date of origin is one of the principal items among these. On the evening of 7 August 1872 the Rev William Charles Plenderleath rose to present a paper to the Wiltshire Archaeological & Natural History Society at Court Hall, Trowbridge, Wiltshire. Two years later, Plenderleath's paper (1874) was published in the Society's magazine.

Plenderleath was not the first to commit Wiltshire's equine hill figures to print, but he did uniquely group them together as subjects worthy of investigation, albeit identifying just six of the county's eight chalk horses and, of those six horses, the Bratton (Westbury) horse drew his particular attention due mainly to its claimed antiquity. Of the remaining five horses, two (Hackpen and Broad Town) were mentioned in name only and the other three (Cherhill, Marlborough and Alton Barnes) attracted relatively detailed examination, the result of memories collected from various local informants. Having delivered his paper at the *conversazione*, Plenderleath was informed by Dr John Thurnam, a member of the audience, that he had failed to include the Devizes horse; an oversight Plenderleath rectified in the printed version of his paper. Two more, broadly contemporary, chalk horses (Rockley and Ham) were presumably unknown to Plenderleath at this time.

Over the next 20 years, Plenderleath produced further publications (1880; 1885; 1891; 1892) featuring Wiltshire's chalk horses, allowing him the opportunity to add new information as he collected it, or relegate some horses to anonymity due to their claimed ephemeral nature; their creation the result of mere leisurely activity. With the exception of Marples (1949), who notes twentieth-century updates (discovery of the Rockley horse and cutting of the Pewsey horse), the long shadow of Plenderleath's origin stories have persisted unchallenged for 150 years and continue to provide the sole basis for a popular understanding of Wiltshire's chalk horses and the country's chalk hill figures. Until Plenderleath's stories are rigorously examined, they will continue to potentially obscure key questions relating to Wiltshire's chalk horses, namely: why and by whom each horse was created, and why multiple chalk horses were uniquely created in a small geographic area?

Antiquity: inspiration for the modern?

Uffington's revels were hugely popular two-day events, which in the eighteenth century could attract up to 40,000 visitors. The revels centred on scouring the horse but there were also numerous sports and pastimes on the hill to entertain and thrill and, doubtless, many wagers were placed on the organised horse races, wrestling and backsword matches. Announcing the forthcoming revels at Uffington in the coronation year of 1838, the *Devizes & Wiltshire Gazette* (15 September 1838) stated the Uffington Horse was the ancient progenitor of Wiltshire's equine hill figures. Plenderleath (1874, 16) concurred, claiming the Uffington Horse was the 'great sire and prototype of them all'. Marples (1949, 17) writes the Uffington Horse was 'father and prototype' of Wiltshire's horses whilst, more recently, Miles (2019, 231) extends the notion to claim the possibility of a bloodline, suggesting it was the Uffington revels that acted as motivation for the creation of no fewer than six Wiltshire hill figures via a network of local Georgian elites who, having attended the revels, were moved to create their own eye-catching marks on Wiltshire's landscape. In a related twist, Edwards (2005, 110) suggests the creation of Wiltshire's chalk horses and revels at Uffington were both closely tied to celebrations marking major patriotic events.

Confirming dates for the creation of Wiltshire's chalk horses is fraught with uncertainty. Plenderleath, like Hughes (who wrote *The Scouring of the White Horse*, 1859 about Uffington), relied on information he was able to harvest from the local community, often relying on informants who volunteered details of events that had taken place up to 80 years previously and who, at times, recounted their stories third hand. If the collective memory of elderly locals (e.g., Joe, William and Thomas), who each provided dates for Uffington's revels as quoted by Hughes, are found to be fallible, and if Plenderleath's informants are similarly shown to be inaccurate, then claims that events at Uffington acted as motivation for, or dovetailed with, the creation of Wiltshire's equine hill figures must be called into question. Equally, if the guiding hand of Wiltshire's landowning elites in the creation of the county's chalk horses can't be established, the Uffington/Wiltshire hill figure link is also called into question.

Rethinking Wiltshire's chalk horses: two case studies

Cherhill's chalk horse

Plenderleath's Creation Story	
Informant:	'A very intelligent old man'
Background of informant:	Born 1786, a resident of Cherhill who had often heard the whole circumstances of the cutting discussed by men who had taken part in it
Date information collected:	c. 1866
Date hill figure cut:	1780
Person(s) responsible for cutting the figure:	Dr Christopher Allsup
Background:	Surgeon and apothecary of Calne
Method of cutting hill figure:	Shape of the horse was defined on the hillside by means of a thick rope under the direction of Allsup standing some distance away on Labour-in-Vain Hill. Allsup communicated instructions to a dozen workmen by means of a 'speaking-trumpet'. Having defined the shape, turf was removed and the area levelled with chalk
References:	Plenderleath (1874, 26–7; 1880, August; 1892, 28–31)
Notes:	Upturned glass bottles, provided by local landowner Mr Angell of Studley, filled the inner part of the hill figure's eye

Recent research has highlighted how the Cherhill horse story developed prior to Plenderleath's published accounts, especially around Allsup's involvement in creating the horse and the 'specious' suggestion it was cut in 1780 (Edwards 2014, 204–6). Of particular note is a piece in the *Monthly Magazine* (Anon. 1799), which Edwards (2014, 204–5) brings to light:

> It is not commonly known that the much admired White Horse on Cher Hill, near Chippenham, was cut only about fifteen years since by three or four gentlemen of Calne, assisted by about a dozen labourers … Mr. Webb, now a respectable surgeon, of Melksham, was one of the persons who assisted in laying it out, and who contributed to the expense.

Rather than Allsup acting as sole protagonist in the enterprise, Edwards' discovery isolates a group of local gentlemen who collectively contributed to the creation of the Cherhill horse. Edwards (2014, 206–7) continues by highlighting Allsup's relationship with the Earl of Shelbourne, whose family seat was located at nearby Bowood, and argues the Doctor and his close circle of friends cut the horse to celebrate Shelbourne's deserved elevation to Marquis of Lansdowne in 1784. Linking the Marquis' elevation, in recognition of his role in ending the American War of Independence, with cutting the Cherhill horse fits comfortably within a framework that argues the creation of Wiltshire's chalk horses marked times of local/national importance. Further, Edwards' later date of 1784 for the creation of the Cherhill horse aligns with the report quoted above, stating the horse 'was cut about fifteen years ago'.

Egan (1819, 26), says of Cherhill's chalk horse:

> On the completion of the (Cherhill) horse, it was celebrated on the spot by a sort of fair, which was kept up on a certain day in every year; but latterly it has been given up.

Egan's report is the earliest mention of revels taking place on the site of Cherhill's chalk horse, which is located below Oldborough (now Oldbury) hillfort. Featured in the *Bath Journal* (27 August 1781), a recently discovered announcement of revels at the horse represents the earliest mention of the Cherhill chalk horse. Supporting Egan's report a fair was 'kept up on a certain day on every year', a similar announcement for activities at Oldborough was published the following year in the *Bath Chronicle & Western Gazette* (12 September 1782).

Richard Webb was a resident of Calne until at least 1781 when the hill figure is first identified in the press. Webb subsequently lived in Melksham for almost 50 years by the time of his death in 1826 so it is possible to place the creation of Cherhill's horse, with some degree of certainty, to the four-year period between 1777 and 1780. Whilst Edwards makes a strong case for connecting the fortunes of Shelbourne and the cutting of Cherhill's horse in 1784, the two recently discovered press reports date the 'Oldborough White-Horse' to 1780 at the latest. The 1781 press announcement indicates the horse was a known landmark, suggesting the Cherhill horse was likely to have been cut in the late 1770s. Any involvement of Shelbourne in the cutting of the Cherhill horse is not evident; in contrast, it appears likely 'three or four gentlemen of Calne' were directly involved in its creation, two of whom were medical men (Christopher Allsup, and co-creator Richard Webb) who also had family ties.

The cutting of Cherhill's horse, now dated to the late 1770s, breaks any link between it and the 1780 Uffington revels that Miles claims provided the impetus for the creation of Cherhill's chalk horse (2019, 231). Edwards' discovery of the direct involvement of Allsup and Webb (and others) in cutting the Cherhill horse suggests future pathways of social history research to illuminate professional and familial links that connect all the 'gentlemen of Calne'. Shelbourne's claimed involvement in the creation of the horse is now all but nullified, as is the reason for cutting the horse, namely, in celebration of Shelbourne's elevation to Marquis of Lansdowne in 1784.

Devizes' chalk horse

Plenderleath's Creation Story	
Informants:	Dr John Thurnam; Mr Barrey
Background of informants:	Superintendent, Wiltshire Asylum, Devizes
Date information collected:	1872; c. 1890
Date hill figure cut:	1845; later, Whitsun 1845
Person(s) responsible for cutting the figure:	Shoe-makers of Devizes
Background:	Not known
Method of cutting hill figure:	Not given
References:	Plenderleath (1874, 28; 1880, November; 1885, 31; 1892, 34–5)
Notes:	The hill figure was cut in outline only. Known as Snob's Horse, Plenderleath took the late eighteenth century dialect meaning of snob to be 'cobbler'.

Devizes' first chalk horse had disappeared by 1870, but attempts to restore it were raised at various times throughout the twentieth century. A local reporter for the *Devizes & Wilts Gazette* (22 July 1909) interviewed a then elderly resident, George Smith, about the horse on Roundway Down. Smith (a former pupil of Richard Biggs' school, Devizes) replied,

> I also remember well, as a boy, when at Dr Biggs' school, how the Doctor took me with all the other boys, and we removed the turf and scoured the horse, which had become overgrown and neglected.

This account was supported by a further press report (*Devizes & Wilts Gazette*, 5 August 1909) from two other elderly residents of Devizes, also former pupils at Biggs' school, Edward Kite and Joseph Marler Sloper, both of whom recalled seeing the horse when they were boys and both reporting unsuccessful attempts to scour the hill figure. Kite's credibility as a witness lay in his observational skills, prolific output, and deep appreciation of Wiltshire's history that raised him from Devizes grocer to a skilled antiquarian (Bradby 1984).

As all three informants attended Biggs' school, it is possible to put approximate dates to the various scouring events they each recalled. Analysing the age of students boarding at Biggs' school in the 1841 and 1851 census returns it is likely that George Smith attended the school between 1833 and 1836; Joseph Marler Sloper between 1840 and 1843; and Edward Kite between 1843 and 1846. Working within these timeframes, Smith was a part of the successful scouring of the horse in the mid-1830s. If the age profile of day pupils was younger than for boarders, the scouring recalled by Smith may have extended back into the late 1820s. The 1830s scouring was followed by a number of unsuccessful attempts made in the early 1840s. Plenderleath claimed the Devizes horse was cut at Whitsun 1845 but it is likely the horse was in fact scoured at this date following a period when the horse greened over. Smith's father had claimed the

TABLE 20.1. History of Devizes White Horse

Event	Date
Hill figure created	pre-1830
Scoured (Smith)	mid-1830s
Not scoured (Sloper & Kite)	c. 1843
Possibly scoured (Plenderleath)	c. 1845
Horse visible (Thurnam)	c. 1850
Horse lost	1850–70

Devizes horse was created at the same time as Wiltshire's other horses, which the timeframe outlined above seems to support (Table 20.1).

Press reports and social history research have demonstrated their joint value in extending knowledge of one of Wiltshire's 'lost' chalk horses. The horse can reliably be considered a Georgian, rather than Victorian, hill figure, placing it firmly alongside a number of its Wiltshire cousins. A longstanding relationship between a local Devizes school and the chalk horse is established, especially as the horse seems to have disappeared from view when the school closed in the 1860s (Gibbons 2021, 280). Successfully scouring the horse in the mid-1840s when doubtless the horse had greened over following previous failed scouring attempts, can account for the hill figure's mistaken Victorian provenance.

Record and preserve: recent fieldwork

Discovering the Devizes chalk horse: 2020

Generally referred to as the Devizes chalk horse, the former hill figure stands some 1.6 km west of its modern cousin, the Millennium Horse (cut 1999). Located within the western boundary of the North Wessex Downs NL and on the Roundway Down and Covert SSSI, the hill figure is not scheduled. Heritage Gateway record (Heritage Gateway 1256630) summarises Plenderleath (1874), but does not include results of latest research. Historic Environment Record (Historic Environment Record ST96SEU04) summary leads with Plenderleath (1874), but includes limited results of new investigations (Horne 2020; Gibbons 2021) alongside Marples (1949); confusingly labelled 'Snob's Horse'. The site of the Devizes hill figure is on the south-facing side of Roundway Hill (NGR SU000645; 139 OD), immediately below a univallate hillfort of Oliver's Camp on an average slope of 26°. The hill figure has been overgrown since the 1860s.

Little reliable information relating to the first Devizes horse was recorded by Plenderleath, not least with respect to his suggestion the horse was cut in outline only. To test this assertion, a photogrammetric topographic survey was conducted in 2020 using a DJI Phantom 4 Advanced Drone over the slope of Roundway Hill below Oliver's Camp in order to locate and describe the hill figure. The UAV undertook a total of three survey flights following a double grid flight path covering sections of Oliver's Camp and slope immediately below the hillfort's earthworks (Horne 2020). Natural features on the slope of Roundway Hill partially obscured parts of the site, exacerbated by soil slip and scrub growth, but a shady outline inscribed into the hillside can readily be identified as a horse (Fig. 20.2). Only anatomical elements to the

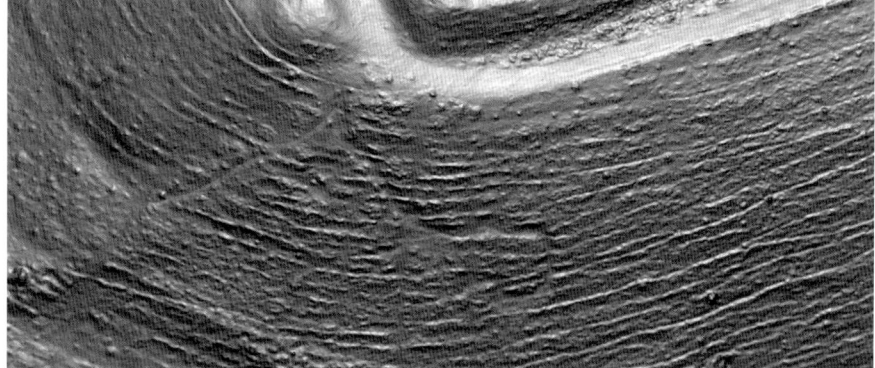

FIGURE 20.2. Devizes chalk horse, UAV topographic survey. Image: Donald Horne 2020

FIGURE 20.3. Devizes chalk horse, partial outline and dimensions; length 33.6 m, height 24.1 m (26.8 m), head length 7.2 m (7.9 m), chest to rump 24.1 m, and neck (length) 10 m (10.8 m). Dimensions in parentheses are ground distances if they differ

front and top of the hill figure's shape have, with any certainty, been picked up by the survey. The head, neck, leading foreleg, back and rump are clearly represented in a variety of visualisations based on off-site 3D modelling by manipulating angles of illumination that enhanced the hill figure's features, allowing a partial delineation of the horse and a good approximation of its dimensions (Fig. 20.3). Survey results demonstrate the hill figure conformed to Wiltshire's other chalk horses, namely, left facing and of naturalistic shape, rather than constructed in outline.

Delineating the Marlborough chalk horse: 2015

Generally referred to as the Marlborough horse, the hill figure is also known as the Preshute horse. Located within the North Wessex Downs AONB, it is not scheduled. The Heritage Gateway record (Marlborough Horse, id 1248725) summary is limited and follows Plenderleath (1974), and does not, as yet, include the results of recent research. The Historic England NMR Report (SU 16 NE 134; 27/10/2015) and Historic Environment Record (White Horse, Record No SU16NE528) summaries usefully includes results of new investigation (Horne *et al.* 2015; Gibbons 2017), but are confusingly labelled 'White Horse'. The hill figure is located on the north-facing side of Granham Hill (NGR SU 183682; 155 m OD) on an average slope of 27°, overlooking the Marlborough Mound (Castle Mound) some 430 m to the north and ½ km north of Marlborough College. The horse measures *c.* 17.5 m × 21.5 m in fenced area of 0.03 ha. The hill figure is maintained by staff at the College.

The earliest visual references for the Marlborough White Horse, two photographs, show a major development of the hill figure across the 1860s. Figure 20.4 (left), dated to 1860, represents the earliest photograph of the Marlborough horse or, indeed, any of the county's chalk horses. It is assumed to be a true likeness of the original hill figure, featuring a horse with two legs. Some six years later the horse has transformed into a four-legged creature (Fig. 20.4, right). See Gibbons (2017, 205–8) for an extended discussion.

A tape and off-set survey, undertaken in 2015 by Rob Read and Garry Gibbons, delineated the hill figure's outline and recorded its immediate topography (Fig. 20.5a). A pole-mounted GPS survey (Horne 2015) of the site was subsequently carried out creating a data set of the hill figure's co-ordinates and mapping the site's topography (Fig. 20.5b). The hill figure revealed through survey is very different in shape and character to the horse photographed in the late nineteenth century; the horse having contracted to primitive abstraction. The most startling elements of this process is evidenced by the current head and neck, yet the whole figure has, over the past century, lost all trace of its original equine identity.

Concluding remarks

Fieldwork investigations on Roundway Hill succeeded in locating and recording Devizes' lost Georgian horse and, in so doing, confirmed its shape and overall dimensions. Survey work at the Marlborough horse has highlighted a process of unintentional degrading at the site. For sure, Wiltshire's extant chalk horse monuments would doubtlessly have disappeared long ago without the ongoing care of the local community, however, results of fieldwork at the Marlborough horse have clearly demonstrated an ongoing process of radical change to the hill figure's once natural equine shape. Crucially, at both sites positive earthworks

294 *Garry Gibbons*

FIGURE 20.4. (top) Photograph of the Marlborough chalk horse c. 1860; (bottom) c. 1866. With permission of Marlborough College

associated with the hill figures features serve to retain the outline of their original shape.

Recent detailed research can question the direct links between the revels at the Uffington White Horse and the creation of six Wiltshire chalk horses as claimed by Miles (2019). An earlier date (late 1770s) for the cutting of the Cherhill horse untethers it from the reported 1780 revels at Uffington. Further, there is no evidence of aristocratic prompting behind the cutting of Cherhill's horse, rather, professional and/or familial networks appear to provide the financial

20. *Wiltshire's chalk equine hill figures: what's the problem?*

FIGURE 20.5. (top) Marlborough chalk horse, tape and off-set survey, 2015. Image: Rob Read 2015; (bottom) Marlborough chalk horse, contour survey with generic slope removed. Image: Donald Horne 2015

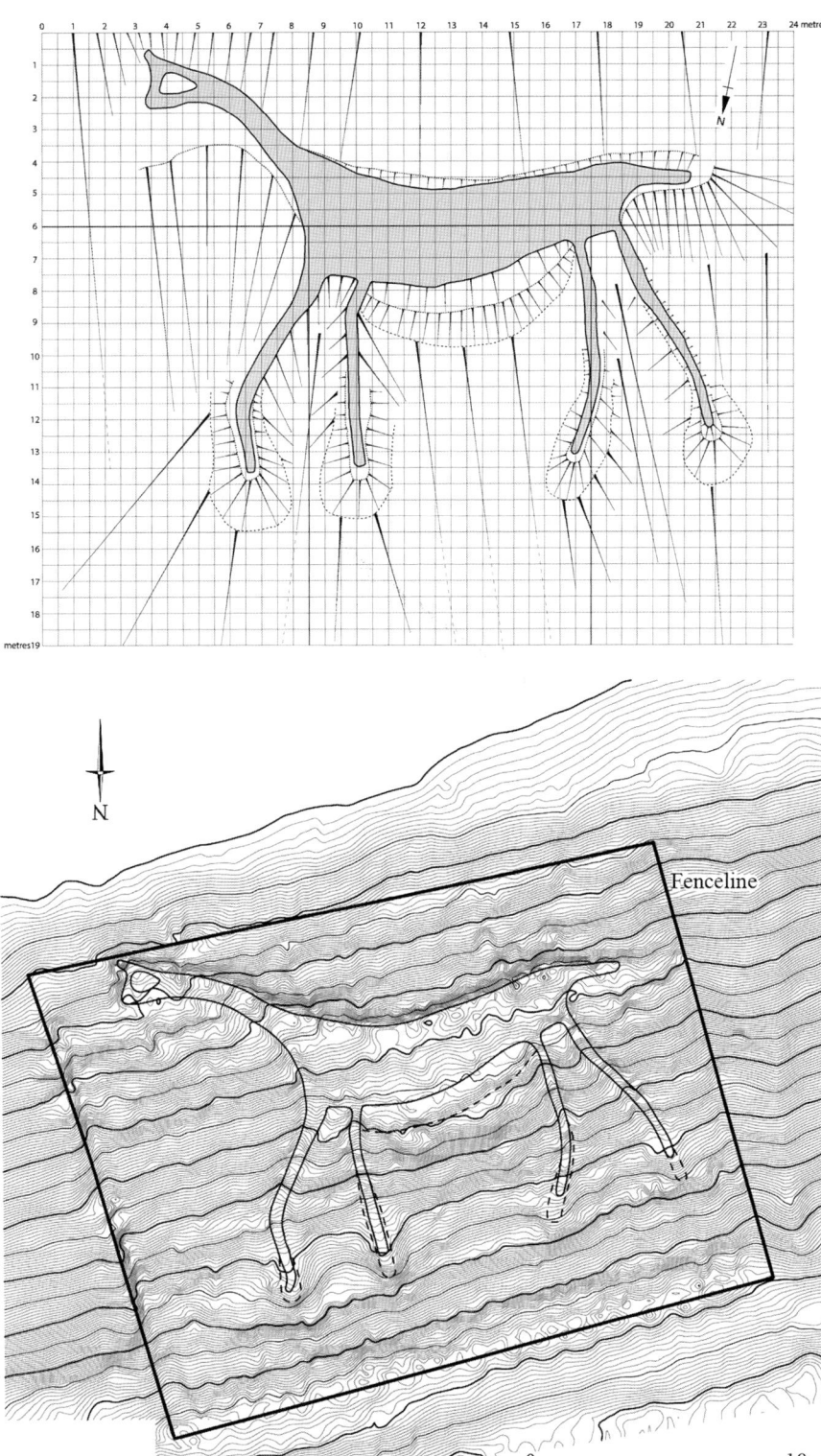

and organisational impetus. At Devizes a local school is closely linked with the ongoing maintenance of a horse whose disappearance coincided with the school's closure. A similar direct connection between school and chalk horse is also evident at Marlborough (Gibbons 2017, 219–20) and, again, there is no evidence for aristocratic meddling. Setting aside its claimed antiquity, the Bratton (Westbury) horse has attracted little research although, as at Devizes and Marlborough, the Bratton horse was maintained by a local school in the eighteenth century (Jennings 1894, 193; Reeves & Morrison 1989, xxxiii).

This chapter opened with an appeal by Flinders Petrie made a hundred years ago; the chapter closes by echoing Petrie's call to action for all Wiltshire's equine hill figures by a systematic programme of (appropriately funded) survey and fieldwork for each of the county's chalk horses, past and present, and promoting evidence-based research designed to challenge the wide-spread circulation of origin stories that have shown to be based on historic guesswork.

Acknowledgements

My sincere thanks are extended to Mike Allen for his gentle guidance, to Gráinne Lenehan (Archivist, Marlborough College), and to David Dawson (Director, Wiltshire Museum). Thanks also go to Steve Cheshire, illustrator (Fig. 20.1), Rob Read, illustrator and surveyor (Fig. 20.5), and Donald Horne, surveyor (Figs 20.2 & 20.6). Figures 20.4 (left and right) are reproduced by kind permission of Marlborough College. Investigations at the Marlborough chalk horse (2015) generously funded by Marlborough College and North Wessex Downs National Landscape. Special thanks go to Emma Burns for reading an early draft of this text, providing thoughtful feedback, and exhibiting unlimited patience. Finally, my gratitude to Brian Edwards whose energetic intellect and generous support remain unparalleled.

CHAPTER TWENTY-ONE

Hill figures in the landscape: contexts, survival and function

Tom Williamson

Recent archaeological investigations into England's three most important hill figures have, for the first time, provided firm evidence for their age. Yet they have also served to reaffirm the extended period of time over which these features were created, confirming a prehistoric date for the Uffington White Horse; moving the Cerne Abbas Giant back from its widely accepted post-medieval date, to the early Middle Ages; while doing the reverse for the Long Man of Wilmington. It would be surprising if these were the only such figures to have been created before the widespread cutting of white horses and other features by local landowners in the later eighteenth and nineteenth centuries. Indeed, as many readers will be aware, there are at least three other extant figures for which a pre-1750 date seems probable or at least possible, the crosses at Bledlow and Whiteleaf in Buckinghamshire and at Ditchling in Sussex. There are also a significant number of lost ones: the horses at Tysoe in Warwickshire and probably Pitstone in Buckinghamshire; the first horse at Westbury in Wiltshire; and the 'giants' at Plymouth, Shotover Hill near Oxford, Wandlebury near Cambridgeshire, and Clifton near Bristol (Marples 1949; Newman 1997b). The famous figures at Uffington, Cerne Abbas and Wilmington need to be considered as part of, and within the context of, this rather wider group.

Documentary evidence makes it clear that the giant at Plymouth was in place by 1480 (Rous 1716, 15; Worth 1893, 93); the probable lost horse at Pitstone by 1581 (Davis 1990); while the single apparent reference to the Clifton figure dates to 1480 (Neale 2000, 32–5: Clark 2016, 110). The Wandlebury giant is first mentioned in 1605 (Marples 1949, 205), the Red Horse at Tysoe in 1607 (Camden 1607, 424), while Aubrey in the late seventeenth century believed the Oxford giant had disappeared soon after 1640 (Marples 1949, 206). All three may have been of early post-medieval date, but could well be earlier. In contrast Francis Wise, writing shortly before 1742, was told that the first Westbury horse had been cut within living memory, so perhaps in the late seventeenth century (Wise 1742, 48). Against this dozen or so figures evidently created over many hundreds of years preceding *c*. 1750 we need to place the very similar number known to have been cut in the single century between 1750 and 1850, and the 20 or more created between 1850 and 1950, as white horses were joined by military badges, kiwis, spitfires and a host of eccentricities. It is often assumed that this chronological pattern simply reflects a growth in the popularity of hill figures over time but it must also be a function of attrition. Less than half of the figures known to have been in existence in 1600 still remain, compared with around two thirds of those extant in 1800 and three quarters of those present in 1850. The older the figure, the less likely it is to have survived, extant or in memory, to the present.

For hill figures are ephemeral, eventually disappearing if not regularly maintained, albeit at a rate which varies with the mode of their formation. The survival of the Uffington Horse, in particular, is remarkable given the number of comparatively recent examples, such as the Inkpen Horse (1868) or the Detling Star (1952), that have vanished without trace (http://www.hillfigures.co.uk/). Most of the early, pre-1750 figures that we know about were located on common land and their maintenance was, in some cases at least, accompanied by the kinds of festivities described, in the case of Uffington, by Thomas Hughes in *The Scouring of the Horse* (Hughes 1859). Such activities might be considered aspects of long-lived 'folk cultures' but they were usually under-pinned by systems of property ownership. The scouring of the Uffington Horse was the responsibility of 'some that dwell hereabouts [who] have an obligation upon their lands to repair and cleanse this Land marke' (Wise 1738, 57). In 1742 the Whiteleaf Cross was not being regularly scoured and Wise remarked that 'if any estates have been formerly charged with the expense, time has long made void the obligation' (Wise 1742, 36). At Tysoe, the maintenance of the red horse was the formal responsibility of 'certain neighbouring landowners who held their land by that tenure', and who paid for the cakes and ale consumed at the festivities (Plenderleath 1892, 37; Marples 1949, 112). It is perhaps noteworthy in this context that the majority of known pre-1750 figures were associated in some way with permanent institutions. This may indicate the particular involvement of corporate bodies (or their members) in the initial creation of hill figures, or the fact that they were more likely to maintain them over extended periods than private estates, subject to changes in ownership or outright dissolution. The figures at Cerne Abbas and Wilmington both lay beside, and on property owned by, monastic houses; those at Uffington and perhaps Tysoe were located on outlying monastic manors; Plymouth Corporation owned the land occupied by the giant cut on Plymouth Hoe; Oxbridge colleges had some involvement with the figures at Wandlebury, Shotover, and reputedly Whiteleaf; while the most recent scouring of the cross at Bledlow, according to Baker in 1856, had been paid for by Eton College, to which the principal manor in the parish was attached (Wise 1742, 36; Baker 1856, 225; Page & Ditchfield 1924, 544–9; Marples 1949, 173–4, 199, 204–12; Salzman 1949, 177–80). Yet if changes in ownership could lead to the gradual disappearance of a hill figure, change of land-use could lead to more immediate destruction and the seventeenth and eighteenth centuries saw much enclosure, and ploughing, of common land on steep slopes in the chalklands of southern England and elsewhere. The Red

Horse of Tysoe, ploughed out following enclosure in 1798, is not the only hill figure to have disappeared in this way (Johnson 1867, 326).

The three celebrated hill figures with which this volume is primarily concerned do not, therefore, seem to represent idiosyncratic and isolated endeavours but surviving examples of a practice which, if never common, was perhaps reasonably frequent at various times in the past. Even in the seventeenth century there may have been significantly more hill figures in existence than we are aware of today. It might be argued that, if this was the case, more of them would have been noted by early antiquaries and topographers. But as the new dates now available for the Cerne Giant, in particular, make clear, such negative evidence is not to be trusted and the silence of antiquaries is perhaps quite easily explained. Many figures were not particularly old when they encountered them and the regular scouring of those that were, if not confused by elite observers with actual creation, at the very least served to associate them with contemporary rural culture, rather than with antiquity. Hill figures may often have been regarded, in effect, as a form of graffiti, not worthy of serious academic attention. Camden simply described that at Tysoe as 'forma equii in colle russescente iuxta *Pillerton* a rusticis incisa' ('the shape of a horse cut by country people on a red hill near Pillerton') (Camden 1607, 424). He only appears to refer to the figure in order to explain why the Vale of the Red Horse was so named; it is a moot point whether it would have been mentioned otherwise.

However common hill figures may have been in the medieval and post-medieval landscape, it is likely that they were created for a wide variety of reasons. In the case of the better documented post-1750 examples, these evidently embraced the display of ownership, humour, the assertion of group identity, the commemoration of important events, remembrance of the dead, commercial advertising, and the enjoyable exercise of artistic creativity, alone or in combination (Marples 1949; http://www.hillfigures.co.uk/). The traditional emphasis in the literature on mythical or religious iconography in the interpretation of, in particular, the Cerne Abbas Giant and the Long Man of Wilmington, may not help very much in elucidating a figure's meaning or purpose, for symbols may come to be used in multiple ways, many quite practical and with only a tenuous connection with any original mythic or religious significance. To take an obvious example, the symbol of the cross, originating with the crucifixion, has developed diverse meanings in modern society, variously marking a place of burial, a retail outlet where pharmaceuticals can be obtained or a location where medical care is provided. It can also, quite independently, signify an approaching junction of roads, or simply draw attention to a particular point in space – 'X marks the spot'. To understand why particular early hill figures were created we therefore also need to examine their landscape context, including not only associations with other archaeological features and aspects of human geography but also relationships to the wider topography, and in particular where they were intended to be viewed from. For hill figures were clearly not, as some have suggested, 'messages to the gods', directed skywards. They were not cut on level and seldom on gently sloping ground. Like advertising hoardings, motorway signs or the 'Angel of the North', visibility was fundamental to their purpose, whatever that might have been in particular cases. What follows is a brief exploration of these ideas, taking as an example the hill figures of the Chiltern Hills in Buckinghamshire, Oxfordshire and Bedfordshire.

The Chiltern hill figures

Of the four extant, or near-extant, figures cut into the steep chalk escarpment of the Chilterns, two are relatively modern. There is no real doubt that the 'White Mark' at Watlington in Oxfordshire – an 82 m long elongated triangle – was created in 1764 by Edward Horne, a local landowner, while the White Lion at Whipsnade in Bedfordshire was unquestionably cut in 1933 by the proprietors of the adjacent zoo (Marples 1949, 146, 217). Of more interest are the two chalk crosses, at Whiteleaf and Bledlow in Buckinghamshire. The former, in the parish of Monks Risborough, is the more impressive of the two (Fig. 21.1). Cut to a depth of 0.5–1 m into a west-facing slope, it comprises a triangular base of bare chalk, traditionally referred to as 'the globe', measuring 40 m in height and some 120 m along the base; which is surmounted by a cross, 24 m high and the same across (Buckinghamshire HER 0164900000). The dimensions of the monument appear to have changed over time, for Wise in 1742 recorded that the 'globe' measured only 57.6 m (189 ft) along the base, and only around 21.6 m (71 ft) in height, while the cross was then around 30 m (100 ft) high and only 21 m across (Wise 1742, 35). How far these changes were the consequence of maintenance and natural erosion, how far the result of a systematic recutting that occurred in 1826 under the direction of the Earl of Buckinghamshire, is

FIGURE 21.1. Whiteleaf Cross, viewed from the air

unclear (Payne 1897, 566–7). The original tapering 'shaft' of the cross, as first described and illustrated by Wise in 1742, has also been changed to its current, parallel-sided form (Fig. 21.2).

It has been suggested that the cross originated as an ancient phallic symbol, Christianised by the addition of the arms (Clinch 1905, 189). This has been supported by a reference, in the boundary clause of a charter for Risborough drawn up in 903, to 'Weland's stocc', interpreted as 'Weland's phallus' (Massingham 1935, 293–5). But to judge from Baines's meticulous reconstruction of the charter bounds, this feature (whatever it may have been) lay some 800 m to the south-east of the cross (Baines 1981, 89), while the cross itself lies some 600 m from the boundary described in the charter, which here coincided with the later parish boundary between Monks and Princes Risborough (Baines 1981, 84). There is likewise no hard evidence for the Anglo-Saxon date, and commemorative function, proposed by some early antiquaries (Wise 1742; Baker 1856, 222–3). On the other hand, there are good grounds for questioning the post-medieval date for the monument proposed, *inter alia*, by Scott, and by the Buckinghamshire Historic Environment Record (Scott 1937; Buckinghamshire HER 0164900000).

It is true that the earliest reference to the cross comes as late as 1742, when it was included by Francis Wise in his pioneering discussion of hill figures more generally (Wise 1742). But for the reasons just noted, not much can be read into such silence. The cross takes its name from the nearby hamlet of Whiteleaf, almost certainly a corruption of White Cliff; the hamlet was sporadically called 'Whitecliff' as late as 1876 (Centre for Buckinghamshire Studies D-X5 18/1) (the cross itself is labelled Whitcliffe Cross on Jeffrey's 1766 map of Buckinghamshire).

FIGURE 21.2. Whiteleaf Cross, as illustrated by Francis Wise in 1742. Looking north-east along the escarpment, with the Risborough Gap in the middle distance

This is clearly a reference to the steep expanse of exposed chalk making up the 'globe', whether entirely man made or partly natural, indicating that this at least was already in existence when the settlement was named. Unfortunately, the name is unrecorded before the eighteenth century but is unlikely to have been new then. More importantly, various aspects of Wise's description, such as his reference to local traditions that 'some of the colleges in Oxford', nearly 30 km away, had once contributed to the costs of maintenance, are hard to square with a recent origin (Wise 1742, 36). Above all, a large cross, visible for many miles, would be more at home in a pre- than a post-Reformation context, while the monument's original resemblance to a medieval stone wayside cross with a tapering base is suggestive. A medieval date for the cross seems, on balance, more probable than a post-medieval one; the base may have been adapted from a natural feature but is likely to be contemporary.

Bledlow Cross, in contrast, is not mentioned by Francis Wise in his article of 1742, even though he specifically discusses Bledlow itself as the possible site of the battle with the Danes that Whiteleaf Cross supposedly commemorated. Putting aside an ambiguous reference in the 1350 Patent Rolls to an individual named 'Henry atte Crouche of Bledelowe', the first mention of the monument comes in an article in the *Gentleman's Magazine* in 1827 (Mawer & Stenton 1925, 168; Anon. 1827). Here it is described as a simple equal-armed cross, '30 feet long in both lines, and of the width of six feet'. Like Whiteleaf it has grown over time, now measuring more than 20 m in both directions, and with a width of more than 4 m, although today overgrown and mainly visible as a cross-shaped depression (Fig. 21.3). Not surprisingly, several writers have suggested that the cross is a late eighteenth-century folly, cut to the orders of a local landowner. But against this we need to note that it was already decayed and overgrown in 1827, and its origins forgotten. Such an interpretation also fails to explain why, in the nineteenth century, the maintenance of the cross appears to have been paid for by Eton College, an institution that had held the

FIGURE 21.3. Bledlow Cross in the early twentieth century, before its condition deteriorated

principal manor in the parish since 1462 but which, by the eighteenth century, was not a major landowner there, and was anyway located more than 30 km to the south-east. Again, a large cross might seem more at home in a medieval than a post-medieval context.

To these extant Chiltern hill figures we can add one probable lost example and possible hints of another. The former, on Pitstone Hill, Pitstone – in Buckinghamshire but close to the county boundary with Hertfordshire – was first noted by Jean Davis in 1990. A map of the parish surveyed around 1809 depicts three small fields on the north-western side of Pitstone Hill called the First, Second and Third White Horse Pieces. They lay close to a lane that crossed the county boundary from the adjacent parish of Aldbury, very probably the 'White horse waye' referred to in a will of 1610. A document from 1581 records the sale of property in Aldbury, which included three roods of land in Nokeden Furlong, in this same area, lying adjacent to 'Whight Horse'. Davis plausibly suggested that 'there might at some time have been a physical feature to link up the references in the two parishes, and a white horse cut into the chalk on Pitstone Hill and dating before 1580 seems a possibility' (Davis 1990). No trace of such a feature, which would have occupied the steep west-facing slope of the hill, survives on the ground today, or can be convincingly discerned on Lidar images, but Davis's suggestion seems convincing.

Much less certain is the possible figure on Aston Hill in Aston Clinton. A contributor to the 1854 edition of *Records of Buckinghamshire* described how:

> Tradition states that a shepherd named Faithful, delighted with the panorama, used to make this spot his common resting place, while attending his master's flock. Becoming at length so attached to it, he exacted a promise from his fellow shepherds that at his death they would bury him here. This promise they fulfilled, and cut in the turf the following epitaph :-
>
> Faithful lived and Faithful died,
> Faithful shepherd on the hill side;
> The field so wide, the hill so round,
> In the day of judgement he'll be found.
>
> The rustics of the neighbourhood used carefully to keep the letters clear; but, having for some time ceased to do so, the word 'Faithful' alone was legible when I saw it ... This was about 1847; and I am afraid the ground has since been ploughed over (Anon. 1854).

The phrase 'faithful shepherd on the hill side' might conceivably be a garbled memory of a hill figure that originally accompanied the text, but such speculation should obviously be treated with extreme caution.

The context and purpose of the Chiltern figures

The Chiltern Hills can thus boast one hill figure of pre-eighteenth-century, perhaps medieval, date, Whiteleaf Cross; two probable examples, at Bledlow and Pitstone; and possible hints of a fourth, at Aston Clinton. What is striking is the proximity of these sites. The Chiltern Hills, strictly defined and ignoring their muted north-easterly continuation as the 'East Anglian Heights', extend for some 75 km from the Thames near Goring to Hitchin in Hertfordshire. Yet the four sites are clustered within 22 km of each other, in the central section of the hills, with only the eighteenth-century 'White Mark' at Watlington, and the twentieth-century lion at Whipsnade, as outliers. Even a cursory examination of a map of physical geography, or a digital surface terrain model, immediately suggests an explanation.

The steep escarpment of the Chilterns constituted a significant obstacle to travellers approaching London from the north-west but fortunately the hills are cut by a number of 'through valleys', which are today followed by main roads and railway lines. For the first *c*. 25 km, moving north-east along the escarpment from the vicinity of Goring, there are no through valleys – and no early hill figures. The escarpment is then interrupted by the Risborough Gap, which is followed by the A4010 and the rail line from London to Bicester and Banbury. Its wide opening is flanked to the south-west, at a distance of *c*. 600 m, by the Bledlow Cross and to the north-east, at roughly the same distance, by that at Whiteleaf (Fig. 21.4) (the latter's position on the edge of the 'gap' is particularly clear on Wise's engraving of 1742: Fig. 21.2). Some 7 km further to the north-east comes the Wendover Gap, followed by the A413 and the rail line from London to Aylesbury, with the site of the possible Aston Clinton figure around 1600 m from its eastern edge. After a further 7 km or so a third passage through the hills opens up, the Tring Gap, today

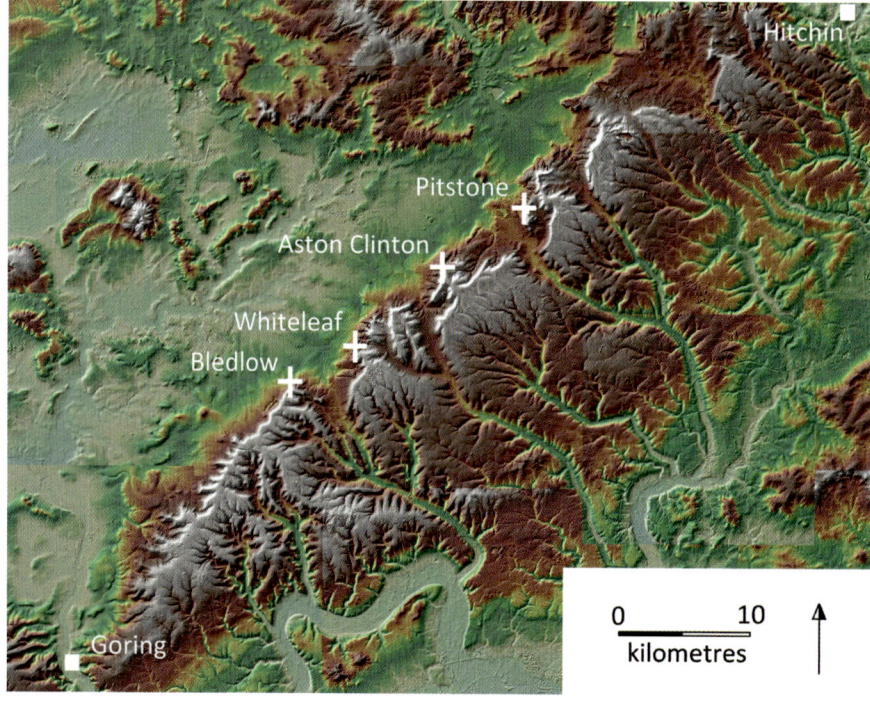

FIGURE 21.4. The relationship of the Chiltern hill figures to the natural topography

followed by the A411 and the main railway line from London to Birmingham. Overlooking the latter, on the west-facing slope forming the eastern edge of the entrance to the gap, is the site of the probable lost horse on Pitstone Hill. It thus seems possible that the main purpose of the two crosses and the Pitstone horse (and perhaps of the Aston Clinton figure, if it existed) was to direct travellers approaching from the west and north-west, across the wide and relatively level claylands of the Vale of Aylesbury and south Oxfordshire, to the main routes cutting through the hills and continuing on to London. This is not, of course, to deny that the crosses may also have fulfilled a devotional function. Medieval wayside crosses, many of which served as important land marks, provide an obvious parallel.

The idea that the Chiltern crosses served as guides for travellers is not in itself new. Marples suggested an association with the Icknield Way, which runs along the escarpment, and a function as 'landmarks for the assistance of the traveller', providing direction to the track (Marples 1949, 147). Payne in 1897 suggested that the two crosses marked routes through the hills to be taken by troops during the Civil War, although strangely *not* through the Risborough Gap (Payne 1897, 565–6). It is true that the crosses, as noted, lie a few hundred metres from the 'mouth' of the gap but this may have been to maximise visibility. In this context it is worth noting that while overall the Chiltern escarpment faces north-west, three of the sites are associated with slopes facing west and this was surely to ensure maximum visibility across the clay vale lying below the escarpment, which extends westward, to Oxford and beyond, and less far in a northerly direction (Fig. 21.4). In the Middle Ages, and well into the post-medieval period, this was an area of open-field farming on heavy clay soils, crossed by a maze of poor local roads and without much in the way of signage, in the form of direction posts or the like. As late as 1813 Priest, writing about Buckinghamshire, described how:

> In the *passages* through lands under the open-field culture, not only the roads are bad, but the difficulty of discerning public roads from mere drift-ways, or from passages to lands of different proprietors, is so great, that without a guide, some of them cannot be travelled by a stranger with safety (Priest 1813, 339–40).

In such circumstances it is easy to see how the charitable impulse of individuals or institutions might have turned to the provision of signs bringing the traveller to within sight of the main gaps through the hills. It is true that there are two further 'gaps' passing through the Chilterns in the area between Tring and Hitchin, which have no record of early hill figures placed close to their openings. Some may once have existed, of course, but as Figure 21.4 clearly shows the topography is more complex here, with lines of chalk hills standing out from the main escarpment, and the clay vale is narrower, the land rising again, after less than 10 km, in the hills of the Greensand ridge. In these circumstances, hill figures cut on the escarpment would have had limited visibility from the lower ground, but more importantly they were less necessary as guides for travellers than on the wide, disorienting expanse of the clay vale lying further to the west.

Discussion

Other early hill figures, elsewhere in the country, may have served a similar function to that suggested here for those in the Chilterns. The cross at Ditchling in Sussex, now in a poor condition, is one example. The steep north-facing escarpment of the South Downs has no 'gaps' like those in the Chilterns but it is crossed by a number of deeply incised tracks or 'bostalls', which run obliquely across the steep slope, providing a comparatively easy ascent to the relatively level ground above. The Ditchling Cross lies next to one of these – Plumpton Bostal. This is now no different from its many neighbours in that it simply provides access to the summit of the Downs. But it may have once had a superior status that it was important to signal to travellers approaching from the north, for three fields lying beside its first stages, on the lower slopes of the escarpment, are named 'Brighton Lane' on the tithe map of 1839, suggesting that it once served as a route to that place, which lies some 9 km to the south west (The National Archive, IR 30/35/80). Attention might also be drawn, in this context, to the lost horse at Tysoe in Warwickshire. This was located (like the Chiltern examples) on a forbiddingly steep escarpment overlooking an extensive clay vale characterised by open-field farming, and it was positioned (in Newman's words) 'where the Banbury-Stratford road descends the Edgehill escarpment' (Newman 1997b, 47). Here, as at Pitstone, a horse might have seemed a particularly appropriate choice for a signal to travellers. The giant cut into Plymouth Hoe might represent a slightly different kind of direction sign, functioning as a 'sea mark', visible to ships passing the mouth of Plymouth Sound and signalling the presence of the town and harbour of Plymouth.

Yet I do not mean to suggest that all, or even most, early hill figures served the function, in Marples' words, of 'road signs' (Marples 1949, 148). They must have been cut for a variety of reasons but these need, for the most part, to be sought within the context of relatively recent societies. Popular studies tend to focus – to quote the cover notes of what is in many ways an important treatment of the subject – on 'druid massacres, conjectured human sacrifice and strange phallic and pagan rites' (Newman 1997b). We somehow want our hill figures to be older, more esoteric, than they really are. In reality, their landscape associations, not least locations close to important urban centres like Plymouth, Oxford, Cambridge and Bristol, strongly suggests that only the white horse at Uffington has any very great antiquity. This, as I suggested earlier, is most unlikely to have been unique; time and chance have served to leave it a sole survivor of a wider population. The other 'early' figures, including the Cerne Abbas Giant and the Long Man of Wilmington, are likewise perhaps best understood as survivors from rather larger populations, their greater numbers reflecting the shorter periods of time for which they have existed. Most were probably created for reasons that were – at least in part – practical, even humdrum, in nature. But to understand what these may have been we need to consider the landscape context, as much as the form and iconography, of these seemingly mysterious features.

CHAPTER TWENTY-TWO

Hill figures: retrospective and a national research agenda

Michael J. Allen and Win Scutt

Scattered across the English countryside, hill figures or 'geoglyphs' stand as distinct, enigmatic emblems etched into the slopes. They are large-scale depictions including symbols, designs or motifs, often in animal form, created by cutting away turf and subsoil to create a visual contrast between the exposed fresh geology (often chalk) with the surrounding grassland. Though unified by their topographical setting and method of creation, each figure boasts a unique identity, history, and genesis. This shared form 'hill figure' belies the diversity within the category. They are not linked by any overarching chronological narrative or social association. Instead, each monument arises from local, individual, or national impulses, often commissioned and crafted by the very communities they inhabit. Many figures serve as responses to immediate contexts, commemorating local groups and events. Regimental badges at Fovant and the Anzac badge at Codford speak to military presence, while the Ditchling Cross memorialises the Battle of Lewes in 1264. National occasions, like coronations or VE Day, inspire fewer figures, suggesting a focus on the local and particular. Some medieval geoglyphs are located near monastic establishments (Lewes Priory; Wilmington Priory; Cerne Abbey). This proximity raises intriguing questions about potential connections and motivations for their creation.

The hill figure 'horizon' in archaeology

Michael J. Allen

Although archaeologists know that the chalk hill figures are of different ages, especially after the OSL dating of Uffington (Miles *et al.* 2003; Chapter 16) and Wilmington (Bell & Butler 2014; Chapter 15), there seems to be an unwritten consensus that they belong to the same monument group and a belief that they share more affinities than they do in reality. As other monument-types tend to share a communal function and age, the same attributes seem to be unconsciously carried over to hill figures. This short piece reminds us that each hill figure has its own specific local identity, specific reason and chronology; they are all individuals and can only be grouped in a single class on topographical and graphical grounds.

Despite archaeologists realising the chronological differences between the hill figures and now more so, with the results from the Cerne Giant showing that the 'big three' (Cerne, Wilmington and Uffington) all occupy completely different spaces in time and reasons for their presence, nevertheless, there is a continued subconscious connection made by archaeologists: they are commonly classed, discussed, and considered together. Even as recently as September 2023, in discussions with a former senior member of English Heritage we discussed academically the values and differences of the 'big three' almost as if they were chronologically a part of the same 'team'; yet knowing each had its own separate identity and that there was no social or chronological connection between them. The fact that we class and discuss them together is exemplified by the number of books on 'hill figures' (e.g., Marples 1949; 1981; Newman 1987; Castleden 2000; Bergamar 1997, etc.). However, they only share a common topography and effect; they are not chronologically, functionally, socially or iconographically similar as, for instance, are barrows, hillforts and churches. This apparent unity means that sometimes they are artificially or erroneously afforded some social, chronological or other functional commonality, when all that actually relates them is one of form and creation. No engraving, painting or graffiti is subconsciously ascribed this same unification; they are all individual pieces, individual statements and of, and at, their time. The same is true of hill figures and that is their fascination for the general public, local residents, archaeologists and historians alike. Each has its own individual time, or times, and story that needs to be teased out by historical and documentary research, archaeological fieldwork survey, ground truthing (survey, geophysical research and augering), and potentially, as at the 'big three', formal clinically targeted archaeological excavations to define construction, processes and chronology. These data provide the narratives of each individual site, the story of their being, part of which may also be shrouded in mystery and folklore, but amongst which some erroneous myths can be (publicly) debunked by sound research. Such research does not remove emotive elements from the figures, nor all of their mystery; it never can, but provides a core of 'truism'. This also conveys community ownership, and value, and in so doing enhances heritage value among the local community, if not (yet) the heritage community (see Gibbons, Chapter 20, and below).

In an attempt to demonstrate their individuality, we have listed 37 hill figures and shown their differences in date, commemoration and style (Table 22.1). In some cases they are signposts to venues in the proximity (e.g., Whipsnade Zoo) or commemoration; military as in the Fovant badges; war time events (e.g., Codford Anzac badge); or regal (coronation or jubilee; e.g., Wye Crown coronation of King Edward VII in 1902), although now these are more frequently celebrated by more ephemeral actions such as lighting community beacons, which leave a mark on the mind, not the land. The compilation of well-researched historical data, archaeological research and the collation of a robust individual narrative for each site will demonstrate their own individuality and heritage value, and greater social and community value.

TABLE 22.1. Hill figures in England, their description and designation. All horses are left facing unless otherwise stated

Hill figure	Location	Image	Popular hill figure date	NHLE number	Date first listed/ scheduled	Key monuments in proximity
Main excavated figures						
Cerne Abbas Giant	Dorset	Naked man		SM1003202	15 Oct 1924	The Trendle
The Long Man, Wilmington	East Sussex	Man with two staves		SM 1002293	1 May 1951	(long barrow)
Uffington White Horse	Oxfordshire	Stylised horse		SM 1008413	13 Dec 1959	Hillfort, The Manger
Other geoglyphs						
Gog Magog (Wandlebury)	Cambridgeshire	Two giants no longer extant	AD 200–300	–	–	Causewayed enclosure, hillfort
Gogmagog and Corineus? (site of)	Plymouth, Devon	Two men, one larger, holding clubs, no longer extant	Pre-1602	–	–	Royal Citadel (17th century)
Horses						
Westbury/Bratton White Horse	Bratton/Westbury, Wiltshire	Iron Age hillfort & white horse	18th century (1778)	SM 1013399	14 May 1934	Bratton Camp hillfort (and long barrow)
Cherhill/Oldbury White Horse	Calne/Cherhill, Wiltshire		1780	–	–	Oldbury Castle hillfort
Marlborough White Horse	Preshute, Wiltshire		1804	–	–	Marlborough Mound
Osmington White Horse	Dorset	Right facing & rider	1808	SM 1005574	11 June 1969	
Alton Barnes White Horse	Alton Barnes, Wiltshire		1812	–	–	Adam's Grave long barrow
Hackpen White Horse	Broad Hinton, Wiltshire		1838	–	–	Ridgeway
Kilburn White Horse	Yorkshire		1857	–	–	
Ham/Inkpen White Horse	Ham, Wiltshire	Left facing according to 1872 OS map	1860s	–	–	
Broad Town White Horse	Broad Town, Wiltshire		1863	–	–	
Pewsey White Horse	Pewsey, Wiltshire	Possibly with rider	1778/?1785	–	–	
Pewsey White Horse (new)	Pewsey, Wiltshire	Left facing	1937	–	–	1937 in chalk above horse
Devizes White Horse	Wiltshire		1845	–	–	Oliver's Castle hillfort
Devizes White Horse (new)	Wiltshire	Right facing	1999	–	–	
Tysoe Red Horse	Warwickshire	Left facing with saddle	1607	–	–	
Litlington White Horse	East Sussex	Right facing	1899	–	–	

(Continued)

TABLE 22.1. Hill figures in England, their description and designation. All horses are left facing unless otherwise stated (Continued)

Hill figure	Location	Image	Popular hill figure date	NHLE number	Date first listed/scheduled	Key monuments in proximity
Folkestone White Horse	Kent	Left facing, prancing	2012	–	–	
Penleigh White Horse	Dilton Marsh, Wiltshire	Left facing, lost	–	–	–	On postcard c. Victorian or Edwardian. Short-lived
Mormond Hill White Horse	Aberdeenshire	Left facing	1794	Canmore ID 111210	–	Ruined hunting lodge
Military				–	–	
Fovant Badges	Wiltshire	Eight military badges	1916	SM 1020132	1 June 2001	
Bulford Kiwi	Wiltshire	–	1919	SM 1443438	12 June 2017	
Codford Anzac Badge	Wiltshire	Military badge	1916–17	–	–	
Royal Warwickshire Regiment	Sutton Mandeville	Military badge	1916	SM 1020134 (1)	1 June 2001	
City of London Regiment	Sutton Mandeville	Military badge		SM 1020134 (2)	1 June 2001	
Map of Australia	Compton Chamberlayne	–	1917	SM 1020133	1 June 2001	
Venue-related				–	–	
Whipsnade Lion	Bedfordshire	Left facing	1933	–	–	
Other animals				–	–	
Laverstock Panda	Wiltshire	Panda face	1969			
Crosses						
Whiteleaf Cross	Buckinghamshire	–	Pre-1827	SM 1014597	26 June 1924	
Bledlow Cross	Buckinghamshire	–	Pre-1350, pre-1827	SM 1006950	?1932	
Lenham Memorial Cross	Kent	–	1922	II* 1438738	5 Dec 2017	
Shoreham Memorial Cross		–	1921	SM 1474978	21 Oct 2021	
Ditchling Cross	Plumpton, Lewes (Battle of Lewes 1264)	Cross – grassed over	Pre-1893	–	–	
Coronation				–	–	
Wye Crown, Edward VII	Kent	Crown	1902	–	–	

NHLE = National Heritage List for England

Hill figures as archaeological monuments

Michael J. Allen, Garry Gibbons and Win Scutt

Thirty-seven hill figure sites are listed in Table 22.1, of which 35 are extant. Fourteen (i.e., 39%) are protected; 13 as Scheduled Monuments and one as Listed. Two (Uffington and Westbury) are considered of such exceptional significance that they are also in the Guardianship of the Secretary of State and in the care of English Heritage. There is, however, a real problem, as Gibbons has alluded to (Chapter 20), with the archaeological perception, curation and care of hill figures. Stakeholders of specific sites have in the past attempted to liaise and share knowledge. In 1992 *The Chiltern Society* took the initiative and wrote to a number of stakeholders in an attempt to set a common 'standard of practice and maintenance', but little seems to have come of it. This perhaps exemplifies a lack of unity and commonality between hill figure stakeholders and the heritage management and curation sector. If we are to consider and treat these sites as archaeological monuments, as we may do for those of perceived antiquity (i.e., Uffington, Cerne, and Wilmington), then why is the same consideration not afforded to white horses and other hill figures? Further, hill figure monuments of antiquity are treated with as much care, and record, as we would on other archaeological sites, except for their scouring events, revels, replenishment and rechalking activities. These are often dealt with as volunteer-based purely maintenance activities with little regard to any archaeological record or sensitivity. The scouring at Cerne, for instance, has never been formally recorded. Scouring removes not only recent but also ancient chalking, which is a part of the history of the monument (e.g., see Chapter 1; and Miles *et al.* 2003, figs 5.7, 5.8 & 5.9). Scouring is ruthless and can, and has, removed certain amounts of *in-situ* archaeological deposits adjacent to the chalked outline and possibly potentially key contextual relationships, without any documentation; not even a basic photographic record. For Cerne, we could not find a single record or photograph of the removal process of scouring, of the chalk infill deposits being dug out, the depth of removal, or the trenches being emptied. The few images that do exist in the 'formal' archive of scouring are of the bosses, directors and scouring team resting on the Giant (Keithley *et al.* 1999, fig. 10). Even published photographs of the more general work (e.g., Keithley *op. cit.*, fig. 12) are not archived. The Cerne Giant is just an example and not the exception.

If the curation and preservation of the monument during these scouring activities are poorly monitored even on ancient and scheduled sites, it is no wonder that both the stakeholders and the archaeological community have less regard or respect for historic figures as a part of the preserved heritage, as we discuss below. And here we also have a dilemma: all of these monuments, ancient, historical and modern, have a history; a history that involves change to, and evolution of, the monuments' shape, nature and form. The change of the Marlborough horse from one with two to one with four legs (Fig. 20.4) is now seen as part of the evolutionary history of that figure: so too, in time, might be the addition of a saddle, blanket or rider, or even change from a mare to stallion on others. But such changes conducted today may now be seen as an act of vandalism or desecration. If so, have we fossilised the monuments and prevented any evolution of change? Something we have specifically rejoiced in, and documented, from the same monuments in the past.

It is, therefore, important and critical to survey and record both the outline of monuments and their physical topographical features (see Fig. 20.5). Surveying the earthworks associated with a hill figure, its shape, outline, form and earthworks, and micro-topography is essential for making a baseline record. None exists for the Cerne Giant for instance, nor indeed the micro-earthworks associated with Uffington. Such surveys would record the nature and condition of the monument and may go some way to helping to curate and manage their original, or extant, shape. It would also create an empirical record against which future changes can be monitored; whether natural erosion or deliberately planned and managed modifications, whether official or unofficial. Moreover they can assess the condition of the monument: the measurement, monitoring and management of weathering such as damage by natural agencies such as animals, vegetation (grass and scrub invasion) and climatic conditions.

That record does not necessarily enshrine the figure, ensure fossilisation, nor prevent change, but enables changes to be measured, monitored and managed. Here is not the place to discuss what the boundaries are, or should be, nor what, who, how or even whether they should be managed or policed. After all, some of the temporary changes and additions to the monument are fleeting, but are just as important a part of the monument's history as they are important pieces of social comment (see Fry, Chapter 13).

This is all directly relevant to their heritage value as perceived by professionals (archaeologists, curators and heritage managers) and the community at large, and which may differ between, and within, each group. We note for instance, the changes made to the 86-year-old Pewsey horse, probably in the belief that they were tidying it up. Former earthworks accentuating the head and neck which had grassed over, were partially dug out thereby destroying the evidence of its original shape and form. No formal record existed before or was made after, although the original 1937 design fortunately survives (Marples 1949, 114). However, the 'shadow' of the original horse can still be made out in a 2020 aerial photograph, underlining the dangers of maintenance when the micro-earthworks are dug out. The original designs for most sites, however, do not exist. Scouring of hill figures is a robust activity, often involving many people, sometimes untrained but willing volunteers of all ages, sometimes with mechanical assistance in the form of mechanical excavators. In all cases, if the outline is not accurately marked out under archaeological supervision, then restoration can ultimately irreversibly alter the shape and form of the figure. This is clearly seen in all the trenches at the Cerne Giant where the 2008 and 2019 rechalking was both offline and downhill of the original cut (Figs 3.14 & 3.16). At Cerne this is only a slight, almost imperceptible change. In other cases more radical changes or loss may have occurred.

Elsewhere in Wiltshire, at Cherhill a committee was formed to care for, and tend the figure, weeding, trimming and undertaking significant rechalking renovations and repairs. In contrast the care of the Pewsey horse has largely been coordinated since the mid-1960s by one man without whom it may have grassed-over and disappeared. The Codford villagers clean their ANZAC badge overlooking the A36 on an annual and communal basis. Many, if not most, figures are taken-in and adopted by their resident community. In recent years (2008 and 2019) the main scouring activity at the Cerne Giant has been undertaken by a troop of National Trust volunteers (including National Trust staff), other interested volunteers, and overseen by the National Trust area rangers. The archaeologist would visit once or possibly twice during the whole scouring process, and although the rangers overseeing the rechalking would inform them 'if they found anything of interest', they are not archaeologists and wouldn't comprehend what is archaeologically significant, nor necessarily recognise *in-situ* stratigraphy and deposits that could be cut away during robust scouring activity. At Westbury in 2023 the cleaning, pointing of gaps between the concrete slabs, repairs to the edging stones and painting were all undertaken over a 6–8 week period by specialist contractors Sally Strachey Historic Conservation, on behalf of English Heritage whose small team of dedicated volunteers explained the work to the many visitors interested in the site and its renovation.

Some sites, and particularly the Cerne Giant, attract official and unofficial attention in the form of additions, modifications and alterations – often with a social message (Fry, Chapter 13). These themselves contribute to the history and narrative of the site, and some, albeit unintentionally, have the potential to create lasting damage. However, the National Trust does not have a formal archive of all of the alterations and additions to the Giant as a record for his appropriate management and curation, and for the Trust's own historical and publicity record. As this is a Scheduled Monument (as inscribed by the Secretary of State, via their agent Historic England) one might have expected that as a part of the care, curation, and management of the hill figure, some form of 'condition' survey would have been made after any prank, or modification to identify any potential damage, and to propose or provide mitigation actions for both themselves as owners and for Historic England. This is the normal course for identified damage to any scheduled monument, and should lead to some form of evaluation to assess and determine the nature and extent of that potential damage. So whilst on the one hand there is community engagement and ownership of the monuments, on the other hand there is, at times, a lapse and *laissez faire* attitude to its monitoring of approved physical restorative activities on the monument, and both the official and unofficial activities they attract. The attitude and outlook seems distinctly different for hill figures *per se* than to, for instance, a Neolithic long barrow or Iron Age hillfort, where curation, preservation and research are far more in the foreground and are accompanied with a stronger sense of curatorial responsibility, than that which seems to be afforded to most hill figures of all ages and forms.

Archaeological record

These differing attitudes towards different monument types can also be seen in the level of archaeological research activity, and the wish, or perceived ability, to fund such operations. With the exception of the 'big three', detailed archaeological survey and background historical research is rare. Why, when the monuments are so obviously visible and so much a part of communities' value and identity, are so few a part of research projects examining and recording them in the way that many other archaeological monument types are? Why do so few have modern records of their present form (outline and topography), let alone state?

None of the 'big three' (all scheduled monuments) have been archaeologically surveyed by English Heritage/Historic England with an analytical earthwork survey, background research and the production of a *Research Report*. Few others have been treated to any formal survey and research. One exception is the scheduled monument of the Osmington White Horse, Dorset, which was surveyed by Stuart Ainsworth and Jon Horgan resulting in an English Heritage Research Report (2013) of typically high standard. This was a collaborative undertaking with the *Osmington Project* to inform the restoration of the 1808 Purbeck Limestone-cut hill figure. This included the removal of the inappropriate Portland scalpings that had been added to the monument in 1989 as a result of an attempted restoration for the *Challenge Anneka* television programme. A few trial trenches had been excavated in 2002 by AC Archaeology on the advice of English Heritage to evaluate any potential damage (Valentin 2002) prior to restoration in 2012 (in advance of the Olympics).

The only other hill figure to be examined by Historic England in this way was the 'lost' Australia badge at Compton Chamberlayne, Wiltshire. A survey was conducted by Historic England and three small trial trenches were excavated across the mainland coastline in 2018 by two Heritage at Risk project officers (Soutar 2018, 9–11, fig. 10). The survey was a part of a project to study/find the 'lost' chalk map of Australia, which was added to the Heritage at Risk Register in 2017. The map was carved on the scarp by members of the volunteer First Australian Imperial Force between 1916 and 1919. The *Rediscovering Australia Project* combined archaeological earthwork survey with aerial investigation and mapping and small-scale trial excavations to provide essential understanding of the map, its history and landscape context. This research informed the conservation of the site in 2018 by local volunteers, and will contribute positively to its future management, and enabled the site to be removed from the 2019 Heritage at Risk Register (Soutar 2018). One other site for which an outline plan with topographical survey, and documentary work has been conducted is that of the Marlborough White Horse (Gibbons 2017; Chapter 20). This is an example of a site that is still visible, while the Devizes White Horse is an example of what might be achieved at a site that is 'lost' (Gibbons 2021; Chapter 20). These comprise only a small proportion of the hill figures sites and only one of the extant Wiltshire white horses (see Table 22.1).

We can only conclude that the paucity of engagement with these well-known sites is a subconscious failure of their acceptance as true archaeological monuments and of the research potential they hold. As briefly discussed in Chapter 7, this is perhaps because these works are seen more as historic building maintenance rather than the restoration of an archaeological site with an extant and intact physical archaeological record. Whilst the completion of works on a restored building can be self-evident, that of an archaeological site requires detailed record, documentation and reporting. As such they, potentially, have just as much inherent archaeological and heritage value.

Heritage value or valued heritage?

Although clearly of great interest to the public at large and forming key destinations on the tourist map, hill figures are not, perhaps, given the same respect that is bestowed on other archaeological sites. Only very limited archaeological survey or research has been conducted on this monument type. Many are poorly served by sound documentary research. Despite this, hill figures are nearly always a living part of the locale and local community. They often abound with tales, theories, myths and legends and these have been summarised in a wide range of books over the past 100 years (e.g., Petrie 1926; Marples 1949; 1981; Morris 1981; Bergamar 1968; 1986; 1997; Newman 1987; Goodman 1998; Castleden 2000; Askew 2002). The search for creation and origin stories has been undertaken by, for instance Plenderleath (1885) nearly 150 years ago. These are important records, but are statements of their time and rife with inaccuracies.

As monuments in the public eye, their interest in terms of historical and archaeological research in general seems to be poor, save the three excavated examples, and even here this is questionable. This is admirably demonstrated by Gibbons' review of white horses (2017). The fact that archaeologists seem not to consider them, or care for them, as sites in the way they do others is exemplified by the lack of recording. Even on scheduled sites (excepting Uffington and Osmington) there is often a lack of recording during major scouring and restorations and what seems to be a lack of requirement for that record by not just the curators and owners of the monument, but of the archaeological community as a whole. Hill figures seem to be considered by many to be more like buildings that can be refurbished, rewhitened and repainted; rather than archaeological sites with intrusive and invasive potentially destructive activities which need to be recorded and reported. Many are sited in SSSIs for which a permit for any work is also required from Natural England, and their monitoring of activities and appropriate reportage required, seems greater than that for invasive archaeology.

The Cerne Giant is an example of this. From the industrial-scale scouring (digging out many tonnes of chalk and inadvertently damaging archaeological layers) of the 1956 and 1979 refurbishment to those undertaken by volunteers (1995, 2008, 2019) there are no records and reports of the invasive activity. This begs the question, when does the old weathered chalk infill itself not represent a contaminated layer set with soils and seeds, but a *bone fide* archaeological layer as carefully recorded in sections at both Cerne and Uffington? Although a specification was drawn up in 1978 for the commercial scouring at Cerne in 1979 (Appendix 3), there are no archaeological records of any of the scouring, of the removal depth, of removal of the chalk infills; no photographs, sections, descriptions, drawings, or plans that you might expect in the work on any archaeological investigation or repair of a monument such as henge banks, hill-fort ramparts, barrow mounds and ditches. Hill figures are in plain view and are some of our most visibly displayed monuments, yet clearly do not seem to be afforded the same care and record as others. Our own research at Cerne Giant to attempt to relate the archaeological record (i.e., the sections, Fig. 3.14) and in particular the chalk infills with the historically recorded scouring events, was clearly hampered/impeded by the lack of detailed records of previous actions of cutting, shoring, removing and replacing of chalk, allowing us only to make crude estimates of which deposits might relate to which events; despite the fact that most of these interventions have occurred since 1945.

The fact that these sites contain stratified deposits and valuable archaeological information could not be better exemplified than by the excellent research, excavation, analyses and publication of the Uffington White Horse (Miles *et al.* 2003 and see Miles & Palmer, Chapter 16), and also shown at Wilmington (Bell & Butler 2014; Chapter 15). Perhaps it is time that archaeologists as a community accept these as archaeological sites, as much as other monuments, and collectively ensure that records of invasive and restorative work are more carefully recorded and monitored, providing a firmer and better legacy for archaeological research in the deep future, than we have today.

National hill figure research agenda

Michael J. Allen and Win Scutt

Research agendas are particularly uncommon for this whole class of monument, despite in some cases being Scheduled Monuments or within designated landscapes (Table 22.1). The archaeological approach to these monuments seems almost as individual as the monuments themselves, with perhaps the exception for the three archaeologically excavated and dated by OSL (Uffington White Horse, Oxfordshire; Long Man of Wilmington, East Sussex; Cerne Abbas Giant, Dorset) for which similar approaches had been adopted. Although research proposals were provided for all proposed (e.g., Oxford Archaeology Unit 1988) and excavations undertaken, none, until now, have resulted in a research agenda. Admittedly many hill figures may have been considered too recent (<100 years) to warrant any field archaeological investigation but few of the early figures have had any concerted thought paid to them let alone a research agenda considered. Here we outline just seven key national research areas applicable to most hill figures; each individual monument however also needs its own specific agenda addressing local points.

1. Historical research and social history: inscription, history and maintenance

Unlike many other archaeological sites, their history, if not their origin, is often, but not always, recorded in local documentation. A detailed examination of local records will provide an outline history of the reason for their inscriptions, history of care, curation, maintenance and changes.

Many hill figures are taken for granted, and although obvious in the landscape remain unseen in terms of comprehension and history; many are under-researched and consequently remain under-valued as culturally significant monuments worthy of protection and investigation. Even small-scale investigations at a few of the Wiltshire chalk horse sites (Gibbons, Chapter 20) have demonstrated the potential of wider, systematic investigations in providing evidence against which older origin claims can be measured and by which a more nuanced understanding of chalk horses might eventually be constructed. In particular the examination of how the terms 'creation', 'loss' and 'restoration' serve to both shape and constrict familiar hill figure histories. Further, little interest has been demonstrated by the heritage sector to critique and refamiliarise the persistence of many long-standing series of creation stories that have circulated uncontested. This may call for the creation of a national database.

Some of the main research questions posed of any figure relate to:

- Date and origin stories
- Reason (who, when, why, where)
- Design
- Redesign
- Maintenance and curation, and
- Ownership (formal and societal)
- Myth, folklore and legend
- Social history of the figures

2. Accurate survey and record

2a. National physical or photogrammetric/laser survey programme

Many monuments have no accurate surveys, or if surveys do exist they may be of considerable age. The basic outline of the Cerne Giant used for the majority of publications was that accurately surveyed by Sir Flinders Petrie and published in 1926. Accurate surveys do not survive, if they ever existed, for many others.

A national programme of physical (outline and micro-topography) and condition survey should be conducted for every extant (and lost) hill figure monument to provide the basis for research, conservation, management and curation. This provides the basis for identifying and interpreting figure shapes and any associated earthworks, and provides baseline data for future repairs/maintenance. Accurate geolocation is necessary in order to create benchmarks against which future movements of the entire figure downslope can be measured.

Survey should also include a thorough review or photographic records including ground and aerial photographs, and Lidar etc., such as those from local and national photographic records (some of which may come to light in historical records and writings), so that any parchmarks and earthworks can be identified and the full context of the figures understood.

Identifying edges of the figure
Unless the figure has been replaced in concrete (Westbury White Horse, Wiltshire) or outlined in solid blocks (Long Man of Wilmington, East Sussex), then the current edges may often be unclear. Colonising vegetation and growth over the figure obscures the edges and are not always recut on the previous or original line during maintenance, scouring and rechalking. Many maintenance activities are too bold and result in lines drifting and changing shape (as seen with the Cerne Giant since 1979 whose lines have increased in places by 13–20 cm; 30–50% increase, and drifted downslope by 15 cm), while too timid maintenance and not cutting back to original line edges can result in the change in shape and overall appearance of the figure, as seen at Uffington where the horse's belly is now much narrower, giving a much sleeker figure than the original (Miles pers. comm.; Miles & Palmer, Chapter 16). Survey may, therefore, require investigative work within the confines of any heritage (Scheduled Monument) or ecological (e.g., SSSI) designations.

2b. Geophysical survey
Whilst surprisingly few figures have an accurate survey, fewer still have had any non-invasive geophysical survey. Admittedly some surveys in the past have performed very poorly (e.g., Cerne Giant, Gale 1999), which may have been due to poor/incorrect instrument set up for the conditions (Cheetham pers. comm.; Chapter 3, 2023 GPR survey). The results from targeted resistivity survey at the Long Man of Wilmington were ambiguous, but this was conducted over 50 years ago (Holden 1971, 50–1), and survey equipment, methods and interpretation have significantly improved since then. Better results were obtained from the Uffington White Horse in 1990 where total resistivity survey and magnetometry were employed across the whole figure (Payne 2003). Although the magnetometry survey was 'found to be of very limited value' (Payne 2003, 66, fig. 5.5 top), resistivity was capable of detecting difference in soil compaction and accumulation, however, it still did not provide a good plan of the figure. All three techniques (resistivity, magnetometry and GPR) should be undertaken or considered. The GPR results from the Cerne Giant clearly show great value and promise. At Westbury, for instance, a GPR survey to examine whether the former figure survives beneath the concrete veneer could be a particularly valuable asset.

2c. Whole monument survey
Surveys of whole monuments are less common, simply because of the large area they represent, and often the steep terrain in which they are situated. Nevertheless with the appropriate conditions, equipment and set up, this non-invasive survey method can be exceptionally valuable, especially in areas where the figure is not enclosed in fencing. Lidar is clearly an under-used resource in its application to hill figures.

2d. Targeted survey

Targeted survey is more common; usually addressing the presence or absence of features no longer present such as the cloak/lion skin and severed head at the Cerne Giant, or a helmet or horns, and trident/scythes on the staves of the Wilmington Long Man (Holden 1971).

Geophysical surveys can, therefore, clearly indicate lost or changed elements of the figures, but are commonly unsuccessful, poor, or unreliable. It is clear that different methods may be appropriate to different soil conditions (both between sites and moisture content within the year), and that different methods (resistivity, magnetometry/gradiometry and ground penetrating radar) may each have its own success or failure rate at individual sites. The marking of figures can vary and the equipment employed needs to clearly take this into account. Ground truthing with simple hand augering may also be advantageous (again dependent upon any necessary permissions or consents).

3. Date of creation and phases of modification

If historical research cannot define the date of inscription of the hill figure, then excavation should be considered either to date the monument directly (unlikely) or indirectly through the dating of deposits *closely* associated with the monument. Typically these deposits fall into two categories: deposits accumulating against and around the monument (local hillwash and local spoil) as at the Cerne Giant and Uffington White Horse (this volume and Miles & Palmer, Chapter 16; Miles *et al.* 2003); and colluvium at the footslope, the erosion and deposition of which may be related to the preparation of the land and construction of the figures, as at the Long Man of Wilmington (Bell & Butler 2014; Chapter 15).

Date of significant modification

Some monuments may be changed by design, whilst others are modified as result of maintenance, or lack of it. The date of deliberate alterations may be recorded in documented decisions and so too might be non-deliberate modifications such as the loss of the navel and extension of the phallus at the Cerne Giant (Grinsell 1980a).

4. Characterisation and setting

One of the only ways to obtain absolute scientific dates is through the sampling of excavated stratified deposits relating to the hill figure. These can provide detailed stratified contextual information, palaeo-environmental land-use histories as well as the scientific date. Although there is a place for small keyhole excavation, on their own they are limited and myopic in their scope. There is an obvious need for more work on the landscape context of nearly every hill figure, and associated investigation of local, potentially contemporaneous, sites.

5. Lost figures

A number of figures, particularly white horses in Wiltshire (Gibbons, Chapter 20) have been lost; mainly though neglect, which may represent or reflect complacency, fashion or change of land ownership. Figures are rarely deliberately or wantonly abandoned. Ground survey and research of historical maps, illustrations and photographs can identify the location and nature of these. More detailed investigation such as aerial photography, parchmark investigation and whole site geophysical survey may bring these back to light, or refine them where they are clearly known (e.g., Cambridgeshire Giants, Gog Magog Hills, Wandlebury).

6. Reinstating the image or lost parts

Knowing the historical record of the hill figures informs discussions of reinstating the image, or restoring lost or removed elements of the figure. Such discussions should address ethical considerations and the wishes of the local community. The changes and evolution of a hill figure are part of its history and so returning a figure to its former stance needs careful prior consideration. Moreover, if one element is reinstated, as we have seen to the Giant (Chapter 14), why not others? As we suggested for the Giant (Chapter 14), establishing a full physical (ground and topographical plan) and historic record and then the careful and considered curation and maintenance is probably a better solution.

7. A protocol for archaeological recording and monitoring during scouring, revels and maintenance events

It is clear that hill figures require maintenance and curation and many have an informal and irregular programme of scouring, revels, replenishment, rewhitening and top dressing that is both historical and individual to each monument. These can be large-scale and costly affairs, and in many cases more recently include volunteer labour and helpers. The latter shows the value and esteem that those volunteers (from the local community, monument-specific organisation or wider conservation organisation such as The National Trust) hold 'their' monument. It is clear, however, that these activities are often seen as acts of repair and refreshment, almost like a building, and that in some, not all, cases they are not treated as archaeological sites or archaeological monuments, nor monitored and supervised from an archaeological perspective, but are dealt with like other countryside activities, such as scrub clearances, with little understanding or regard for the archaeological sensitivity. This has led, in some cases, to loss of information and even of (damage to) *in-situ* archaeological deposits.

What is clearly needed is for the volunteers to be imbued with a (greater) understanding of the archaeological importance, and for these activities to be under more rigorous archaeological supervision. A protocol, or guidelines, to assist organisations managing, curating and maintaining these monuments, and written by those involved from a curatorial, archaeological and practical perspective, should be a requirement and would clearly go some way to mediating these observed and perceived problems.

APPENDICES

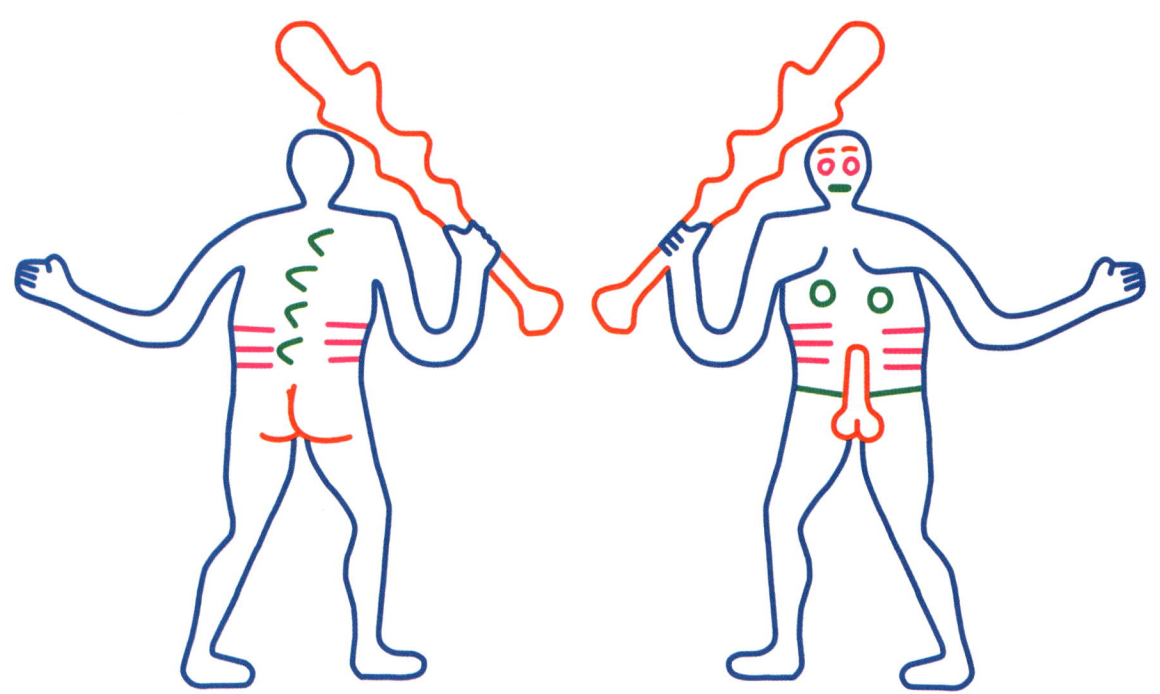

Cerne Giant 1 and Cerne Giant 2 © James Deller

APPENDIX I

Description of Giant Hill and chalk grassland vegetation; loose insert in National Trust Management Plan November 1974

The spur is largely of an open chalk grassland character in its southern parts including the Giant, with very little or no woody cover. Increasingly northwards towards the cross-dyke and banks complex, some scrub clumps and scattered woody shrubs occur. The site is bordered by taller hedges associated with the adjacent field enclosures including a thicket of emergent trees with associated shrub near the main entry point at the southernmost tip.

The site as a whole has been subjected to much surface disturbance, and is rich in archaeological artefacts, especially on the level ground of the plateau. Some chalk-quarrying has taken place along the lower slopes, principally in the SE sector.

The chalk grassland of the spur, in the immediate vicinity of the Giant, is in good condition, with as low dense sward, supporting a wide range of species. The Giant is fenced-in, with a two-strand barbed wire fence, there being insufficient space below the lower strand for sheep to move into the enclosure without obstruction. Evidence of both sheep and cattle grazing was noted throughout, and also the abundant droppings of rabbit, particularly in the areas with scrub near the cross-dykes *[may be seen everywhere]*.

In general, the vegetation and the SW–W–NW facing slope appeared to contain a greater proportion of grasses than the dryer and warmer S and SE slopes. In both however, the main species are *Festuca* spp., *Agrostis* sp., *Anthoxanthum odoratum*, *Briza mediam* and *Holcus lanatus*, with a little *Dactylis glomerata*, in places. The Dwarf Thistle (*Cirsium acaule*), is particularly abundant in dryer and steeper places, along with *Poterium sanguisorba*. The associated flora includes *Hippocrepis comosa*, *Lotus corniculatus*, *Thymus drucei*, *Trifolium patense*, *Carlina vulgaris*, *Leontodon autumnale*, *Plantago lanceolata*, *Polygala calcarea*, *Stachys betonica*, *Campanula glomerata*, and *C. rotundifolia*, *Linum cartharticum*, *Asperula cyanchica*, *Helianthemum nummularium*, *Scabiosa columbaria*, *Prunella vulgaris*, *Galium verum*, and *G. erectum*, *Plantago media*, *Euphasia offinalis*, *Carex flacca*, *Carduus lanceolatus*, and *Cirsium arvense*, *Achillea millefolia*, and some scattered *Senecio jaocbea*.

The plateau grassland, although in general, less rich in species, and possibly having been under the plough in some distant time, is not coarse, and carried a high proportion of the smaller less vigorous grasses. The general lack (or ?absence) of the two chalk grassland invaders, *Brachypodium pinnatum*, and *Bromus erectus* is a particularly noteworthy feature of the site.

APPENDIX 2

Placing Cerne Abbas 'On the Map'; Stuart Piggott's 1946 BBC radio broadcast (21 June 1946) on the theme of the Giant

Jan Lewis

When, at midsummer in 1946, archaeologist Stuart Piggott presented a 10-minute radio talk about the Cerne Abbas Giant, he was already an experienced broadcaster. Having been a regular radio contributor since 1935, Piggott was very familiar with the British Broadcasting Corporation's requirement for their expert speakers to attempt to paint pictures in sound, and to take listeners with them on an imaginative journey. Piggott was by now well-versed in the techniques broadcasters could use in order to create an informative and enjoyable radio talk. As far as BBC producers were concerned, the task of collaborating experts was not merely to describe aspects of their chosen topic, but to write and deliver vivid and evocative scripts that would bring the listener with them in their imagination, marshalling radio's nature as an intimate medium, and evoking the power of 'the mind's eye' to respond to the speaker's words. A close analysis of Piggott's talk, which is preserved on microfilm at the BBC Written Archives, provides one brief glimpse of the portrayal of archaeology to the British public at this time when radio was still the predominant form of mass media, a role that all too soon would be overtaken by television. Importantly, it also records some detail about the place of the Cerne Abbas Giant in British popular culture during the immediate post-war period, and infills a forgotten fragment of radio history.

The broadcast, transmitted on the West of England Home Service, commenced at 6:50 pm on Friday 21 June 1946. Following a brief studio rehearsal a little more than an hour beforehand, Piggott would have spoken 'live' to his early evening audience, reading a script that he had written himself, and that would have been carefully checked for duration and suitability by a Talks department producer. The opening announcement reported that this was the 12th programme in a series entitled 'On the Map', and that Piggott had recently been appointed Professor of Archaeology at Edinburgh University.

Forming part of the newly constituted post-war organisation of the BBC, the Home Service was intended to provide a diet of 'light, but informative' radio.

In common with other archaeological radio presentations of this period, the talk in question was in certain aspects designed to tap in to narratives of national identity evoking the scenic beauty and timeless appeal of the English landscape, with the factual archaeology content of the broadcast being in some ways secondary to this appeal to the cultural tropes of the 'countryside turn' and their connection to national pride. This strand of programming had first become prominent on BBC radio during the 1930s, and in addition to informing the expanding radio audience of the glories of the British countryside, set out to encourage listeners to experience the open air for themselves, an aspiration that had become more feasible due to the increased prosperity, leisure and transport opportunities that became available during the interwar period. Now, with the return of peace, listeners could return to countryside excursions, and resume engagement with aspects of the 'Deep England' that the recent conflict had sought to preserve and protect.

The topic of archaeology had regularly featured on the airwaves since the inception of regular BBC broadcasting in 1922, forming part of its diet of science communication, propaganda and light entertainment. By the early 1930s, BBC managers were on the look-out for expert contributors who could write and present the type of archaeology-themed radio content that would paint pictures in sound, and could therefore help to create the type of vivid radio talk that would appeal to listeners. In possession of vocal qualities that were regarded as pleasant enough to satisfy the demands of the microphone, and the requisite energy to combine a busy professional life with regular trips to the radio studios, Stuart Piggott amply fulfilled this brief, and had for a number of years proved a reliable source of entertaining and informative archaeology-themed broadcast content.

Starting by describing its position in the landscape, and moving on to a lyrical description of the village itself, Piggott informed listeners that:

> I remember my first sight of the village many years ago, at the end of a long day's walking from Shaftesbury along the crest of the downs, when I came over the hill eastwards of Cerne and looked down on the grey stone roofs and the great fifteenth century tower of the church. It seemed then extraordinarily withdrawn and closed in by an atmosphere of its own, in that quiet level light of a late summer's evening which picked out in bars and strips of shadow the grass-grown outlines of the prehistoric field systems on the hill slopes around.

Far from being a dispassionate and factual account of the Cerne Abbas Giant, the talk reveals something of Piggott's personal relationship with the village itself, as a location he had often enjoyed visiting over the years:

> There is a small monument in the church of which I am very fond – a brass tablet of the early eighteenth century with rather amateurish lettering recording the death of a small boy with a simple, mis-spelt quatrain – 'A little time did blast my prime / And brought me hether / The fairest flower within an hower / May fade and wether'. There is something of the haunting cadence of De La Mare in these lines of the village poet of two centuries ago.

Piggott's evocation of the popular contemporary poet Walter De La Mare, demonstrates his imperative to represent the expert archaeological contributor as possessing an eclectic appreciation of high culture, as befitted the role of the well-educated public intellectual and cultural commentator. Such touches also served to add additional resonance by anchoring the talk in the present day. It was this type of skilful harvesting of nuggets of interest that elevated a radio talk into more than merely an opportunity to relay information. This ability to forge a personal connection with the listener and to present a talk that would resonate in the imagination was the mark of the talented broadcaster, as well as helping to fulfil the established remit of the BBC to 'inform, educate and entertain'.

Having spent some time describing the village of Cerne Abbas's aesthetic and architectural appeal, Piggott now moved on to bring the listener further detail of the Giant's physical situation, leading them in their imagination past 'a range of remarkable half-timbered houses of the later fifteenth century', and up 'a lane to the Manor, to tremendous elms, huddled close together by a long-grassed graveyard with cypresses, and to a great bluff of downland which dominates the whole scene. This is the real centre of Cerne, though it lies on the edge of the village'. There is a sense that Piggott is by now almost leading the listener by the hand, towards the real nub of the matter. On he goes, briefly describing the remains of the Benedictine abbey, and progressing past the graveyard near which lies 'the Holy Well, which tradition associates with Saint Augustine', finally approaching the Giant itself and remarking that:

> It may surprise you to know that this quiet village has an international reputation among those scholars who study the ancient religions of pre-Christian Europe from the archaeological evidence and the testimony of the classical writers – to them, the Giant of Cerne ranks as one of the most astonishing survivals of primitive religion in the whole of Western Europe. What is this Giant, then? On that dominating bluff of the downs immediately above the Abbey ruins there is cut in outline through the turf and down into the solid chalk rock the representation of an enormous man, whose virile qualities are emphasised with frank realism and who brandishes over his head a huge knotted club. The slope of the hill is here very steep, and from across the valley the figure can be seen almost as if chalked on a wall, but to a fabulous scale – nearly two hundred feet from his toes to the top of his club. It is terrific, and without a parallel elsewhere. When one looks at the Giant of Cerne it is as if the centuries had been ruthlessly torn aside and one was given a sudden disconcerting glimpse of the insistent barbaric fertility cults of the prehistoric world.

Informing his audience that the Giant had been allowed to become overgrown by vegetation during the war, in case he should become a landmark for enemy aircraft, Piggott went on to refer to his recent work to restore the figure, in liaison with the National Trust. Disappointingly for the purposes of the present volume, despite noting that he was in charge of this work, and 'came to know the Giant very well as I spent day after day on the great figure', he furnishes no further detail of the exact processes carried out during this restoration work.

Having spent the majority of his talk in describing the Giant's topographical setting and physical form, towards the end of his broadcast Piggott moved on to contemplate his historical meaning and significance. He betrayed no hesitation in assigning it to the Romano-British period, and to its constituting a unique survival of pre-Christian, pagan religion. Acknowledging that 'From such a figure it is impossible to get direct archaeological evidence by excavation' he turns to historical sources to inform his analysis of the Giant's origins. In his view the figure suggests the classical god Hercules, who was a popular object of worship in the Roman provinces. Noting that 'The Celtic peoples did not as a rule, in their own art tradition, represent the figure of their gods', he interprets the Giant as a 'Celtic sky-god' assimilated into the Roman cult of Hercules, which he notes flourished in Britain during the time of the Emperor Commodus, around AD 190:

> The clumsy realism of the Cerne Giant is that of Romano-British art; the modelling of the face recalls small stone carvings of the same period. I think it reasonable to assume that the Giant represents a native, Celtic, god assimilated to Hercules after the Roman conquest. The Durotriges were the tribe occupying that part of Wessex, and at Cerne may have been one of their great cult-centres.

The broadcast concludes with a nod to the medieval legend that St Augustine had visited Cerne to 'root out a centre of paganism', meeting with a hostile reception from the locals. Piggott suggests that by substituting St Aldhelm of Sherborne for Augustus, this account may acquire historical validity. He also recounts that the medieval chronicler Walter of Coventry, having made mention of Cerne Abbas and other Dorset monasteries, remarks 'In which county they used to worship the god Helith', adding that the eighteenth-century antiquary William Stukeley recorded that the locals called the Giant 'Helis', both of which variants are related philologically to diminutives of the Hercules name that were current in the Middle Ages and referred to 'a ghostly club-bearing giant who was to be seen leading a troup [sic] of souls from the underworld'.

Having regaled his early-evening radio audience for long enough, Piggott now drew his talk to a conclusion by emphasising the lucky preservation of the Giant:

> It is an odd survival of an ancient god into Christian folklore, but scarcely less remarkable is the survival of the figure of the god himself at Cerne Abbas, still intact despite the recurrent dangers of obliteration from zealous missionaries or puritanical bigots over nearly two thousand years. Today he is a national monument and an impressive reminder of the ancient faiths of prehistoric Britain.

Acknowledgements

The author notes that all script excerpts quoted here are the copyright of the BBC, and is most grateful for the excellent help and guidance of the staff of the BBC Written Archives at Caversham near Reading, Berkshire in undertaking her research.

APPENDIX 3

National Trust Management Plan November 1974; Appendix 2, The Cerne Giant: Schedule of Works

The following shows the level of detail and care the National Trust exercised when putting the scouring and rewhitening of the Giant out to tender; points B2–8.

B2. Under the explicit instructions of the Agent of his representative cut out all loose chalk from the parts of the Giant to be indicated on the site. This cutting down to hard chalk is to be carried out as to leave a horizontal or stepped or serrated edges to form a key for the new chalk. The first operation will be trim back all the growth etc., to the outline to enable scars of non-historic portions of the outline to be seen clearly and instructions given for repair.

B3. All cutting and subsequent work must be carried out so that the outlines/ditches are not widened or damaged, and the ditch when cut out to the satisfaction of the Agent must be strictly preserved and protected from damage. Where scars and broken edges are to be seen, hardboard linings are to be provided for the purpose of rebuilding up in earth and turf to the correct outline as the same time as the new chalk is being infilled into the form of the hill figure. These scars are not expected to be many.

B4. None of the scoured chalk from the excavations of the Giant is to be reused in the new work, and it must be removed from these and disposed of in a place and in a manner to be arranged with the owner of the surrounding land.

Strong barricades of hurdles or their materials are to be fitted beneath the Giant, as directed to prevent soil from falling or washing down the hillside. At the time of tendering the Contractor will be informed whether the scoured chalk and spoil from the repairs may be spread or otherwise disposed.

B5. The new chalk which must be clean, hard and free from dirt, must be broken down into pieces roughly the size of an egg. It will then be spread in suitable areas in layers not more than 150 mm in thickness and well sprinkled with water. Each layer must be rammed with a heavy wooden punner until the whole layer is one homogenous mass free of any air pockets. Each portion other than the top layer is to be left with a very rough surface and allowed to dry before the next layer is applied. The final surfacing of the new chalk shall be rammed in very hard and finished off with a smooth surface and left to harden. It shall be suitably protected from earth from the surrounding edges and the whole area kept clean. The final surface of the chalk will be between 100–150 mm below the level of the turf according to the various parts of the hill figure under repair.

B6. The Giant is the property of the National Trust, and is a Scheduled Ancient Monument. All instructions will be given in the first place by the Trust. The staff of the Department of the Environment (Ancient Monuments Division) will visit the site during the contract and the advice issued by the Department will be complied with so far as possible and reported to the Agent.

B7. For of Tender will be as agreed with the Agent. It is essential that all items to be charged are properly set out in and covered by the tender, and no extra charges will be made except for further work carried out on the written instructions of the Agent.

B8. Upon completion. All equipment etc. shall be cleared away and the hill figure left clean and tidy and any chalk scars or overspill shall be removed from the surface of the grass. Any fencing which it is required to remove for access to the hill figure for repair shall be made good by the Contractor.

APPENDIX 4

Location of OSL sample 1

There is some ambiguity over the location of one of the OSL samples advanced for dating and its precise location makes a significant and critical difference to the date of the construction, and history, of the Giant. The sample was taken from the south section of Trench C, which was not drawn and the location projected onto the opposite section drawing (Figs 3.14 & 3.16). The location

FIGURE A4.1. a) Location of OSL sample 1 in Trench C (south side, not drawn), and b) Close-up of OSL sample 1 in the basal deposit next to the chalk infill. Images © P. Toms 2020

Context	Original wt	>4 mm	>2 mm	>1 mm	>0.5 mm	<0.5 mm
227	2000 g	1584 g, 79%	289 g, 14%	114 g, 6%	7 g, <0.5%	6 g, <0.5%

TABLE A4.1. Fractions of the sieved 2 kg bulk sample of the basal chalk rubble (227) in Trench C

of OSL sample 1 in Trench C is either from the lower chalk rubble fill (223) of the sharp edged chalk fill cut, or from the adjacent chalky downslope colluvial soil (226; Fig. A4.1a & b). Two land snail samples were also taken from this section from the colluvial soil lower down the slope (away from the area being sampled for OSL), and from the lowest upslope colluvium (227). All three of these samples were taken from the southern (north-facing) section, which unfortunately was not drawn in the field, nor was a detailed record of the OSL sample made on the context sheets, so samples had been transposed onto equivalent locations on the north (south-facing) section drawing without their *precise* location being physically recorded.

The ambiguity lies in the fact that the OSL sample was taken either in the edge of the chalky rubble of the chalked infill, or in the chalky downslope colluvial soils and weathered Cw, developed against the cut for which there is no formal archaeological record. Phil Toms feels he took it from the edge of the chalk rubble fill (223), but on-site geoarchaeological notes suggest that it was from the chalky colluvial soil immediately adjacent at 76 cm depth (226), in what was subsequently identified as a fill and chalky fill of Giant 1 and described as 'yellowish brown (10YR 4/2) silt loam with locally small and common small and subrounded stones … base of colluvial soil'. The chalk rubble fills (223), described as well-packed abundant small with common medium subangular and subrounded chalk rubble in a calcareous marl, with some voids, although loose was impossible to get a metal or plastic kubiena tin into, even cutting or forcible tapping collapsed the section or distorted the tin, while samples for snails and soil micromorphology were successfully taken from the adjacent basal deposits (226 and 227).

A 2 kg bulk sample of the chalk rubble (227), taken primarily for the acquisition of the chalk to compare with kibbled chalk (208), was sieved (no flot), and when fractioned it produced little material <0.5mm (<0.5%; (Table A4.1). The larger chalk fraction was predominantly subangular and all the residue was chalk ($CaCO_3$), which may suggest that silica/quartz would be extremely sparse. The OSL sample tube can carry about 150 g of medium chalk rubble, <8% of the bulk sample.

If OSL 1 was taken from the chalk fill, then Giant 1 is AD 700–1100, and Giant 2a existed as a trenched outline (AD 650–1310), then Giant 2b as a chalk-filled outline. Colluvium building up against the chalk stack the dates to AD 990–1510 and AD 1080–1400. Alternatively if OSL 1 was taken from the cut of Giant 1 adjacent to the chalk stack then Giant 1 is AD 650–1310 and AD 700–1100, then Giant 2 existed as trenched outline (AD 650–1310), then Giant 2b as chalk-filled outline in AD 990–1510 (see Table A4.2).

Appendix 4. Location of OSL sample 1

TABLE A4.2. The OSL results in relation to the two sample location options

Option 1	Option 2
Giant 1 128: AD 700–1100 (**910**) B, OSL 5	Giant 1 128: AD 700–1100 (**910**) B, OSL 5 226: AD 650–1310 (**980**) C, OSL 1
Post G2a post trenched Giant colluvium, 127: AD 1080–1400 (**1240**) B, OSL 6 216: AD 990–1510 (**1250**) C, OSL 3	Post G2a post trenched Giant colluvium, 127: AD 1080–1400 (**1240**) B, OSL 6 216: AD 990–1510 (**1250**) C, OSL 3
Giant 2b Chalk stack (in fill of open trench) 223: AD 650–1310 (**980**) C, OSL 1	

The following is a list of stratigraphic events indicating which have been dated OSL:

Stratigraphic sequence of events	*OSL samples*
Colluvial soil	
G1 cut G1 fills up	OSL 1 & 5
Initial colluvium	
Trench cut and open	
colluvium	
Trench scoured and open	
Colluvium	
Trench scoured and initial chalk but trench open	
Colluvium	OSL 3 & 6
Trench soured and filled with chalk	[OSL 1]

Bibliography

Unless otherwise indicated, sources will be found in the National Trust Cerne Giant files 1945–1992 (NT CGa) and the British Newspaper Library (BNL). Only Newspaper (and similar) references that comprise a titled article are cited in the bibliography, others are only cited in text with full date of publication, or page number.

National Trust files (archive)

NTf 1. Letters Leslie Grinsell to Andrew Saunders, 14 August 1979; Gerald Pitman to Martin Papworth, 20 September 1994

NTf 2. Memo Knollys to Simon Buxton, 6 October 1948

National Trust, Dorset Cerne Abbas, Archaeology Reports and Correspondence Files, Tisbury Hub, Court Street, Tisbury, Wiltshire

Bibliography

Able Smith, J. 2015. *Wilmington Priory History Album*. Shottesbrooke: Landmark Trust. Available online: https://www.landmarktrust.org.uk/globalassets/3.-images-and-documents-to-keep/history-albums/wilmington-priory-history-album.pdf

Adamiec, G. & Aitken, M.J. 1998. Dose-rate conversion factors: new data. *Ancient TL* 16, 37–50

Ainsworth, S. & Horgan, J. 2013. *A regal restoration: surveyed and historical evidence for the re-established outline*. English Heritage Research Report Series no. 27–2013. Portsmouth: English Heritage

Aitken, M.J. 1985. *Thermoluminescence Dating*. New York: Academic Press

Aitken, M.J. 1998. *An Introduction to Optical Dating: The Dating of Quaternary Sediments by the Use of Photon-stimulated Luminescence*. Oxford: Oxford University Press

Allen, M.J. 1988. Archaeological and environmental aspects of colluviation in south-east England, in Groenman-van Waateringe, W. & Robinson, M. (eds), *Man-Made Soils*, 67–92. British Archaeological Report S410

Allen, M.J. 1991. Analysing the landscape: a geographical approach to archaeological problems, in Schofield, A.J. (ed.), *Interpreting Artefact Scatters: Contributions to Ploughzone Archaeology*, 39–57. Oxford: Oxbow Monograph 4

Allen, M.J. 1992. Products of erosion and the prehistoric land-use of the Wessex chalk, in Bell, M.G. & Boardman, J. (eds), *Past and Present Soil Erosion: Archaeological and Geographical Perspectives*, 37–52. Oxford: Oxbow Books

Allen, M.J. 2005. Beaker occupation and development of the downland landscape at Ashcombe Bottom, nr Lewes, East Sussex. *Sussex Archaeological Collections* 143, 7–33

Allen, M.J. 2007. Evidence of the prehistoric and medieval environment of Old Town, Eastbourne: studies of hillwash in the Bourne valley, Star Brewery Site. *Sussex Archaeological Collections* 145, 33–66

Allen, M.J. 2017a. Land snails in archaeology, in Allen, M.J. (ed.), *Molluscs in Archaeology: Methods, Approaches and Applications*, 6–29. Studying Scientific Archaeology 3. Oxford: Oxbow Books

Allen, M.J. 2017b. The geoarchaeology of context: sampling for land snails, in Allen, M.J. (ed.), *Molluscs in Archaeology: Methods, Approaches and Applications*, 30–47. Studying Scientific Archaeology 3. Oxford: Oxbow Books

Allen, M.J. 2019. Cerne Abbas Giant, Dorset; Dating the Giant and other research 1: notes on walkover survey and preliminary research framework. Unpubl. report version AEA 414.01.01, dated 15 September 2019. National Trust Archive, Tisbury Wiltshire

Allen, M.J. 2020a. Cerne Abbas Giant, Dorset; Dating the Giant and other Research 2: Project Design and Written Scheme of Investigation, Unpubl. report for The National Trust and Historic England. Report AEA 414.02.03, dated 24 February 2020, revised 2 March 2020. National Trust Archive, Tisbury Wiltshire

Allen, M.J. 2020b. Cerne Abbas (AEA 414): geoarchaeology, OSL and palaeo-environmental sampling – the geoarchaeology and OSL sampling of the Giant. Unpublished report for the National Trust AEA 414.05.12, dated 19 May 2020, revised 22 May 2020. National Trust Archive, Tisbury Wiltshire

Allen, M.J. 2023a. Cerne Abbas (AEA 414): Soils of Giant Hill. Nature of the soils and soil depth on the scarp slope of Giant Hill adjacent to the Cerne Giant Scheduled Monument. Report for David Charman, Natural England and Martin Papworth, National Trust, version AEA 414.12.1 dated 25 September 2023

Allen, M.J. 2023b. Cerne Abbas Giant, Dorset: Geophysical Survey 2023 (GPR and resistivity): preliminary report of survey activity. Unpubl. AEA report version AEA 414.12.02 dated 12 December 2023, revised 13 December 2023. Available at the National Trust, Tisbury

Allen, M.J. 2024. Cerne Abbas Giant, Dorset (SM 1003202): Investigations of the putative belt; preliminary geophysical survey results, and full auger survey results. Unpubl. AEA report version AEA 414.13.02, dated 16 January 2024. Available at Historic England and National Trust Archive, Tisbury, Wiltshire

Anderson, O.S. 1939. *The English Hundred-Names: The South-Eastern Counties*. Lunds Universitets Arsskrift 37(1). Lund: Hakan Ohlsson

Anderson, R. 2005. An annotated list of the non-marine Mollusca of Britain and Ireland. *Journal of Conchology* 38 (6), 607–37

Anon. [Swift, J.] 1726. *Travels into Several Remote Nations of the World*. London: Benj. Motte

Anon. 1763a. [Anonymous letter]. *Royal Magazine* 9 (September), 140–1

Anon. 1763b. [Anonymous letter]. *St. James's Chronicle* (Tuesday 4 October–Thursday 6 October 1763), 404

Anon. [possibly Stukeley] 1764. Description of a gigantic figure. *Gentleman's Magazine* 34 (August), 335–6

Anon. 1799. *Monthly Magazine* 1, 418

Anon. 1827. Untitled note on the Bledlow Cross. *Gentleman's Magazine* 97, pt. 2, 29

Anon. ('PPP'). 1854. The Shepherd's Grave. *Records of Buckinghamshire* 1, 124

Anon. 1924. The Giant of Cerne Abbas. *Country Life* (18 October 1924), 612

Anon. 1937. The Cerne Giant. *Nature* 139, 876

Anon. 1956. Untitled article in *Cerne and Up Cerne, Godmanston and Nether Cerne Parish Magazine*, June 1956, new edition no. 103. NT Box 12979

ApSimon A.M. 1964. The Roman Temple on Brean Down, Somerset. *Proceedings of University of Bristol Spelaeological Society* 10 (3), 195–258

Archer, H.G. 1903. White Horses. *Good Words* 44, 195 (photograph by W. Pouncy, c. 1902–3)

Ashley, S. 2007. The lay intellectual in Anglo-Saxon England: Ealdorman Æthelweard and the politics of history, in Wormald, P. & Nelson, J. (eds), *Lay Intellectuals in the Carolingian World*, 218–45. Cambridge: Cambridge University Press

Askew, M. 2002. *Hill Figures of England*. Ramsbury: Crowood Press

Assize Rolls – for etymology and historical mentions: https://epns.nottingham.ac.uk/browse/Dorset/Cerne+Abbas/532852f9b47fc4099d002af7-Cerne+Abbas

Attenborough, F.L. 1922. *The Laws of the Earliest English Kings*. Cambridge: Cambridge University Press

Attree, F.W.T. & Booker, J.H.L. 1904. The Sussex Colepepers. *Sussex Archaeological Collections* 47, 47–81

Bacon, F. 1620. Novum Organum. Reprinted in Burtt, E.A. (ed.), 1939. *The English Philosophers from Bacon to Mill*, 24–87. New York: Random House

Baines, A.H.J. 1981. The boundaries of Monks Risborough. *Records of Buckinghamshire* 23, 76–101

Baker, A. 1856. Ancient crosses incised on the Chiltern Hills at Monks Risborough and Bledlow. *Records of Buckinghamshire* 1, 219–24

Baker, J. 2019. Meeting in the shadow of heroes? Personal names and assembly places, in Carroll, J., Reynolds, A. & Yorke, B.A.E. (eds), *Power and Place in Europe in the Early Middle Ages*, 37–63. Oxford: British Academy

Baker, J. & Brookes, S. 2013. Monumentalising the political landscape: a special class of Anglo-Saxon assembly-sites. *Antiquaries Journal* 94, 147–62

Baker, J. & Brookes, S. 2015a. Explaining Anglo-Saxon military efficiency: the landscape of mobilisation. *Anglo-Saxon England* 44, 221–58

Baker, J. & Brookes, S. 2015b. Identifying outdoor assembly sites in early medieval England. *Journal of Field Archaeology* 40.1, 3–21

Barbados Agricultural Reporter 1905. Vanishing Giant: a relic of prehistoric England. *Barbados Agricultural Reporter*, 28 September 1905, 3

Barker, K. 1988a. Introduction, in Barker, K., (ed.), *The Cerne Abbey Millennium Lectures*, vii–ix. Cerne Abbas: The Cerne Abbey Millennium Committee

Barker, K. 1988b. Ælfric the Mass-priest and the Anglo-Saxon Estates of Cerne abbey, in Barker, K. (ed.), *The Cerne Abbey Millennium Lectures*, 27–42. Cerne Abbas: The Cerne Abbey Millennium Committee

Barker, K. (ed.), 1988c. *The Cerne Abbey Millennium Lectures*. Cerne Abbas: The Cerne Abbey Millennium Committee

Barker, K. 1997. Brief encounter: the Cerne Abbas Giantess project, summer 1997. *Proceedings of the Dorset Natural History and Archaeological Society* 119, 179–82

Barker, K. 1999. Medieval Giants: bible, history, heritage and landscape, in Darvill, T., Barker, K., Bender, B. & Hutton, R., *The Cerne Giant: An Antiquity on Trial*, 88–108. Oxford: Oxbow Books

Barker, K. 2005a. *Anni Domini Computati*: the Sherborne Benedictine Millennium, in Barker, K., Hinton, D.A. & Hunt, A. (eds), *St Wulfsige of Sherborne: Essays to Celebrate the Millennium of the Benedictine Abbey*, 40–52. Oxford: Oxbow Books

Barker, K. 2005b. Bishop Wulfsige's lifetime: Viking campaigns recorded in the Anglo-Saxon Chronicles for southern England, in Barker, K., Hinton, D.A. & Hunt, A. (eds), *St Wulfsige of Sherborne: Essays to Celebrate the Millennium of the Benedictine Abbey*, 124–32. Oxford: Oxbow Books

Barker, K. & Darvill, T. 1999. The Giant on trial and the assessors' report, in Darvill, T., Barker, K., Bender, B. & Hutton, R., *The Cerne Giant: An Antiquity on Trial*, 162–5. Oxford: Oxbow Books

Barnard, J. 2015. Dryden's Virgil (1697): gatherings and politics. *The Papers of the Bibliographical Society of America* 109:1, 131–9

Barraclough, E.M.C. 1969. The flags of the Bayeux Tapestry. *Armi Antichi* 1, 117–24

Baxter, S.B. 1992. William III as Hercules: the political implications of court culture, in Schwoerer, L.G. (ed.), *The Revolution of 1688–1689: Changing Perspectives*, 95–106. Cambridge: Cambridge University Press

BBC News 2013. Cerne Abbas Giant sports moustache for Movember. *BBC News*, 1 November 2013. Available at https://www.bbc.co.uk/news/uk-england-dorset-24772626 (Accessed 14 November 2023)

BBC News 2019. Cerne Abbas Giant's manhood given floral makeover. *BBC News*, 8 March 2019, https://www.bbc.co.uk/news/uk-england-dorset-47493907 (Accessed 14 November 2023)

BBC TV 1996. *The Trial of the Cerne Abbas Giant*, Close-Up West. Broadcast on BBC 2 May 1996

Beeston, H. 1999. Making 'the Trial of the Cerne Abbas Giant', in Darvill, T., Barker, K., Bender, B. & Hutton, R., *The Cerne Giant: An Antiquity on Trial*, 166–72. Oxford: Oxbow Books

Bell, M.G. 1977. Excavations at Bishopstone. *Sussex Archaeological Collections* 115

Bell, M.G. 1983. Valley sediments as evidence of prehistoric land-use on the South Downs. *Proceedings of the Prehistoric Society* 49, 119–50

Bell, M. & Butler, C. 2014. The Long Man of Wilmington: and interim report on a giant conundrum, in Allen, M.J. (ed.), *Eastbourne: Aspects of Archaeology, History and Heritage*, 18–30. Eastbourne: Eastbourne Natural History & Archaeological Society

Bell, M. & Hutton, R. 2004. Not so long ago. *British Archaeology* 77, 17–21

Bender, B. 1999. A living Giant, in Darvill, T., Barker, K., Bender, B. & Hutton, R., *The Cerne Giant: An Antiquity on Trial*, 126–9. Oxford: Oxbow Books

Bergamar, K. 1968. *Discovering Hill Figures*. Aylesbury: Shire Publications

Bergamar, K. 1986. *Discovering Hill Figures*. 3rd edn. Dyfed: C.I. Thomas & Sons

Bergamar, K. 1997. *Discovering Hill Figures*. 4th edn. Princes Risborough: Shire Books

Berger, G.W., Mulhern, P.J. & Huntley, D.J. 1980. Isolation of silt-sized quartz from sediments. *Ancient TL* 11, 147–52

Berry, S. 2021. *Country Houses of the South Downs*. Lewes: Sussex Archaeological Society

Beswick, M. 2001. *Brickmaking in Sussex: A History and Gazetteer*. Midhurst: Middleton Press

Bettey, J.H. 1981. The Cerne Abbas Giant: the documentary evidence. *Antiquity* 55, (July) 118–21

Bettey, J.H. 1988. The Dissolution and after at Cerne Abbas, in Barker, K. (ed.), *The Cerne Abbey Millennium Lectures*, 43–53. Cerne Abbas: The Cerne Abbey Millennium Committee

Bettey, J. 1999. The Cerne Giant revisited, in Darvill, T., Barker, K., Bender, B. & Hutton, R., *The Cerne Giant: An Antiquity on Trial*, 75–88. Oxford: Oxbow Books

Billingsley, D. 2016. Instances and contexts of the head motif in Britain, in Hutton, R. (ed.), *Physical Evidence for Ritual Acts, Sorcery and Witchcraft in Christian Britain*, 68–90. London: Palgrave Macmillan

Bird, E. 1995. *Geology and Scenery of Dorset*. Bradford upon Avon: Ex Libris Press

Bishop, G. 2021. In search of an Abbey, *Cerne Historical Society Magazine*, Spring 2021, 4–9

Bolton, T. 2017. *Cnut the Great*. New Haven and London: Yale University Press

Bond, D. 1982. An examination of a scheduled area and fields at Black Hill, Cerne Abbas undated earthworks. *Proceedings of the Dorset Natural History & Archaeological Society* 104, 67–70

Booth, W. 2021. Scientists unravel a mystery about a naked giant carved into an English hill. *Washington Post* 17 May 2021. Available at https://www.

washingtonpost.com/world/europe/cerne-giant-hill-figure/2021/05/16/4203c15a-b407-11eb-bc96-fdf55de43bef_story.html (Accessed 11 August 2023)

Bøtter-Jensen, L., McKeever, S.W.S. & Wintle, A.G. 2003. *Optically Stimulated Luminescence Dosimetry*. Amsterdam: Elsevier

Bøtter-Jensen, L., Mejdahl, V. & Murray, A.S. 1999. New light on OSL. *Quaternary Science Reviews* 18, 303–10

Bouchardon, E. 1741. *L'anatomie nécessaire pour l'usage du dessein*. Paris: Gabriel Huquier

Bournemouth University 2014. Site of Benedictine Abbey, Cerne Abbas, Dorset. Archaeological Geophysics. Unpublished Centre for Archaeology, Anthropology and Heritage Report No. 0071

Bowden, M. 1991. *Pitt Rivers: The Life and Archaeological Work of Lieutenant-General August Henry Lane Fox Pitt Rivers, DCL, FRS, FSA*. Cambridge: Cambridge University Press

Boyer, T.M. 2016. *The Giant Hero in Medieval Literature*. Leiden: Brill

Bradby, E. 1984. Edward Kite, antiquary of Devizes (1832–1930). *Wiltshire Archaeological & Natural History Magazine* 78, 82–6

Breeze, A. 2005. Where were Bede's Uilfaresdun and Paegnalaech? *Northern History* 42, 189–91

Bridge, M. 2021. Naughty noble 'dropped Cerne Abbas giant's trousers for a laugh in 1600s'. *The Times* 12 May 2021. Available at https://www.thetimes.co.uk/article/naughty-noble-dropped-cerne-abbas-giant-s-trousers-for-a-laugh-in-1600s-fbn9tllss (Accessed 23 November 2023)

Bridport News 1868a. The Giant, *Bridport News*, 14 March 1868, 3

Bridport News 1868b. Cerne Abbas and its antiquities: The Abbey, The Giant, etc. *Bridport News* four-part serial 28 March–18 April 1868

Bridport News 1886. The Giant, *Bridport News* 23 July 1886, 5

Bridport News 1888. The Giant Refreshed, *Bridport News* 18 October 1888, 5

Bridport News 1899. Cerne Echoes, *Bridport News* 15 November 1899, 5

Britton, J. & Brayley, E.B. 1803. *The Beauties of England and Wales*, Vol. IV of IV, 483. London: Vernor & Hood, Longman & Rees, Cuthell & Martin, J. & A. Arch, W.J. & J. Richardson, J. Harris, and E. Crosby

Brookes, S. 2020. Domesday Shires and Hundreds of England. Available at https://doi.org/10.5284/1058999 (Accessed December 2023)

Brookes, S. 2023. Assembly practices in 10th-century England: continuities and innovations in military mobilisation, in Santos Salazar, I. & Tente, C. (eds), *The 10th Century in Western Europe. Change and Continuity*, 53–63. Oxford: Archaeopress

Brown, M.P. 1996. *The Book of Cerne: Prayer, Patronage and Power in Ninth-century England*. London: The British Library

Brown, M. 2021. Man of the moment. *Archaeology*, September/October 2021. https://www.archaeology.org/issues/439-2109/digs/9916-digs-england-cerne-abbas-giant

Brundle, L.M. 2019. *Image and Performance, Agency and Ideology: Representations of the Human Figure in Funerary Contexts in Anglo-Saxon Art, AD 400–680*. Oxford: British Archaeological Reports (British Series) 645

Brundle, L. 2020. Exposed and concealed bodies: exploring the body as subject in metalwork and text in 7th–century Anglo-Saxon England. *Medieval Archaeology* 62.2, 193–225

Bryson, B. 2000. *The English Landscape*. London: Profile Books

Budgen, W. 1928. Wilmington Priory: historical notes. *Sussex Archaeological Collections* 69, 29–52

Camden, W. 1607. *Britannia*. London

Camden, W. 1637. *Britannia*. London

Camden, W. 1789a. *Camden's Britannia*, Vol. 1 (of 3), ed. R. Gough. London: John Nichols for T. Payne & Son

Camden, W. 1789b. *Camden's Britannia*, 2nd edn, Vol. 1 (of 4), ed. G.G.J. Robinson & J. Robinson. London: John Stockdale for J. Nichols & Son

Cameron, R.A.D. & Morgan-Huws, D.I. 1975. Snail faunas in the early stages of a chalk grassland succession. *Biological Journal of the Linnean Society* 7, 215–29

Campbell, A. 1962. *Chronicon Æthelweardi. The Chronicle of Æthelweard*. London: Thomas Nelson & Sons

Campion, N. 2005. 'Thousand is a perfect number …' quoth Aelfric of Cerne, in Barker, K., Hinton, D.A. & Hunt, A. (eds), *St Wulfsige of Sherborne: Essays to Celebrate the Millennium of the Benedictine Abbey*, 33–9. Oxford: Oxbow Books

Carpenter, E., Barber, M. & Small, F. 2013. South Downs, Beachy Head to the River Ouse: Aerial investigation and mapping project. Swindon: English Heritage Research Report Series 22–2013

Cartwright, J.J. 1888. *The Travels through England of Dr Richard Pococke, Vol. II*, 143–4. London: Camden Society, new series XLIV

Castleden, R. 1983. *The Wilmington Giant: The Quest for a Lost Myth*. Wellingborough: Turnstone Press

Castleden, R. 1993. Report on the geophysical survey of the Cerne Giant 1989–90. Unpubl. ms, dated September 1993. Available at Historic England (the Inspectorate) and the National Trust, Tisbury

Castleden, R. 1995a. The Cerne Giant Project, Phase 2: surveys undertaken in 1995. Unpubl. ms, dated December 1995. Available at Historic England (the Inspectorate) and the National Trust, Tisbury

Castleden, R. 1995b. The Cerne Giant Project, Phase 3: Proposals for 1996. Research Proposals for English Heritage and the National Trust. Unpubl. ms, dated December 1995. Available at Historic England (the Inspectorate) and the National Trust, Tisbury

Castleden, R. 1996. *The Cerne Giant.* Wincanton: Dorset Publishing Company

Castleden, R. 1999. Iconography and the identity of the Giant, in Darvill, T., Barker, K., Bender, B. & Hutton, R., *The Cerne Giant: An Antiquity on Trial*, 43–51. Oxford: Oxbow Books

Castleden, R. 2000. *Ancient British Hill Figures*. Seaford: SB Publications

Castleden, R. 2002. Shape-shifting: the changing outline of the Long Man of Wilmington. *Sussex Archaeological Collections* 140, 81–95

Castleden, R. 2021. The Long Man of Wilmington: memorialising a Sussex martyr? *Current Archaeology* 379, 47–9

Chambers, R.W. 1967. *Beowulf. An Introduction to the Study of the Poem.* Cambridge: Cambridge University Press

Chandler, J. 1993. *Leland's Itinerary; Travels in Tudor England.* Stroud: Alan Sutton

Chappell, H.G., Ainsworth, J.F., Cameron, R.A.D. & Redfern, M. 971. The effect of trampling on a chalk grassland ecosystem. *Journal of Applied Ecology* 8, 869–82

Charman, J. 2020. The Geology of Cerne and its impact on village life. Available at (Accessed 5 July 2023)

Chartrand, J. 1999. The Cerne Giant: a place in space, in Darvill, T., Barker, K., Bender, B. & Hutton, R., *The Cerne Giant: An Antiquity on Trial*, 2–6. Oxford: Oxbow Books

Chase, C. 1981. *The Dating of Beowulf*. Toronto: University of Toronto Press

Cheetham, P.N. 2001. A symmetrical response, high resolution and reduced near-surface noise multiple potential electrode earth resistivity array for archaeological area survey, in Doneus, M., Edler-Hinterleitner, A. & Neubauer, W. (eds), *Archaeological Prospection: Fourth International Conference on Archaeological Prospection*, 79–80. Vienna: Austrian Academy of Sciences

Cheetham, P. 2003. A practical and effective, high-resolution, symmetrical response and reduced near-surface noise multiple potential electrode resistivity array configuration for the detection of clandestine graves, in Pye, K. & Croft, D. (eds), *Forensic Geoscience: Principles, Techniques and Applications*, 17. Conference Abstracts, The Geological Society, London, 3–4 March 2003

Cheetham, P. 2014. Mary, Mary, quite unwary, what could your garden geophysical survey show?, in *Recent Work in Archaeological Geophysics*, 40–2. Conference Abstracts, English Heritage & The Geological Society, London, 2 December 2014

Cheetham, P. 2024. Geophysical Surveys of Aspects of the Cerne Abbas Giant undertaken in December 2023. Unpubl. Section 42 licence report for Historic England dated 1st July 2024. Bournemouth University

Childe, V.G.C. 1938. The Orient and Europe. *American Journal of Archaeology* 43(1), 10–26

Churchill, N. 2007. Giant opportunity? *Bournemouth Echo*, 18 July 2007. Available at https://www.bournemouthecho.co.uk/news/1552290.giant-opportunity/ (Accessed 14 November 2023)

Clark, A.J. 1980. Archaeological detection by resistivity, Unpubl. PhD thesis, University of Southampton

Clark, A.J. 1983. The Cerne Giant. *Archaeological Journal* 140, 29–30

Clark, G. 1940. *Prehistoric England*. London: Batsford

Clark, J. 2016. Trojans at Totnes and giants on the Hoe: Geoffrey of Monmouth, historical fiction and geographic reality. *Report and Transactions of the Devonshire Association for the Advancement of Science* 148, 89–130

Clark, M. 2020. Another mystery. *Cerne Historical Society Magazine*, 12 June, 12–14

Clemoes, P. 1997. *Ælfric's Catholic Homilies. The First Series: Text*. Early English Text Society. Supplementary series 17. London: Oxford University Press

Clinch, G. 1905. Early Man, in Page, W. (ed.), *Victoria History of the County of Buckinghamshire,* 177–94. London: Institute of Historical Research

Coates, R. 1978. The linguistic status of the Wandlebury Giants. *Folklore* 89.1, 75–8

Coates, R. 2011. Some local place-names in medieval and early-modern Bristol. *Transactions of the Bristol & Gloucestershire Archaeological Society* 129, 155–96

Coker, J. 1732. *Survey of Dorsetshire; containing the antiquities and natural history of the county with a particular description of all the places of note and ancient seats …* London: J. Wilson

Cohen, G.J. 1999. Of Giants: sex, monsters, and the Middle Ages. *Medieval Cultures* 17, 29–61. Minneapolis and London: University of Minnesota Press

Colgrave, B. & Mynors, R.A.B. (eds), 1969. *Bede's Ecclesiastical History of the English People*. Oxford: Oxford University Press

Colley, L. 2010. Little Englander Histories. *London Review of Books* 32(14), 12–14

Cooke, G.A. 1807. *Topographical and Statistical Description of the County of Dorset*. London: G.A. Cooke

Cooper, G.M. 1851. Wilmington Priory and Church. *Sussex Archaeological Collections* 4, 37–63

Copson, C. 1988. Giant conundrum. *Dorset County Magazine* 2 (June), 13–15

Copson, C. 2022. Medieval encaustic tiles from the site of Cerne Abbey. *Cerne Historical Society Magazine*, Autumn 2022, 15–16

Coulthard, S. 2020. *A Short History of the World According to Sheep*. London: Head of Zeus

Country Life 1924. The Giant of Cerne Abbas. *Country Life*, 56, no. 1450, 18 September 1924, 612

Crawford, O.G.S. 1922. Former white horse at Ham. *Wiltshire Archaeological & Natural History Magazine* 42, 73

Crawford, O.G.S. 1929. The Giant of Cerne and other hill-figures. *Antiquity* 3, 277–82

Creighton, J. 2000. *Coins and Power in Late Iron Age Britain*. Cambridge: Cambridge University Press

Cressy, D. 1992. The fifth of November remembered, in Porter, R. (ed.) *Myths of the English*, 68–90. Oxford: Polity Press

Cubitt, C. 2009. Ælfric's lay patrons, in Magennis, H. & Swan, M. (eds), *A Companion to Ælfric*, 165–92. Leiden: Brill

Curwen, E.C. 1928. The antiquities of Windover Hill. *Sussex Archaeological Collections* 69, 92–101

Curwen, E.C. 1954. *The Archaeology of Sussex*. 2nd edn. London: Methuen

Daily Mail 1905. Cerne Abbas. *Daily Mail*, 22 August 1905

Daily Mirror 1989. Private Parts. *Daily Mirror*, Tuesday 29 August 1989, 13

Daily Telegraph 2023. Trust says sorry as chalk giant hidden by grass. *Daily Telegraph*, 7 October 2023

Damon, R.F. 1884. *Geology of Weymouth, Portland and the Cast of Dorsetshire, from Swanage to Bridport-on-the-sea; With Natural History and Archaeology Notes*. 2nd edn. Weymouth: R.F. Damon.

Daniel, G. 1976. Editorial. *Antiquity* 50 (March), 93–4

Darton, F.J.H. 1922. *Marches of Wessex: A Chronicle of England*. 2nd edn. London: Nisbet & Co.

Darton, F.J.H. 1935. Appendix, The Giant of Cerne, in *English Fabric: A Study of Village Life*, 319–31. London: G: Newnes

Darvill, T. 1999a. The Cerne Giant: an English hill-figure, in Darvill, T., Barker, K., Bender, B. & Hutton, R., *The Cerne Giant: An Antiquity on Trial*, 7–17. Oxford: Oxbow Books

Darvill, T. 1999b. A prehistoric warrior-god?, in Darvill, T., Barker, K., Bender, B. & Hutton, R., *The Cerne Giant: An Antiquity on Trial*. 29–31. Oxford: Oxbow Books

Darvill, T. & Barker, K. 1999. Preface, in Darvill, T., Barker, K., Bender, B. & Hutton, R., *The Cerne Giant: An Antiquity on Trial*, v–xi. Oxford: Oxbow Books

Darvill, T., Barker, K., Bender, B. & Hutton, R. 1999. *The Cerne Giant: An Antiquity on Trial*. Oxford: Oxbow Books

Davidson, A. 2010. Henry Compton, in Thrush, A. & Ferris, J.P. (eds), *History of Parliament: The House of Commons 1604–1629*. Cambridge: Cambridge University Press

Davidson, H.E. 1958. Weland the smith. *Folklore* 69, 145–59

Davis, D.J. 2015. *From Icons to Idols: Documents on the Image Debate in Reformation England*, 4. Cambridge: James Clarke & Co.

Davis, J. 1990. A lost white horse on Pitstone Hill? *Records of Buckinghamshire* 32, 151

Dearsley, W.A. St John 1890. Wilmington Giant. Report of the Rev. W.A. St John Dearsley on Mr F. Russell's work in restoring a portion of the above. Manuscript in Sussex Archaeological Society Library Security Cupboard

Dewar, H.S.L. 1968. *The Giant of Cerne Abbas*. Toucan: West Country Folklore Series no. 1

Doel, F. & Doel, G. 2007. *Folklore of Dorset*. Stroud: Tempus

Dorset County Chronicle 1835. Dorsetshire: Cricketing. *Dorset County Chronicle*, Thursday 25 June 1835, 3. col 1

Dorset County Chronicle 1920. The Cerne Park Races. *Dorset County Chronicle*, 23 September 1920, 2, col 5

Dorset County Express and Agricultural Gazette 1867. Forester's Fete at Cerne Abbas. *Dorset County Express*, 16 July 1867, 2, col 3

Dorset Echo 2007. Homer sets up a 'quickie mart' in Cerne Abbas. *Dorset Echo*, 17 July 2007. Available at https://www.dorsetecho.co.uk/news/1549960.homer-sets-up-a-quickie-mart-in-cerne-abbas/ (Accessed 23 June 2023)

Dorset History Centre 1768a. Survey by Benjamin Pryce with image of Giant. D-PIT/P/5

Dorset History Centre 1768b. Map of Cerne Abbas based on Benjamin Pryce 1768 but with cartouche c. 1798 and a different but deteriorating image of Giant. D-PIT/P/6

Dorset History Centre 1884. Map of Cerne Abbas 1884 by A.H. Green with image of Giant without genitalia. D-PIT/P/8

Dorset History Centre: RON/2/2/Cerne Abbas/149, Newspaper article re: Giant clothed in camouflage of vegetation by Home Guard to prevent German bombers using it as guideline for Bristol

Douglas, D.C. & Greenaway, G.W. 1981. *English Historical Documents 1042–1189*. London: Eyre Methuen

Downey, R., King, A. & Soffe, G. 1979. *The Hayling Island Temple; third interim report on the excavation of the Iron Age and Roman temple 1976–78*. Privately printed

Drower, M.S. 1995. *Flinders Petrie: A Life in Archaeology*. Madison, WI: University of Wisconsin Press

Dryden, J. 1697. *The works of Virgil containing his Pastorals, Georgics, and Aeneas: adorn'd with a hundred sculptures translated into English verse by Mr. Dryden*. London: for Jacob Tonson

Duell, M. 2020. National Trust blasts PR team behind new Borat film for 'defacing' protected Cerne Abbas Giant chalk figure with a MANKINI. *MailOnline*, 23 October 2020. Available at https://www.dailymail.co.uk/news/article-8872185/National-Trust-blasts-Borat-PR-team-defacing-Cerne-Abbas-Giant.html (Accessed 14 November 2023)

Duller, G.A.T. 2003. Distinguishing quartz and feldspar in single grain luminescence measurements. *Radiation Measurements* 37, 161–5

Duller, G.A.T., Bøtter-Jensen, L., Kohsiek, P. & Murray, A.S. 1999. A high sensitivity optically stimulated luminescence scanning system for measurement of single sand-sized grains. *Radiation Protection Dosimetry* 84, 325–30

Eagles, B. 1994. Archaeological evidence for settlement in the fifth to seventh centuries AD, in Aston, M. & Lewis, C. (eds), *The Medieval Landscape of Wessex*, 15–32. Oxford: Oxbow Books

Edwards, B. 2005. The scouring of the White Horse Country. *Wiltshire Archaeological & Natural History Magazine* 98, 90–127

Edwards, B. 2007. A seventeenth century Wiltshire white horse? Some thoughts on narrowing the field at the start line. *Regional Historian* 17, 17–21

Edwards, B. 2014. Contemporary recollections and later traditions of the Cherhill White Horse. *Wiltshire Archaeological & Natural History Magazine* 107, 204–18

Edwards, B. 2020. The Giant's story revisited: identifying the Cerne Abbas hill figure as 'The Choice of Hercules' *Current Archaeology* 365 (August), 41–45. Available at https://the-past.com/feature/the-giants-story-revisited-identifying-the-cerne-abbas-hill-figure-as-the-choice-of-hercules/ (Accessed 23 November 2023)

Edwards, R. 1992. Letter Rachel Edwards (of Yetminster) to Ian Smith, dated 16 December 1992. NT Cerne Abbas archive box 1297

Egan, P. 1819. *Walks Through Bath, Describing Everything Worthy of Interest*. Bath: Meyler & Son

Ekwall, E. 1947. *Early London Personal Names*. Lund: C.W.K. Gleerup

Emery, P. 2021. The Cerne Giant – the possibility of Roman connection. Unpublished report. National Trust ref SNA 690231, HBSMR 110511

Enenkel, K. 2018. From Chivalric Family Tree to National Gallery: The Portrait Series of the Counts of Holland, ca. 1490–1650, in Enenkel, K.A.E. & Konrad A.O. (eds), *The Quest for an Appropriate Past in Literature, Art and Architecture*, 233–301. Leiden: Brill

Entwistle, R. & Bowden, M. 1991. Cranborne Chase; the molluscan evidence, in Barrett, J., Bradley, R. & Hall, M. (eds), *Papers on the Prehistoric Archaeology of Cranborne Chase*, 20–48. Oxford: Oxbow Monograph 11

Eppinger, A. 2021. The Early Christian Heracles, in Ogden, D. (ed.), *Oxford Handbook of Heracles*, 522–39. Oxford: Oxford University Press

Evans, J. 1956. *A History of the Society of Antiquaries*. Oxford: Oxford University Press

Evans, J.G. 1972. *Land Snails in Archaeology*. London: Seminar Press

Evans, J.G. 1984. Stonehenge – the environment in the Late Neolithic and Early Bronze Age and a Beaker burial. *Wiltshire Archaeological & Natural History Magazine* 78, 7–30

Fairholt, F.W. 1859. *Gog and Magog. The Giants in Guildhall; Their Real and Legendary History. With an Account of other Civic Giants, at Home and Abroad*. London: John Camden Hotten

Farmer, D.H. 1988. The monastic reform of the tenth century and Cerne Abbas, in Barker, K. (ed.), *The Cerne Abbey Millennium Lectures*, 1–10. Cerne Abbas: The Cerne Abbey Millennium Committee

Farrant, J. 1993. The Long Man of Wilmington, East Sussex: the documentary evidence reviewed. *Sussex Archaeological Collections* 131, 129–38

Farrant, J. 1995. A drawing of the Long Man of Wilmington, East Sussex, by the Revd D.T. Powell. *Sussex Archaeological Collections* 133, 282–4

Faulkner, M. 2008. Ælfric, St Edmund, and St Edwold of Cerne. *Medium Ævum* 77 (1), 1–9

Finberg, H.P.R. 1953. Sherborne, Glastonbury, and the Expansion of Wessex. *Transactions of the Royal Historical Society*, 5th Ser., 3, 101–24

Finberg, H.P.R. 1964. *The Early Charters of Wessex*. Leicester: Leicester University Press

Findlay, D.C., Colborne, G.J.N., Cope, D.W., Harrod, T.R., Hogan, D.V. & Staines, S.J. 1983. *Soils of England and Wales. Sheet 5, South West England*. Southampton: Ordnance Survey

Findlay, D.C., Colborne, G.J.N., Cope, D.W., Harrod, T.R., Hogan, D.V. & Staines, S.J. 1984. *Soils and their Use in South West England*. Soil Survey of England and Wales, Bulletin No. 14

Fitter, R. 1947. Letter from R. Fitter (editor *The Countryman*) to H.J.F. Smith, National Trust 29 September 1947. NT Box 12979

Foliot, G., Morey, A. & Brooke, C.N.L. 1948. The Cerne Letters of Gilbert Foliot and the Legation of Imar of Tusculum. *The English Historical Review* 63(249), 523–7. Available at http://www.jstor.org/stable/556005 (Accessed 5 July 2023)

Foster, J.J. 1888. Dorset folk-lore. *The Folk-Lore Journal* 6(2), 115–19

Foulser, B. 2022. The water and drainage systems of Cerne Abbey. *Cerne Historical Society Magazine*, Spring 2022–3, 23–7

Foulser, B. 2023. The Abbey's water and drainage system II. *Cerne Historical Society Magazine*, Winter 2022–3, 26–32

Fowler, P.J. 2000. *Landscape Plotted and Pieced: Landscape History and Local Archaeology in Fyfield and Overton, Wiltshire*. London: Society of Antiquaries of London

Fox, A. 2000. *Oral and Literate Culture in England, 1500–1700*. Oxford: Clarendon Press

Foxe, J. 1563. *Actes and Monuments of these Latter and Perilous Days, Touching Matters of the Church*. London: John Day (Foxe's *Book of Martyrs*)

Freeman, J.P. 1998. Archaeological Evaluation at Alton Lane, Cerne Abbas, Dorset. Unpublished Exeter Archaeology Report number 98.67

French, C. 2015. *A Handbook of Geoarchaeological Approaches to Investigating Landscapes and Settlement Sites*. Studying Scientific Archaeology 1. Oxford: Oxbow Books

Galbraith, R.F. & Green, P.F. 1990. Estimating the component ages in a finite mixture. *Nuclear Tracks and Radiation Measurements* 17, 196–206

Galbraith, R.F. & Laslett, G.M. 1993. Statistical models for mixed fission track ages. *Nuclear Tracks and Radiation Measurements* 21, 459–70

Galbraith, R.F., Roberts, R.G., Laslett, G.M., Yoshida, H. & Olley, J.M. 1999. Optical dating of single and multiple grains of quartz from Jinmium rock shelter (northern Australia): Part I, Experimental design and statistical models. *Archaeometry* 41, 339–64

Gale, J. 1999. The 1996 geophysical survey of the Giant, in Darvill, T., Barker, K., Bender, B. & Hutton, R., *The Cerne Giant: An Antiquity on Trial*, 57–67. Oxford: Oxbow Books

Gameson, F. 1999. Goscelin's Life of Augustine of Canterbury, in Gameson, R. (ed.), *St Augustine and the Conversion of England*, 391–409. Stroud: Sutton

Garmondsway, G.N. (trans.), 1991. *Ælfric's Colloquy*. Exeter Medieval Texts and Studies. Liverpool: Liverpool University Press

Garrison, J. 1975. *Dryden and the Tradition of Panegyric*. Berkeley: University of California Press

Gaze, J. 1988. *Figures in a Landscape: A History of the National Trust*. London: Barrie and Jenkins

Gelling, M. 1993. *Place-Names in the Landscape*. London: J.M. Dent

Geological Survey 1958. *Geological Survey Sheet 328 Dorchester (Solid and Drift)*. London: Geological Survey and Ordnance Survey

Gerrard, T. 1980. *'Coker's' survey of Dorsetshire*, ed. R. Legg. Milborne Port: Dorset Publishing Company

Gibbons, A.O. 1962. *Cerne Abbas: Notes and Speculations on a Dorset Village*. Dorchester: Longmans

Gibbons, G. 2017. Delineating the Marlborough White Horse: shaping the hill-figure c. 1860–1901, and 2015 survey. *Wiltshire Archaeological & Natural History Magazine* 110, 203–21

Gibbons, G. 2021. The Devizes white horse: rethinking a lost hill-figure. *Wiltshire Archaeological & Natural History Magazine* 114, 273–81

Gliganic, L.A., Cohen, T.J., Slack, M. & Feathers, J.K. 2016. Sediment mixing in Aeolian sandsheets identified and quantified using single-grain optically stimulated luminescence. *Quaternary Geochronology* 32, 53–66

Godden, M. 2007. Did King Alfred write anything? *Medium Ævum* 76, 1–23

Godden, M. 2016. *The Old English History of the World: An Anglo-Saxon Rewriting of Orosius*. Cambridge, MA: Dumbarton Oaks Medieval Library

Godden, M. & Irvine, S. 2009. *The Old English Boethius. An Edition of the Old English Versions of Boethius' De Consolatione Philosophiae*. 2 volumes. Oxford: Oxford University Press

Godfrey, I. forthcoming. The Uffington White Horse Complex as a Winter Solstice Observatory. *Oxoniensia*

Godfrey, W.H. 1928. Wilmington Priory: an architectural description. *Sussex Archaeological Collections* 69, 1–28

Gomme, G.L. 1880. *Primitive Folk-Moots; or Open-Air Assemblies in Britain*. London: Sampson Low, Marston, Searle & Rivington

Gomme, G.L. (ed.), 1885. *The Gentleman's Magazine Library: English Traditional Lore*, vii. London: Elliot Stock

Goodman, K. 1998. *Chalk Figures of Wessex*. Salisbury: Wessex Books

Gough, R. (trans.), 1809. *Camden's Britannia: Surrey and Sussex*. London: Hutchinson.

Gover, J.E.B., Mawer, A. & Stenton, F.M. 1938. *The Place-Names of Hertfordshire*. English Place-Name Society vol. 15. Cambridge: Cambridge University Press

Gover, J.E.B., Mawer, A. & Stenton, F.M. 1970. *The Place-Names of Wiltshire*. English Place-Name Society vol. 16. London: English Place-Name Society

Grazier, K. & Cass, S. 2017. *Hollyweird Science: The Next Generation, from Spaceships to Microchips*. New York: Springer Nature

Gretsch, M. 2013. Historiography and literary patronage: Æthelweard's *Chronicon*. *Anglo–Saxon England* 41, 205–48

Griffiths, S. & Stone, A. 2022. Luminescence: optically stimulated luminescence and thermoluminescence, in Griffiths, S. (ed.), *Scientific Dating in Archaeology*, 63–94. Oxford: Oxbow Books

Grimmer, M. 2007. Britons in early Wessex: the evidence of the law code of Ine, in Higham, N. (ed.), *Britons in Anglo-Saxon England*, 102–14. Woodbridge: Boydell Press

Grinsell, L.V. 1939. *White Horse Hill and the Surrounding Country*. London: Saint Catherine Press

Grinsell, L. 1980a. The Cerne Abbas Giant 1764–1980. *Antiquity* 54 (March), 29–33

Grinsell, L.V. 1980b. Thomas Hardy and the Cerne Abbas Giant. *Notes & Queries for Somerset & Dorset* (March 1980) 31, 38

Grün, R. 2001. Trapped charge dating (ESR, TL, OSL), in Brothwell, D.R. & Pollard, A.M. (eds), *Handbook of Archaeological Sciences*, 47–62. Chichester: Wiley

Grundy, G.B. 1937. Dorset Charters. *Proceedings of the Dorset Natural History & Archaeological Society* 59, 95–118

Haigh, J.G.B. 1991. Automatic grid balancing in geophysical survey, in Lock, G. & Moffett, J. (eds), *Computer Applications and Quantitative Methods in Archaeology 1991*, 191–6. Oxford: British Archaeological Report S577

Hall, A. 2007. *Elves in Anglo-Saxon England. Matters of Belief, Health, Gender and Identity*. Woodbridge: Boydell Press

Hall, T.A. 2000. *Minster Churches in the Dorset Landscape*. British Archaeological Reports British Series 304. Oxford: BAR Publishing

Halpern, L.C. 2002. Wrest Park 1686–1730s: exploring Dutch influence. *Garden History* 30 (2), 131–52

Halsall, G. 2003. *Warfare and Society in the Barbarian West, 450–900*. London: Routledge

Hamalainen, P. 2019. *Lakota America: A New History of Indigenous Power*. New Haven: Yale University Press

Hamilton, N.E.S.A. (ed.), 1870. *Willelmi Malmesbiriensis Monachi De Gestis Pontificum Anglorum Libri Quinque*. London: Longman & Co, Rolls Series, No. 51

Hamilton, S. 2014. Under-representation in contemporary archaeology. *Papers from the Institute of Archaeology* 2, article 24. Available at https://doi.org/10.5334/pia.469 (Accessed 21 November 2023)

Hanham, A.A. 2010. Spencer Compton Earl of Wilmington 1674–1743, in *Oxford Dictionary of National Biography*. Oxford: Oxford University Press. Available at https://doi.org/10.1093/ref:odnb/6036 (Accessed 23 March 2024)

Hannah, I.C. & Peckham, W.D. 1928. Brambletye. *Sussex Archaeological Collections* 69, 103–12

Hardy, A., Dodd, A. & Keevill, G.D. 2003. *Ælfric's Abbey. Excavations at Eynsham Abbey, Oxfordshire, 1989–92*. Thames Valley Landscapes 16. Oxford: Oxford Archaeology

Hardy, F. 1925. Letter to Alda Hoare dated 21 September 1925, copy in National Trust SW archives, and Hardy, T. 1925. Letter to Alda Hoare dated 21 September 1925, Stourhead papers WRO 383/954 and copy in National Trust SW archives

Harris, S.J. 2003. Ælfric's Colloquy, in Kline, D.T. (ed.), *Medieval Literature for Children*, 112–30, London: Routledge

Hartley-Parkinson, R. 2019. Volunteers polish giant's erection by hand. *Metro* 29 August 2019. Available at https://metro.co.uk/2019/08/29/volunteers-polish-giants-erection-hand-10648785/ (Accessed 14 November 2023)

Hawes, P. 2015. Sheep grazing and management of chalk grassland. *British Wildlife* 27 (10), 25–30

Hawes, P., Pywell, R. & Ridding, L. 2018. Long-term changes in chalk grassland. *British Wildlife* 59 (3) (Feb 2018), 184–9

Hawes, P., Riddling, L., Walls, R., Bailey, J., Pescott, O., Pilkington, S. & Pywell, R. 2020. Colonisation of exposed chalk surfaces and the restoration of chalk grassland. *Conservation Land Management*, summer 2020, 17 (2), 20–7

Hawes, P., Green, M. & Walls, R. 2024. Restoring chalk grassland from ex-arable fields and bare chalk surfaces. *Conservation Land Management*, spring 2024, 22 (1), 31–6

Hawkes, C. 1965. The Long Man of Wilmington: a clue, in Hawkes, S.C., Ellis Davidson, H.R. & Hawkes, C., The Finglesham Man. *Antiquity* 39, 17–32

Hawkes, J. 1951. *A Guide to the Prehistoric and Roman Monuments in England and Wales*. London: Chatto and Windus

Hawkes, J. 1954. *Man on Earth*. London: Cresset Press

Hayton, D., Cruickshanks, E. & Handley, S. (eds), 2002. Freke, Thomas I (c.1638–1701), of Shroton and Melcombe Horsey, Dorset, in *The History of Parliament: The House of Commons 1690–1715*. Available at https://www.historyofparliamentonline.org/volume/1690-1715/member/freke-thomas-i-1638-1701 (Accessed 23 November 2023)

HE: Bede. 1969. *Historia Ecclesiastica Gentis Anglorum*, ed. and trans. B. Colgrave & R.A.B. Mynors. Oxford Medieval Texts. Oxford: Clarendon Press

Hearne, T. 1774. *John Leland, De Rebus Britannicis Collectanea*. 6 volumes. 3rd edn. London: G. & J. Richardson

Henning, B.D. (ed.), 1983. Freke, Thomas I (c. 1638–1701), in *The House of Commons, 1660–1690: Introductory Survey, Appendices, Constituencies, Members C-L., Volume 2, 365–6*. London: Published for the History of Parliament Trust by Secker & Warburg. Available at https://www.historyofparliamentonline.org/volume/1660-1690/member/freke-thomas-i-1638-1701 (Accessed 23 November 2023)

Herren, M. 1998. The transmission and reception of Graeco-Roman mythology in Anglo-Saxon England 670–800. *Anglo–Saxon England* 27, 87–103

Hilts, C. 2021. Latest dating evidence for the Cerne Abbas hill figure. *Current Archaeology* 376, 14–15

Hindle, B.P. 1993. *Roads, Tracks and their Interpretation*. London: B.T. Batsford

Hinton, D. 1998. *Saxons and Vikings*. Discover Dorset. Wimborne: Dovecote Press

Hoade, B. 1978. Scope for excavation, in Legg, R., Dorset and Cerne's God of the Celts. *Dorset County Magazine*, special issue (April), 28–9

Hodgson, J.M. 1997. *Soil Survey Field Handbook*. Silsoe: Soil Survey and Land Research Centre

Holden, E.W. 1971. Some notes on the Long Man of Wilmington. *Sussex Archaeological Collections* 109, 37–54

Holloway, W. 1808. Giant of Trendle Hill: a legendary tale, in *The Minor Minstrel; or, Poetical Pieces, Chiefly Familiar and Descriptive*. London: W. Suttaby

Horne, D. 2015. The Marlborough Downs White Horse Survey. Unpubl. report for G. Gibbons Marlborough fieldwork

Horne, D. 2020. Roundway White Horse. Unpub. report for G. Gibbons Devizes fieldwork

Horne, D., Britton, D., Herring, V., James, L. & Hutton, K. 2015. The Marlborough Downs White Horse Survey. Unpubl. report for G Gibbons' Marlborough fieldwork

Hughes, T. 1859. *The Scouring of the White Horse*. London: Macmillan & Co.

Huntley, D.J., Godfrey-Smith, D.I. & Thewalt, M.L.W. 1985. Optical dating of sediments. *Nature* 313, 105–7

Hutchins, J. 1774. *The History and Antiquities of the County of Dorset, compiled from the best and most ancient historians, inquisitions post mortem, and other valuable records and mss. in the public offices, and libraries, and in private hands*. 1st edn, 2 volumes. London: W. Bowyer & J. Nichols aka *The History and Antiquities of Dorset*

Hutchins, J. 1860–1. *History and Antiquities of Dorset*, Vol. IV. Westminster: John Bowyer Nichols & Sons

Hutchins, J. 1870. *History of Dorset*, 3rd edn. London: John Bowyer Nichols & Sons

Hutton, R. 1994. *The Rise and Fall of Merry England: The Ritual Year*, 257–60. Oxford: Oxford University Press

Hutton, R. 1999a. Naming places and the role of the landscape, in Darvill, T., Barker, K., Bender, B. & Hutton, R., *The Cerne Giant: An Antiquity on Trial*, 87–8. Oxford: Oxbow Books

Hutton, R. (ed.), 1999b. The case for a post-medieval Giant, in Darvill, T., Barker, K., Bender, B. & Hutton, R., *The Cerne Giant: An Antiquity on Trial*, 79–124. Oxford: Oxbow Books

Hutton, R. 1999c. A 17th century marvel, in Darvill, T., Barker, K., Bender, B. & Hutton, R., *The Cerne Giant: An Antiquity on Trial*, 70. Oxford: Oxbow Books

Hutton, R. 2004. Back to modern culture. *British Archaeology* (July 2004), 20–1

Insley, J. 2003. Pre-Conquest Personal Names. *Reallexikon der Germanischen Altertnmskunde* 23, 367–96

Irvine, S. 2003. Wrestling with Hercules: King Alfred and the classical past, in Cubitt, C. (ed.), *Court Culture in the Early Middle Ages*, 171–88. Turnhout: Brepols

Jackson, M.L., Sayin, M. & Clayton, R.N. 1976. Hexafluorosilicic acid regent modification for quartz isolation. *Soil Science Society of America Journal* 40, 958–60

Jennings, W.T. 1894. The white horse of Westbury. *Wiltshire Notes & Queries* 1, 193–4

Johnson, D.F. 1995. Euhemerisation versus demonisation: the pagan gods and Ælfric's *De Falsis Diis*, in Hofstra, T., Houwen, L.A. & MacDonald, A.A. (eds), *Pagans and Christians: The Interplay between Christian–Latin and Traditional Germanic Cultures in Early Medieval Europe*. Gronngen: Egbert Forsten

Johnson, W. 1867. *Folk-Memory; or, the Continuity of British Archaeology*. Oxford: Clarendon Press

Jones, C. 2004. *Ælfric's Letter to the Monks of Eynsham*. Cambridge: Cambridge University Press

Jones, G. 1998. Penda's footprint? Place-names containing personal names associated with those of early Mercian kings. *Nomina* 21, 29–62

Just, M.M. 2002. The Reception of Gulliver's Travels in Britain and Ireland, France, and Germany, in Carey D. & Boulaire, F. (eds), *Les Voyages De Gulliver: Mondes Lointains Ou Mondes Proches*, 81–100. Caen: Presses Universitaires de Caen

Kalinsky, G.K. 1972. *The Herakles Theme: The Adaption of the Hero in Literature from Homer to the Twentieth Century*. Oxford: Basil Blackwell

Keithley, W., Papworth, M. & Thackray, D. 1999. Owning and managing a giant, in Darvill, T., Barker, K., Bender, B. & Hutton, R., *The Cerne Giant: An Antiquity on Trial*, 18–26. Oxford: Oxbow Books

Kelly, S.E. (ed.), 1996. *Charters of Shaftesbury Abbey*, Anglo-Saxon Charters 5, xiii. Oxford: Oxford University Press

Kemble, J.M. 1839–48. *Codex Diplomaticus Aevi Saxonici*. 6 volumes. London: Sumptibus Societatis

Kerney, M.P. 1966. Snails and man in Britain. *Journal of Conchology* 26, 3–14

Kerney, M.P. 1970. The British distribution of *Monacha cantiana* (Montagu) and *Monacha cartusiana* (Müller). *Journal Conchology* 26, 152–60

Kerney, M.P. 1999. *Atlas of the Land and Freshwater Molluscs of Britain and Ireland*. Colchester: Harley Book

Keynes, S. 1994. Cnut's earls, in Rumble, A.R. (ed.), *The Reign of Cnut. King of England, Denmark and Norway*, 48–88. London: Leicester University Press

Keynes, S. 2005. Wulfsige, monk of Glastonbury, abbot of Westminster (c. 990–3) and bishop of Sherborne (c. 993–1002), in Barker, K., Hinton, D.A. & Hunt, A. (eds), *St Wulfsige of Sherborne: Essays to Celebrate the Millennium of the Benedictine Abbey*, 53–95. Oxford: Oxbow Books

King, A. & Soffe, G. 1994. The Iron Age and Roman temple on Hayling Island, in A.P. Fitzpatrick, & E.L. Morris (eds), *The Iron Age in Wessex: Recent Work*, 114–16. Salisbury: Wessex Archaeology

King, J.E. 1952. *Bede's Historical Works*. London: Heinemann

Kitchen, F. 1997. John Norden (c. 1547–1625): estate surveyor, topographer, county mapmaker and devotional writer. *Imago Mundi* 49, 43–61

Knowles, D. 1966. *The Monastic Order in England, 940–1216*, 2nd edn. Cambridge: Cambridge University Press.

Kuhn, T.S. 1962. *The Structure of Scientific Revolutions*. Chicago, IL: University of Chicago Press

Kwaad, F.J.P.M. 1977. Measurements of rainsplash erosion and the formation of colluvium beneath deciduous woodland in the Luxembourg Ardennes. *Earth Surface Processes* 2, 161–73

Kwaad, F.J.P.M. & Mücher, H.J. 1977. The evolution of soils and slope deposits in the Luxembourg Ardennes near Wiltz. *Geoderma* 17, 1–37

Laidlaw, J. 2014. Ethics, in Boddy, J. & Lambek, M. (eds), *A Companion to the Anthropology of Religion*, 171–88 (esp. 176–7). Chichester: John Wiley and Sons

Lapidge, M. 1982. *Beowulf*, Aldhelm, the *Liber Monstrorum* and Wessex. *Studi Medievali* 3.23, 150–92

Latham, J.A. 2015. Inventing Gregory 'the Great': memory, authority, and the afterlives of the 'Letania Septiformis', *Church History* 84 (1), 1–31. Available at www.jstor.org/stable/24537289 (Accessed 16 March 2021)

Legg, R. 1978. Dorset and Cerne's God of the Celts. *Dorset County Magazine* 66, Special Issue, April

Legg, R. 1980. Afterword, in Gerrard, T. (ed. Legg. R.), 'Coker's' survey of Dorsetshire. Milborne Port: Dorset Publishing Company

Legg, R. 1986. *Cerne's Giant and Village Guide*. Milbourne Port: Dorset Publishing Company

Legg, R. 1990. *Cerne's Giant and Village Guide*. 2nd edn. Wincanton: Dorset Publishing Company

Legg, R. 1992. One Giant and his dog. *Dorset Magazine*, November, 29–30

Legg, R. 1999. Making and re-making the Cerne Giant, in Darvill, T., Barker, K., Bender, B. & Hutton, R., *The Cerne Giant: An Antiquity on Trial*, 129–34. Oxford: Oxbow Books

Leigh-Smith, C. 2018. Cerne Abbey, a Critical Evaluation Using Landscape Archaeology to Discover the Layout and Water Management Including Construction Elements Before its Dissolution in 1539, Compared to its Foundation in 987. Unpublished MA Thesis, Birkbeck, University of London

Leland, J. 1770. *Antiquarii De Rebus Britannicis Collectanea*, 67, ed. T. Hearne, 2nd edn., vol. 4 of 6, London: William and John Richardson [Another edition cited by Faulkner, M. 2008. Ælfric, St Edmund, and St Edwold of Cerne, *Medium Ævum* 77 (1), 1–9. JSTOR, Available at www.jstor.org/stable/43630592 (Accessed 16 March 2021)]

Leneghan, F. 2020. *The Dynastic Drama of Beowulf*. Cambridge: D.S. Brewer

Lethbridge, T.C. 1956. The Wandlebury Giants. *Folklore* 67.4, 193–203

Lethbridge, T.C. 1957. *Gogmagog: The Buried Gods*. London: Routledge and Kegan Paul

Licence, T. 2006. Goscelin of St Bertin and the Life of St Eadwold of Cerne. *Journal of Medieval Latin* 16, 182–207

Licence, T. 2020. *Edward the Confessor: Last of the Royal Blood*. New Haven and London: Yale University Press

Limbrey, S. 1975. *Soil Science and Archaeology*. London: Academic Press

Lloyd, R. 1982. The Cerne Giant: another document. *Antiquity* 56 (March), 51–2

Lower, M.A. 1851. *The Sussex Martyrs: Their Examinations and Cruel Burnings in the Time of Queen Mary*. Lewes: Baxter & Son

Lucas, G. 2017. The paradigm concept in archaeology. *World Archaeology* 49, 260–70

Lukis, W.C. (ed.) 1883. Letters from John Hutchins to William Stukeley, 22 October & 29 November 1763, in *The Family Memoirs of the Rev. William Stukeley*, Vol. II, 129–33. London: The Surtees Society, 76

Lumb, A. 2023 National Trust issue apology for overgrown Cerne Giant, *Dorset Echo*, 6 October 2023. Available at https://www.dorsetecho.co.uk/news/23833919.national-trust-issue-apology-overgrown-cerne-giant/#comments-anchor (Accessed 23 November 2023)

Lyon, H. 2022. *Memory and the Dissolution of the Monasteries in Early Modern England*. Cambridge: Cambridge University Press

March, H.C. 1899. Dorset Folklore Collected in 1897. *Folklore* 10 (4), December, 478–89

March, H.C. 1901. The Giant and the Maypole of Cerne. *Proceedings of the Dorset Natural History and Antiquarian Field Club* 22, 101–18

March, H.C. 1902. *The Giant and the Maypole of Cerne*. Dorchester: Dorset County Chronicle, 8

Margary, I. 1973. *Roman Roads in Britain*, 3rd edn. London: John Baker

Markey, B.G., Bøtter-Jensen, L. & Duller, G.A.T. 1997. A new flexible system for measuring thermally and optically stimulated luminescence. *Radiation Measurements* 27, 83–9

Marples, M. 1949. *White Horses and other Hill Figures*. London: County Life Books

Marples, M. 1981. *White Horses and other Hill Figures*. 3rd edn. Gloucester: Alan Sutton

Marren, P. 2006. *Battles of the Dark Ages: British battlefields AD 410 to 1065*. Barnsley: Pen & Sword

Maslin, E. 2023. Cheese company 'castrates' Dorset's Cerne Abbas Giant. *Dorset Echo*, 2 June 2023. Available at https://www.dorsetecho.co.uk/news/23562157.cheese-company-castrates-dorsets-cerne-abbas-giant/ (Accessed 14 November 2023)

Massingham, H.J. 1926. *Fee, Fi, Fo, Fum, or, The Giants in England*, 54–8. London: Kegan Paul, Trench, Trubener & Co.

Massingham, H.J. 1935. *Through the Wilderness.* London: Cobden-Sanderson

Maton, W.G. 1797. *Observations Relative Chiefly to the Natural History, Picturesque Scenery, and Antiquities, of the Western Counties of England, Made in the Years 1794 and 1796*, 17–18. Salisbury: J. Easton

Mattelaer, J.J. 2010. The phallus tree: a Medieval and Renaissance phenomenon. *The Journal of Sexual Medicine* 7, 633–856

Mawer, A. & Stenton, F.M. 1925. *The Place-Names of Buckinghamshire*. Cambridge: Cambridge University Press

Mawer, A. & Stenton, F.M. 1926. *The Place-Names of Bedfordshire and Huntingdonshire.* English Place-Name Society 3. Cambridge: Cambridge University Press

Mawer, A. & Stenton, F.M. (in collaboration with Houghton, F.T.S.) 1927. *The Place-Names of Worcestershire.* English Place-Name Society 4. Cambridge: Cambridge University Press

Mead, R. 2021. Letter from England: the mysterious origin of the Cerne Abbas Giant. *The New Yorker,* 12 May 2021. Available at https://www.newyorker.com/magazine/2021/05/24/the-mysterious-origins-of-the-cerne-abbas-giant (Accessed 5 July 2022)

Meech, R. 2012. Youngsters recreate Olympic torch on Cerne Abbas's chalk giant. *Dorset Echo* 29 May 2012. Available at https://www.dorsetecho.co.uk/news/9732051.youngsters-recreate-olympic-torch-on-cerne-abbas-chalk-giant/ (Accessed 14 November 2023)

Mejdahl, V. 1979. Thermoluminescence dating: beta-dose attenuation in quartz grains. *Archaeometry* 21, 61–72

Menendez, E. 2019. Monument's massive erection covered with a flower for International Women's Day. *Metro* 8 March 2019. Available at https://metro.co.uk/2019/03/08/monuments-massive-erection-covered-with-a-flower-for-international-womens-day-8868921/ (Accessed 14 November 2023)

Miles, D. 1999. The Uffington Horse and its antiquity, in Darvill, T., Barker, K., Bender, B. & Hutton, R., *The Cerne Giant. An Antiquity on Trial*, 39–43. Oxford: Oxbow Books

Miles, D. 2019. *The Land of the White Horse: Visions of England.* London: Thames & Hudson

Miles, D., Palmer, S., Lock, G., Gosden, C. & Cromarty, A.M. 2003. *Uffington White Horse and its Landscape: Investigations at the White Horse Hill Uffington, 1989–95, and Tower Hill Ashbury, 1993–4.* Thames Valley Landscapes Monograph 18. Oxford: Oxford Archaeology

Mileson, S. & Brookes, S. 2021. *Peasant Perceptions of Landscape: Ewelme Hundred, South Oxfordshire, 500–1650.* Oxford: Oxford University Press

Mills, A.D. 2010. *The Place-Names of Dorset*, vol. 4, English Place-Name Society vols 86–7. London: English Place-Name Society

Milmo, C. 2021. Cerne Abbas Giant: UK's largest chalk figure revealed as medieval art. *iNews*, 12 May 2021. Available at https://inews.co.uk/news/cerne-abbas-giant-uks-largest-chalk-figure-revealed-as-medieval-art-and-not-an-insult-to-oliver-cromwell-996256 (Accessed 25 November 2023)

Milne, J.L. 1945. *The National Trust. A Record of Fifty Years' Achievement.* Norwich: Jarrold & Sons

Minden, M., Bachmann, H. & Curran, J.V. 2002. *Fritz Lang's Metropolis: Cinematic Visions of Technology and Fear.* New York: Camden House

Morcom, T. & Gittos, H. 2021. The curious tale of an Anglo-Saxon giant. *BBC History Extra* podcast, 31 May 2021. Available at https://www.historyextra.com/membership/curious-tale-cerne-abbas-giant-podcast-tom-morcom-helen-gittos/ (Accessed 6 June 2023)

Morcom, T. & Gittos, H. 2024. The Cerne Giant in its early medieval context. *Speculum. A Journal of Medieval Studies* 99 (1), 1–38

Morgan, L.H. 1877. *Ancient Society.* London: McMillan

Morgan Evans, D. 1998. Eighteenth-century descriptions of the Cerne Abbas Giant. *Journal of Antiquaries* 78, 463–71

Morgan Evans, D. 1999. Eighteenth-century descriptions of the Cerne Abbas Giant, in Darvill, T., Barker, K., Bender, B. & Hutton, R., *The Cerne Giant: An Antiquity on Trial*, 108–24. Oxford: Oxbow Books

Morris, M. 1981. *White Horses and other Hill Figures.* Gloucester: Alan Sutton

Morrison, R. 2021. Is the mystery of the Cerne Giant finally solved? UK's largest chalk man was made in Saxon times 'in tribute to the God Helith' and NOT to insult Oliver Cromwell (but his appendage may have been added later for laughs), scientists say. *MailOnline* 12 May 2021. Available at https://www.dailymail.co.uk/sciencetech/article-9566687/Cerne-Giant-constructed-late-Saxon-period-study-shows.html (Accessed 23 November 2023)

Mortimer, J. 2020. Preaching Cross. *Cerne Historical Society Magazine*, 12 June, 5–6

Murray, A.S. & Olley, J.M. 2002. Precision and accuracy in the Optically Stimulated Luminescence dating of sedimentary quartz: a status review. *Geochronometria* 21, 1–16

Murray, A.S. & Wintle, A.G. 2000. Luminescence dating of quartz using an improved single-aliquot regenerative-dose protocol. *Radiation Measurements* 32, 57–73

Murray, A.S. & Wintle, A.G. 2003. The single aliquot regenerative dose protocol: potential for improvements in reliability. *Radiation Measurements* 37, 377–81

Murray, J. 1856. *Murray's Handbook for Travellers in Wiltshire, Dorsetshire, and Somersetshire*. London: John Murray

National Trust 1974a. Land use Description, p2–3, The National Trust Wessex Region management plan Cerne Giant Dorset, November 1974. File Ref. 6215588. Archive management plan (top copy), National Trust, Tisbury

National Trust 1974b. Appendix 2, The National Trust Wessex Region management plan Cerne Giant Dorset, November 1974. File Ref. 6215588. Archive management plan (top copy), National Trust, Tisbury

Nature 1937a. The Cerne Giant a correction. *Nature* 139 (29 May), 920. Available at https://doi.org/10.1038/139920c0 (Accessed 20 November 2023)

Nature 1937b. The Cerne Giant. *Nature* 139 (22 May), 876. Available at https://doi.org/10.1038/139876a0 (Accessed 20 November 2023)

Neale, F. (ed.), 2000. *William of Worcestre: the topography of Medieval Bristol*. Bristol: Bristol Record Society

Nees, L. 1991. *A Tainted Mantle: Hercules and the Classical Tradition at the Carolingian Court*. Philadelphia: University of Pennsylvania Press

Neidorf, L. 2014. *The Dating of Beowulf. A Reassessment*. Cambridge: D.S. Brewer

Newman, P. 1987. *The Gods and Graven Images; and Chalk Hill Figures of Britain*. London: R. Hale

Newman, P. 1997a. *Gods and Graven Images: The Chalk Hill-figures of Britain*. 2nd edn. London: Robert Hale

Newman, P. 1997b. *Lost Gods of Albion: The Chalk Hill-figures of Britain*. Stroud: Sutton Publishing

Newman, P. 1999. In defence of antiquity, in Darvill, T., Barker, K., Bender, B. & Hutton, R., *The Cerne Giant: An Antiquity on Trial*, 31–7. Oxford: Oxbow Books

Newman, P. 2000. *Lost Gods of Albion: The Chalk Hill-figures of Britain*. Barton-under-Needwood: Wrens Park Publishing

Ng, D.Y. & Swetnam-Burland, M. 2018. Introduction: reuse and renovation in Roman material culture, in Ng, D. & Swetnam-Burland, M. (eds), *Reuse and Renovation in Roman Material Culture: Functions, Aesthetics, Interpretations*, 1–23. Cambridge: Cambridge University Press

Nichols, J. 1818. *Illustrations of the Literary History of the Eighteenth Century*. London: J. Nichols

Nickerson, R.S. 1998. Confirmation bias: a ubiquitous phenomenon in many guises. *Review of General Psychology* 2(2), 175–220

Norden, J. (transcribed by Smith-Burns, C.) 1617. *Cerne Abbas, Survey*. Dorset Record Office cat. no. Ph 248/4-9

North, R. 2006. *The Origins of Beowulf from Vergil to Wiglaf*. Oxford: Oxford University Press

O'Donovan, M.A. 1988. *Charters of Sherborne*. British Academy Charters 3. Oxford: Oxford University Press

Ogden, D. (ed.), 2021. *Oxford Handbook of Heracles*. Oxford: Oxford University Press

Orchard, A. 1995. *Pride and Prodigies. Studies in the Monsters of the Beowulf Manuscript*. Toronto: Toronto University Press

Orchard, A. 1997. *Norse Myth and Legend*. London: Cassell

Oxford Archaeology Unit 1998. Cerne Abbas Giant; archaeological research design. Unpubl. report for The National Trust, dated December 1998

Page, W. (ed.) 1908. Houses of Benedictine monks; the abbey of Cerne, in *A History of the County of Dorset Volume 2*, 53–8. London: Victoria County History

Page, W. & Ditchfield, P.H. 1924. *Victoria History of the County of Berkshire, Volume 4*. London: Institute of Historical Research

Pantos, A. 2002. Assembly-Places in the Anglo-Saxon Period: aspects of form and location, 3 vols., unpubl. DPhil thesis, University of Oxford

Papworth, M. 2011. *The Search for the Durotriges; Dorset and the West Country in the Late Iron Age*. Stroud: The History Press

Papworth, M. 2021a. Dating the Cerne Giant Results! *Archaeology National Trust SW* 9 May 2021. Available at archaeologynationaltrustsw.wordpress.com/2021/05/09/dating-the-cerne-giant-results/ (Accessed 5 July 2023)

Papworth, M. 2021b. Cerne Abbas Giant, four excavation trenches and interim report of OSL dates, *Proceedings of the Dorset Natural History & Archaeological Society* 142, 239–4

Papworth, M. & Keighley, W. 2004. *The Cerne Giant and Dorset Hill-forts. Their Past Revealed*. Charminster: The National Trust

Patrologiae (Cursus Completes) Latina 80. 1863. Paris: Apud Garnier Fratres

Payne, A. 2003. Geophysical survey (The White Horse), in Miles, D., Palmer, S., Lock, G., Gosden, C. & Cromarty, A.M., *Uffington White Horse and its landscape: investigations at the White Horse Hill Uffington, 1989–95, and Tower Hill Ashbury, 1993–4*, 66–8. Oxford: Oxford Archaeology

Payne, E.J. 1897. Whiteleaf Cross. *Records of Buckinghamshire* 7, 559–67

Peddie, J. 1989. *Alfred the Good Soldier. His Life & Campaigns*. Bath: Millstream Books

Penn, K.J. 1980. *Historic Towns in Dorset*. Dorchester: Dorset Natural History & Archaeological Society Monograph Series, No. 1

Petrie, W.M.F. 1926. *The Hill-figures of England*, Royal Anthropological Institute Occasional Paper 7. London: Royal Anthropological Institute

Piggott, S. 1931. The Uffington White Horse. *Antiquity* 5 (17), 37–46

Piggott, S. 1932. The name of the Giant of Cerne, *Antiquity* 6 (June), 214–16

Piggott, S. 1938. The Hercules Myth – beginnings and ends. *Antiquity* 12, 323–31

Piggott, S. 1946. On The Map (24 talks about interesting West Country places). BBC Radio (Broadcast 21 June 1946; text in BBC Archives)

Piggott, S. 1985. *William Stukeley. An Eighteenth-century Antiquary*. London: Thames & Hudson

Pitman, G. 1978. Navel, letter published in *Dorset Magazine* 70 (September), 27

Plenderleath, W.C. 1874. On the white horses of Wiltshire and its neighbourhood. *Wiltshire Archaeological & Natural History Magazine* 14, 12–30

Plenderleath, W.C. 1880. The white horses of Wiltshire. *North Wilts Church Magazine* (April–November 1880)

Plenderleath, W.C. 1885. *The White Horses of the West of England with Notices of some other Turf-monuments*. London: Alfred Russell Smith

Plenderleath, W.C. 1891. White horse jottings. *Wiltshire Archaeological & Natural History Magazine* 25, 57–68

Plenderleath, W.C. 1892. *The White Horses of the West of England with Notices of some other Ancient Turf-monuments*. 2nd edn. London: Allen & Storr

Pococke, R. 1754. *The Travels through England of Dr Richard Pococke*, Volume II. London: Camden Society, new series XLIV

Pollard, J. 2017. The Uffington White Horse Geoglyph as Sun-horse. *Antiquity* 91 (356), 406–20

Pope, F.J. 1901. The site of Cerne Abbey. *Somerset and Dorset Notes and Queries* 7, 332–4

Pope, R. 2011. Processual archaeology and gender politics: the loss of innocence. *Archaeological Dialogues* 18(1), 59–86

Porter, L. 1998. Tarts, tampons and tyrants: women and representation in British comedy, in Wragg, S. (ed.), *Because I Tell a Joke or Two: Comedy, Politics and Social Difference*, 65–92. London and New York: Routledge

Pratt, D. 2007. *The Political Thought of King Alfred the Great*. Cambridge: Cambridge University Press

Preest, D. (trans.), 2002. *William of Malmesbury, The Deeds of the Bishops of England* (*Gesta Pontificum Anglorum*), 122–3. Woodbridge: Boydell Press

Prescott, J.R. & Hutton, J.T. 1994. Cosmic ray contributions to dose rates for luminescence and ESR dating: large depths and long-term time variations. *Radiation Measurements* 23, 497–500

Priest, S.-J. 1813. *General View of the Agriculture of Buckinghamshire*. London: Board of Agriculture

Priestly, H. 1976. *The Observer's Book of Ancient and Roman Britain*. London: Frederick Warne

Putnam, B. 1999. The Cerne Valley in prehistoric and Roman times, in Darvill, T., Barker, K., Bender, B. & Hutton, R., *The Cerne Giant: An Antiquity on Trial*, 51–6. Oxford: Oxbow Books

Pywell, R.F., Bullock, J.M., Pakeman, R.J., Munford, J.O., Warman, A.E., Wells, T.C.E. & Walker, K. 1995. *Review of Calcareous Grassland and Heathland Management*. Cambridge: Institute of Terrestrial Ecology (NERC), MAFF/BERC Project T020501

RCHME 1974. *An Inventory of Historical Monument in the County of Dorset Volume 1. West Dorset*. London: HMSO

Rees-Jones. J. & Tite, M.S. 1996. Optical dating of the Uffington White Horse. *Archaeological Sciences 1995*, 171–4

Rees-Jones, J. & Tite, M. 2003. Optically stimulated luminescence (OSL) dating results from the White Horse and linear ditch, in Miles, D., Palmer, S., Lock, G., Gosden, C. & Cromarty, A.M., *Uffington White Horse and its landscape: investigations at the White Horse Hill Uffington, 1989–95, and Tower Hill Ashbury, 1993–4*, 269–71. Thames Valley Landscapes Monograph 18. Oxford: Oxford Archaeology

Reeves, M. & Morrison, J. (eds), 1989. *The Diaries of Jeffery Whitaker Schoolmaster of Bratton, 1739–1741*. Trowbridge: Wiltshire Record Society

Reuter, T. 2005. Introduction: Sherborne and the millennium, in Barker, K., Hinton, D.A. & Hunt, A. (eds), *St Wulfsige of Sherborne: Essays to Celebrate the Millennium of the Benedictine Abbey*, 1–9. Oxford: Oxbow Books

Reynolds, A. & Langlands, A. 2007. Social identities on the macro scale: a maximum view of Wansdyke, in Davies, W., Halsall, G. & Reynolds, A. (eds), *People and Space in the Middle Ages, 300–1300*, 13–44. Turnhout: Brepols

Reynolds, A. & Langlands, A. 2011. Travel as communication: a consideration of overland journeys in Anglo-Saxon England. *World Archaeology* 43.3, 410–27

Rickman, P. 1967. Letter from Phoebe Rickman, 11 September 1967, NT Box 12979

Ridding, L.E., Bullock, J.M., Prescott, O.L., Hawes, P., Walls, R., Pereira, M.G., Thacker, S.A., Keeenan, P.O., Dragosts, U. & Pywell, R.E. 2020. Long-term change in calcareous grassland vegetation and drivers over three time periods between 1970 and 2016. *Plant Ecology* 221, 377–94

Ridding, L.E., Hawes, P., Walls, R. Pilkington, S.L., Pywell, R.F. & Prescott, O.L. 2024. Do functional traits influence primary succession patterns for bryophytes and vascular plants: Evidence from a 33-year chronosequence on bare chalk. *Journal of Ecology* 112, 68–85. DOI: 10.1111/1365-2745.14219

Riley, H. & Wilson-North, R. 1999. From pillow mounds to parterres: a revelation at Cerne Abbas, in Pattison, P., Field, D. & Ainsworth, S. (eds), *Patterns of the Past: Essays in Landscape for Christopher Taylor*, 71–5. Oxford: Oxbow Books

Roach, L. 2016. *Æthelred the Unready*. New Haven and London: Yale University Press

Roach, L. 2018. A tale of two charters: diploma production and political performance in Ætheldredian England, in Naismith, R. & Woodman, D.A. (eds), *Writing, Kingship and Power in Anglo-Saxon England*, 234–56. Cambridge: Cambridge University Press

Robertson, N. & Schofield, J. 2000. Monuments in wartime. Conservation policy in practice, 1939–45, a survey. *Conservation Bulletin* 37, 16–19

Robinson, S. & Valentin J. 2004. A proposed school site, at Simsay, Cerne Abbas, Dorset (centred on NGR ST 6677 0132); results of an archaeological evaluation. Unpubl. AC Archaeology report no 4903/2/0, dated November 2004

Rodnight, H., Duller, G.A.T, Wintle, A.G. & Tooth, S. 2006. Assessing the reproducibility and accuracy of optical dating of fluvial deposits. *Quaternary Geochronology* 1, 109–20

Ross, A. 1967. *Pagan Celtic Britain: Studies in Iconography and Tradition*. London: Routledge & Kegan Paul

Rous, J. 1716. *Joannis Rossi Antiquarii Warwicensis Historia Regum Angliæ*. Oxford: Sheldonian Theatre

Rudling, D. 2001. Chanctonbury Ring re-visited: The excavations of 1988–91. *Sussex Archaeological Collections* 139, 75–121

Sadler, T. (ed.), 1869. *Diary, Reminiscences, and Correspondence of Henry Crabb Robinson*, Vol. 1 (of 2), 7. Boston: Fields, Osgood & Co.

Salway, P. 1981. *Roman Britain*. Oxford: Oxford University Press

Salzman, L.F. 1949. *Victoria History of the County of Warwick, Volume 5: Kingston Hundred*. London: Institute of Historical Research

Sawyer, P.H. 1968. *Anglo-Saxon Charters. An Annotated List and Bibliography*. London: Royal Historical Society

Schwyzer, P. 1999. The scouring of the white horse: archaeology, identity, and heritage. *Representations* 65, 42–62

Scott, W.L. 1937. The Chiltern White Crosses. *Antiquity* 11, 100–4

Seaman, A. 2023. Hillforts in southern Britain: power and place in the late antique landscape, in Pergola, P., Castiglia, G., Hanna, E.E.K., Martinetto & Segura, J.-A. (eds), *Perchement et réalitées fortifiées en Méditerranée et an Europe, 5ème-10ème siècles*, 420–32. Oxford: Archaeopress Publishing

Selbourne, R. 1892. *Ancient Facts and Fictions concerning Churches and Tithes*. London: Macmillan

Sellers, V.B. 2001. *Courtly Gardens in Holland 1600–1650*, 225–227. Woodbridge: Garden Art Press

Semple, S. 2013. *Perceptions of the Prehistoric in Anglo-Saxon England Religion, Ritual, and Rulership in the Landscape*. Oxford: Oxford University Press

Shaw, S. 1791. Excursion from Lewes to Eastbourne in Sussex. *The Topographer* 3, 376

Shippey, T. 2016. *The Cerne Giant*. Cerne Abbas: privately printed

Shoosmith, E. 1938. Letter to *Sussex County Magazine* 12, 281

Simpson, C. 2021. Rebel baron undressed the Cerne Abbas giant to get a rise out of Oliver Cromwell. *The Telegraph* 12 May 2021. Available at https://www.telegraph.co.uk/news/2021/05/12/rebel-baron-undressed-cerne-abbas-giant-get-rise-oliver-cromwell (Accessed 25 November 2023)

Skeat, W.W. 1885. *Ælfric's Lives of the Saints*, Volume II. Old Series 82. London: Early English Text Society

Smart, T.W.W. 1872. The Cerne Giant. *Journal of the British Archaeological Association* 28, 65–70

Smith, B.W., Rhodes, E.J., Stokes, S. & Spooner, N.A. 1990. The optical dating of sediments using quartz. *Radiation Protection Dosimetry* 34, 75–8

Smith, D.M. (ed.), 2001–2008. *Heads of Religious Houses*. 3 vols. Cambridge: Cambridge University Press

Smith, L.T. 1907–10. *The Itinerary of John Leland in or about the Years 1535–1543*, Parts I to XI, London

Society of Antiquaries 1762–1765a. Minute book (volume 9) of The Society of Antiquaries of London. *Minute Book*, 9 February 1764. Ref: SAL/02/009/045. Available at (Accessed 5 July 2023)

Society of Antiquaries 1762–1765b. Minute book (volume 9) of The Society of Antiquaries of London, 204 for 16 February 1764, SAL/02/009/046. Although read on 16 February, Stukeley's note was entered under a significantly different title into the minutes under 15 March 1764, 233–35. SAL/02/009/050. https://collections.sal.org.uk/sal.02.009.046 (Accessed 5 July 2023)

Soutar, S. 2018. *Compton Chamberlayne Wiltshire: Rediscovering Australia - Surveying the 'Lost' Chalk Map*. Historic England Research Report Series no. 80-2018. Swindon: Historic England

Southern Times 1864. Sherborne Park Fete. *Southern Times,* 27 August 1864, 4, col 4

Southern Times and County Herald 1883. The Cerne Giant. *Southern Times and County Herald*, 6 October 1883, 3 col 4

Southern Times and Dorset County Herald 1900. Cricket. *Southern Times and Dorset County Herald*, Saturday 8 September 1900, 7

Southern Times and Dorset County Herald 1908. Cleaning the Giant. *Southern Times and Dorset County Herald*, Saturday 1 February 1908, 8

Spooner, N.A. 1993. The validity of optical dating based on feldspar. Unpublished D.Phil. thesis, Oxford University

Squibb, G.D. 1984. The foundation of Cerne abbey. *Notes and Queries for Somerset and Dorset* 31, 373–6.

Squibb, G.D. 1988. The foundation of Cerne abbey, in Barker, K., (ed.), *The Cerne Abbey Millennium Lectures*, 11–14. Cerne Abbas: The Cerne Abbey Millennium Committee

Stafford, E. 2021. The reception of Heracles, in Ogden, D. (ed.), *Oxford Handbook of Heracles*, 540–56. Oxford: Oxford University Press

Stafford, P. 1978. Church and society in the Age of Ælfric, in Szarmach, P. & Bernard, F. Huppe, B.F. (eds), *The Old English Homily and its Backgrounds*, 11–32. New York: State University of New York Press

Stenton, F.M. 1971. *Anglo-Saxon England*. Oxford: Oxford University Press

Sterckx, C. 1975. Le Géant de Cerne Abbas. *Antiquité Classique* 44: 570–80

Stern, J. 2009. The Orangist myth, 1650–1672, in Cruz, L. & Willem, F., *Myth in History, History in Myth*, 33–52. Leiden: Brill

Stoczkowski, W. 2002. *Explaining Human Origins: Myth, Imagination and Conjecture*. Cambridge: Cambridge University Press

Stubbs, W. (ed.), 1872–73. *Memoriale Fratris Walteri de Coventria, The Historical Collections of Walter of Coventry*, 2 vols, Rolls Series, No. 58. London: Longman & Co.,

Stukeley, W. 1764. [The Cerne Giant] Minute book (volume 9) of the Society of Antiquaries of London, 199–200 (see Society of Antiquaries [of London] Minute book 1764)

Swanton, M. (ed.), 1975. Aelfric's Colloquy, in *Anglo-Saxon Prose*, 105–15. London: J.M. Dent & Sons

Sydenham, J. 1842. *Baal, Durotrigensis; a dissertation on the ancient colossal figure at Cerne, Dorsetshire*. London: W. Pickering

Taylor, I. 1987. *The Giant of Penhill*. York: Northern Lights

Templer, R.H. 1985. The removal of anomalous fading in zircons. *Nuclear Tracks and Radiation Measurements* 10, 531–7

The Bystander 1905. The Vanishing Giant of Cerne. *The Bystander*, 30 August 1905, 46

The Guardian 2007. Homer chalk giant angers pagans. *The Guardian*, 17 July 2007. Available at https://www.theguardian.com/film/2007/jul/17/news1 (Accessed 14 November 2023)

The Observer 1923. The Cerne Giant: ancient turf-cuttings and how to preserve them. *The Observer*, 14 January 1923, 9

The Tatler 1945. Theory. *The Tatler*, 12 September 1945, 14

The Telegraph 2007. Pagans to use 'rain magic' on Homer. *The Telegraph* 17 July 2007. Available at https://www.telegraph.co.uk/news/uknews/1557661/Pagans-to-use-rain-magic-on-Homer.html (Accessed 14 November 2023)

Thomas, A.S. 1960. Changes in vegetation since the advent of myxomatosis. *Journal of Ecology* 48, 287–306

Thomas, A.S. 1963. Further changes in vegetation since the advent of myxomatosis. *Journal of Ecology* 51, 151–83

Thomas, R. 2014. Evidence, archaeology and law: an initial exploration, in Chapman, R. & Wylie, A. (eds), *Material Evidence: Learning from Archaeological Practice,* 255–70. London: Routledge

Thorn, C. & Thorn, F. 1983. *Domesday Book 7: Dorset.* Chichester: Phillimore

Thorn, F.R. 1991. Hundreds and Wapentakes, in Williams, A. & Martin, G.H. (eds), *The Dorset Domesday,* 27–44. London: Alecto Historical Editions

Thurston-Hopkins, R. 1922. *Thomas Hardy's Dorset,* 97. New York: D. Appleton & Co.

Toller, T.N. (ed.) 1898. *An Anglo-Saxon Dictionary Based on the Manuscript Collections of the Late Joseph Bosworth.* Oxford: Clarendon Press

Toms, P.S. & Wood, J.C. 2021. Optical dating of sediments: Cerne Abbas Giant, UK, Unpubl. report University of Gloucestershire Luminescence dating laboratory, dated 5 March 2021

Toulmin-Smith, L. (ed.) 1964. *Leland's Itinerary in England and Wales* 4 vols, i 255, iv 82, 106, 107, 108,109, v 207. London: Centaur Press

Town Towels 2021. We were blasted by the Sun newspaper: 'Tea towel company causes outrage by censoring the Cerne Abbas Giant's penis', 5 October. Available at https://townteatowels.com/blog/we-were-blasted-by-the-sun-newspaper-tea-towel-company-causes-outrage-by-censoring-the-cerne-abbas-giants-penis/ (Accessed 14 November 2023)

Treves, F. 1906. *Highways and Byways of Dorset,* 338–9. London: Macmillan

Tripadvisor 2020. Pragmatic Review 2020, Swindon, 9 October 2020, https://www.tripadvisor.co.uk/ShowUserReviews-g1604560-d210251-r773661760-Cerne_Abbas_Giant-Cerne_Abbas_Dorset_England.html (Accessed 25 November 2023)

Udal, J.S. 1893. Dorsetshire birth, death, and marriage customs and superstitions. *Proceedings of the Dorset Natural History & Antiquarian Field Club* 14, 182–200, esp. 194 (vii)

Udal, J.S. 1922. *Dorsetshire Folk-Lore.* Hertford: Austin & Sons

Vale, V. 1999. Churchwardens and the other God, in Darvill, T., Barker, K., Bender, B. & Hutton, R., *The Cerne Giant: An Antiquity on Trial,* 71–5. Oxford: Oxbow Books

Vale, V. & Vale, P. 2000. *The Parish Book of Cerne Abbas: Abbey and After.* Tiverton: Halsgrove

Valentin, J. 2002. Osmington White Horse, Osmington Parish, Dorset. Results of archaeological evaluation. Unpubl. AC Archaeology Report no. 6602/2/0

Van Houts, E. 1992. A note on *Jezebel* and *Semiramis,* two Latin Norman poems from the early eleventh century. *Journal of Medieval Latin* 2, 18–24

Vandenberghe, D., de Corte, F., Buylaert, J-P., Kucera, J., Van den haute, P. 2008. On the internal radioactivity in quartz. *Radiation Measurements* 43, 771–5

VCH 1908. *Victoria County History of Dorset*, Volume 2. London: Archibald Constable & Co Ltd.

Wallenberg, J.K. 1931. *Kentish Place-Names.* Uppsala: Uppsala Universitets Årsskrift

Walling, P. 2014. *Counting Sheep: A Celebration of the Pastoral Heritage of Britain.* London: Profile Books

Wallis, R.H. 1976. An approach to the space variant restoration and enhancement of images. *Proceedings of the Symposium on Current Mathematical Problems in Image Science.* Montery, 10–12 November 1976

Walsham, A. 2011. *The Reformation of the Landscape,* 8–152, 510. New York: Oxford University Press

Walter of Coventry, 1872–1873. Memoriale fratris Walteri de Coventria: the historical collections of Walter Coventry. *Chronicles and Memorials of Great Britain and Ireland during the Middle Ages* (Rolls Series) 58, 2 vols, ed. W. Stubbs. London: Longman Green

Warminster Herald 1868. Cerne Abbas. *Warminster Herald,* 21 March 1868, 5, col 5

Webster, L. 2012a. *The Franks Casket.* London: British Museum

Webster, L. 2012b. *Anglo-Saxon Art: A New History.* London: The British Museum

Weir, A. & Jerman, J. 1986. *Images of Lust.* London: Batsford

West, A. 2023. 'Where's Willy? Cheese maker causes stink by featuring the Cerne Abbas Giant on its packets – without his huge willy. *The Sun* 1 June 2023, https://www.thesun.co.uk/news/22553266/cheese-maker-cerne-abbas-giant-packets-without-willy/#:~:text=A%20CHEESE%20maker%20has%20caused,famous%20chalk%20figure%20non%2Dbinary. (Accessed 14 November 2023)

Western Chronicle 1888. Dorset Field Club. *Western Chronicle,* Friday 6 July 1888, 3, col 5

Western Chronicle 1905a. Cerne Abbas. *Western Chronicle,* 25 August 1905, 6

Western Chronicle 1905b. The Cerne Giant. *Western Chronicle,* 25 August 1905, 4

Western Daily Mercury 1908. The Cerne Giant. *Western Daily Mercury,* Saturday 8 August 1908, 7, col 7

Western Daily Press 1974. Giant Task. *Western Daily Press*, Wednesday 1 May 1974, 4, col 4

Western Gazette 1863. Cerne Abbas. *Western Gazette*, 19 September 1863, 6

Western Gazette 1905. The Cerne Giant: satisfactory result of a clean out. *Western Gazette*, 6 October 1905, 12, col 6

Wheeler, R.E.M. 1943. *Maiden Castle, Dorset*. Report of the Research Committee of the Society of Antiquaries of London

Whitelock, D. 1961. *The Anglo-Saxon Chronicle*. London: Eyre & Spottiswoode

Whitelock, D. 1979. *English Historical Documents* c. *500–1042*. 2nd edn. London: Eyre and Spottiswoode

Wilkin, R. 2021. Speculations about the Abbey. *Cerne Historical Society Magazine,* Summer 2021, 5–9

Willcox, T. 1988. Hard times for the Cerne Giant: 20th century attitudes to an ancient monument. *Antiquity* 62, (Sept) 524–6

William of Malmesbury 1870. *Willelmesbiriensis Monachi De gestis pontificum Anglorum libri quinque* (Rolls Series) 52, ed. N.E.S.A. Hamilton. London: Longman Green

Williams, H. 2006. *Death and Memory in Early Medieval Britain*. Cambridge: Cambridge University Press

Williams, T.J.T. 2013. The Battle of Ashdown. *Medieval Warfare* 3.5, 16–20

Williams, T.J.T. 2016. The place of slaughter: exploring the West Saxon Battlefield, in Lavelle, R. & Roffey, S. (ed.), *Danes in Wessex*, 35–55. Oxford: Oxbow Books

Willmott, H. 2022. New archaeological survey in Beauvoir Field. *Cerne Historical Society Magazine,* Autumn 2022, 2–3

Willmott, H. 2023. Excavations at Cerne Abbey, a project design for 2023. *Cerne Historical Society Magazine*, Summer 2023, 13–19

Wilson-North, R. & Riley, H. 2003. From pillow mounds to parterres: a revelation at Cerne Abbas, in Wilson-North, R. (ed.), *The Lie of the Land: Aspects of the Archaeology and History of the Designed Landscape in the South West of England, 100–6.* Exeter: Mint Press

Winterbottom, M. (ed. & trans.), 1978. *Gildas. The Ruin of Britain and Other Works.* History from the Sources: Arthurian Period Source Vol. 7. Chichester: Phillimore

Winterbottom, M. (ed.), 2007. *William or Malmesbury, Gesta Pontificum Anglorum (The History of the English Bishops).* Vo.l 1: text and Transaltion ed. R.M. Thomson. Oxford: Clarenden Press

Wintle, A.G. 1973. Anomalous fading of thermoluminescence in mineral sample. *Nature,* 245, 143–4

Wise, F. 1738. *A letter to Dr Mead concerning some antiquities in Berkshire, particularly shewing that the White Horse, which gives name to the Vale, is a Monument of the West-Saxons, made in memory of a great victory obtained over the Danes A.D. 871.* Oxford: Thomas Wood. Available at https://archive.org/details/bim_eighteenth-century_a-letter-to-dr-mead-conc_wise-francis_1738 (Accessed 28 September 2023)

Wise, F. 1742. *Further observations upon the White Horse and other antiquities in Berkshire, with an account of Whiteleaf-Cross in Buckinghamshire. As also of the Red Horse in Warwickshire, and some other monuments of the same kind.* Oxford: Thomas Wood

Withers, B.C. & Wilcox, J. (eds), 2003. *Naked before God: Uncovering the Body in Anglo-Saxon England.* Medieval European Studies Series III. Morgantown: West Virginia University Press

Witney, K.P. 1976. *The Jutish Forest.* Tiptree: Athlone Press

Wood, I. 1994. The Mission of Augustine of Canterbury to the English. *Speculum* 69 (1), 1–17

Woodward, A. 1992. *Shrines and Sacrifice.* London: Basford/English Heritage

Woodward, P.J. & Cox, P.W. 1984. Field survey of the ancient fields and settlement enclosures at Back Hill, Cerne Abbas, Dorset. *Proceedings of the Dorset Natural History and Archaeological Society* 106, 111

Woodward, P.J., Bellamy, P.S. & Cox, P.W. 1988. Field survey of the ancient fields and settlement enclosures at Black Hill, Cerne Abbas, Dorset. *Proceedings of the Dorset Natural History and Archaeological Society* 109, 55–64

Woolner, D. 1965. The White Horse, Uffington. *Transactions of the Newbury District Field Club* 11 (3), 27–44

Woolner, D. 1967. New Light on the White Horse, *Folklore* 78, 90–111

Worsley, L. 1998. *Hardwick Old Hall.* London: English Heritage

Worth, R.N. 1893. *Calendar of the Plymouth Municipal Records.* Plymouth: William Brendon & Son

Yorke, B. 1988. Aethelmaer: the foundation of the Abbey at Cerne and the Politics of the Tenth Century, in Barker, K. (ed.), *The Cerne Abbey*

Millennium Lectures, 15–26. Cerne Abbas: Cerne Abbey Millennium Committee

Yorke, B.A.E. 1995. *Wessex in the Early Middle Ages*. London: Leicester University Press

Yorke, B.A.E. 2015. The fate of otherworldly beings after the conversion of the Anglo-Saxons, in Ruhmann, C. & Brieske, V. (eds), *Dying Gods – Religious Beliefs in Northern and Eastern Europe in the Time of Christianisation.*, 167–76. Neue Studien zur Sachsenforschung 5. Hannover: Niedersächsisches Landesmuseum

Yorke, B. 2001. The origins of Mercia, in Brown, M.P. & Farr, C.A. (eds), *Mercia: An Anglo-Saxon Kingdom in Europe*, 13–22. London: Leicester University Press

Yorke, B. 2017. King Alfred and Weland: Tradition and Transformation at the Court of King Alfred, in Insley, C. & Owen-Crocker, G.R. (eds), *Transformation in Anglo-Saxon Culture. Toller Lectures on Art, Archaeology and Text*, 47–69. Oxford: Oxbow Books

Zimmerman, D.W. 1971. Thermoluminescent dating using fine grains from pottery. *Archaeometry* 13, 29–52

Internet sources

Social media

Dorset Council 2020. [Facebook] 23 July, Available at https://www.facebook.com/share/P3MC5H1ff2eDsnsW/?mibextid=WC7FNe (Accessed 14 November 2023)

Dorset Echo 2023. [Facebook] 6 October, Available at https://www.facebook.com/share/K4Zc6Pekq03xY4EL/?mibextid=WC7FNe (Accessed 14 November 2023)

Have I Got News for You 2020. [Twitter] 8 July. Available at https://x.com/haveigotnews/status/1280849914612809734?s=20 (Accessed 14 November 2023)

National Trust 2019. [Twitter] 30 August. Available at https://x.com/nationaltrust/status/1167341913629962240?s=20 (Accessed 14 November 2023)

Richard Osgood 2020a. [Twitter] 20 April. Available at https://x.com/richardosgood/status/1252252295875694596?s=20 (Accessed 14 November 2023)

Richard Osgood 2020b. [Facebook] 25 April. Available at https://www.facebook.com/richard.osgood.77/posts/pfbid02y73MZ3MgU66f2FaZHatuV4Mrhvmz6APcP3Yn4M1x7b35CHiQ3tniPyQxHH68Ei4Jl (Accessed 14 November 2023)

Stephen Fry 2019. [Twitter] 29 August. Available at https://x.com/stephenfry/status/1167178902428618752?s=20 (Accessed 14 November 2023)

Index

Page numbers in *italic* refer to illustrations, those in **bold** to tables. Places are in Dorset unless stated otherwise.

3-D ground surface model *see* photogrammetric surveys

abbey *see* Cerne Abbey
additions to the Cerne Giant 152, 216–18, **217**, *217*
Ælfric Bata 184
Ælfric of Eynsham 34, 126, 127, 128, 131, 184, 192
Æthelmær 124, 125–6, 127, 135, 183, 191–2
Æthelnoth, Archbishop of Canterbury 127
Æthelred II, King 126, 127
Æthelweard 124, 125, 126, 127, 130, 133, 135, 183, 191
Alfred, King 123, 130
Allsup, Christopher 288–9
ancient giant, arguments for *19*, 197, 203–7, 208–13
Anglo-Saxon Chronicle 123, 124, 133, 275, 276, 277, 280
Anglo-Saxon lawcodes 123
animation *201*, 202
anthropogenic sculpting 54, *54*
antiquarian interest 29–30, 140, 141–3, *142*, 144–5, *144*, 162, 203
archaeological context 12–15, *13–15*
archaeological reasoning 214
Arthur C. Clarke's Mysterious Worlds (Yorkshire TV) 68–9
Ashdown, Battle of 276
assemblages, scholarly 213
assembly places 275–83
 Cerne Giant and 146
 character of 277–81
  boundaries and borderlands 278
  Domesday hundred names 278–9
  'hanging promontories' 279
  naming of 279–81
  routeways 277–8
 corpus of 132, 133, 275–7, *276*
 giants, assembly and wayfinding 281–3
Aston Clinton, Buckinghamshire 303, 304, *304*
Aubrey, John 30, 38
auger surveys
 current project 55, *56*, 103–5, *105*
 previous work *14*, 44–5, 160
Augustine, Saint 268
Aylesford, Kent 275, *276*, 277, 278, 280

Barnett, Cathie 87
barrows 131, 132–3
 see also Wayland's Smithy
Beard & Co. 48, 86
Bede 275
Bell, Martin 45
bell ringing 155
Benedictine Abbey *see* Cerne Abbey
Benedictine reforms 124, 184, 190, 191
Beowulf 129, 133
Bettey, Joseph 31, 196, 198
Beverston, Gloucestershire 275–6, *276*
Bewley, Bob 65
biases 213, 214
Black Hills (Paha Sapa), South Dakota 248–9
Blackmore 35, 192, 193
Bledlow Cross, Buckinghamshire 298, 302–3, *302*, 304, *304*
Boethius, *Consolation of Philosophy* 130–1, 132
Book of Cerne 185

Bournemouth University 24, 199–200, *199–201*, 202
Bratton White Horse *see* Westbury White Horse
bricks 240–1, 242, 259–60
Bryson, Bill 247–8
burial mounds *see* barrows; Wayland's Smithy

Camden, William 299
Castleden, Rodney
 The Cerne Giant 205–6, 210, 211
 geophysical surveys
  interpretation of 20, *21*, 22, 23, 73
  methods 41, 42, *42*, 49
  results 43–4
censorship 146, 150, 216
ceramic building material 240–1, 242, 259–60
Cerne (Cerne Abbas)
 archaeology 15
 estate 123–4
 history 29, 32, 34
 location and setting 11
 in the West Saxon kingdom 122–4
Cerne Abbey 183–9
 abbey church 185–7, *186–7*
 abbey, Saxon 184–5
 archaeology 14, *15*, 187–8, *189*
 Eadwold and 133–4
 and the Giant 189, 272
 history 14, 32, 124–8
  Æthelweard and Æthelmær, family of 126–7
  foundation of 124–6, 183, 191–2
 land management 192–4
 prayer book '*Book of Cerne*' 185
Cerne Giant
 archaeological context 12–15, *13–15*
 archaeology 38–45, **41**, *42–3*
  conclusions 157–8
  two (or more) giants 163, *165*, *167*, *169*
 comparison with Long Man of Wilmington 258
 historical records 16, 27, 29–30, *31–3*, 35, 36–8
  absence from 37–8, 157–8, 162–3, 173–4
 identity 38
 in the landscape 120–1
 location and setting 8, *9*, 11–12
 medieval origin, first suggestion of 146
 post-Dissolution period 147
 previous work 8–10, 20–1, 22
 timeline 175–80
 see also dating of the Cerne Giant
Cerne Giant, anatomy of 15–23, *16*
 below left arm (lion skin, cloak and severed head) 20–2, *21*, 160–1, 166, 168, 170
  auger survey 103–5
  geophysical surveys 42–3, 71–3, *72*, 76, *76*, 160, 264–5
  photogrammetric surveys 64, *67*
 belt? *63*, 65, *67*, *266*, 268
  auger survey 105
  geophysical surveys 73–4, 75–6, *75*
 face and body
  face 18, *63*, 64
  geophysical surveys 70–1, *72*
  nose 64, *67*, 152–3, 156, 170

Index

outline 17–18, *17*, 64, *66–7*
   developmental sequence 105–7, *106*
   geophysical surveys 70–1, *72*
   movement downslope 82–3, *97*, 171
   trenched form *82–3*, 164–6, *165*, *167*, 193–4, *195–6*, *196*
phallus
   drawings *142*, 143, *144–5*, 145–6
   iconography of 269–70, *270*
   and navel 18–20, *19*, 63, 64, 146, 265, *265*, 268
   photographs *139*, 150
   symbols 22–3, *23*, 32, 64, 141, 170
   geophysical surveys 76, *76*
true plan image 67–8, *68*
viewing of 16–17, 24 *see also* visibility of the Cerne Giant
Cerne Giant: an antiquity on trial (1996) 5, 10, 195–202
   background 195–6
   public inquiry event 197–9
      ancient giant 197, 203
      living giant 198
      outcome 198–9, 214
      post-medieval giant 197–8
Cerne Giantess 24, *24*, 199–200, *199–201*, 202
Cerne River 12
chalk fill
   dating of 53–4, 82–3, 89–90, 173
   excavated sequence 80–1, *82–3*, 84–7, **85**, 94–5, *95–6*
   origins of 89
Charles the Bald 128
Charman, John 89
Cherhill White Horse, Wiltshire 285, 288–9, 312
Childe, V. Gordon 214
Chiltern hill figures 300–5, *300–2*, *304*, **310**
Christianisation of monuments 212, 267–8
churchwardens' accounts 154, *154*
Clark, Grahame 252
Clark, Tony 22, 41, 42, *43*, 44, 68, 69, *70*
clubs 212
Cnut, King 127
coins 254
Colley, Linda 247
colluvium *see* geoarchaeology
Commonwealth period *see* Cromwell, Oliver
community engagement 151–2
community interest 138–9, 148
companions 23–4, *24*
Compton Chamberlayne badge, Wiltshire 314
Compton, Sir Henry 244–5
Compton, Spencer (Earl Wilmington) 245
computer animation *201*, 202
condition of the Cerne Giant 46–7, 48, 136
confirmation bias 213, 214
COVID-19 pandemic
   images of the Cerne Giant **217**, 218, *219–20*
   impact of 6–7
Crawford, O.G.S 204, 209, 211
Cromwell, Oliver 10, 31, 38, 159, 196
crosses 298, 300–6, *300–2*, *304*, **310**
Cuckhamsley, Berkshire 132, *276*, 277, 278, 280
cultural evolution 207
cultural significance of the Cerne Giant 219–21
Cwichelm, King 132–3
Cwichelmeshlæw 132–3, 277, 278, 280

dancing 35
Darvill, Timothy 210–11
dating of the Cerne Giant
   aims and objectives 50–1
   current project 6–7
   early medieval dates (7th to 14th cent.)
      7th cent. 157, 158, 270, *270*
      10th cent. 111–12, 157, 159–60
      13th cent. 113, 193–4
   fieldwork methods 55–7, *56*
   historical records 36–8

17th cent. dates 37–8, 153–4, *154*
OSL results 109–13, **110–11**
previous work 5–6, 9–10
problems of OSL dates 271, 274
walkover survey 51, *52–4*, 53–5
Davis, Jean 303
Defoe, Daniel 252
Devizes White Horse, Wiltshire 285, 290–2, **291**, *292*, 314
Diddersley Hill, North Yorkshire 275, *276*, 277
Dissolution of the monasteries 185
Ditchling Cross, East Sussex 306
Domesday Book 123–4, 127, 277, 278–9
Dorset Ooser 35, *36*
drawings 141–5, *142–5*, 163–4
   see also measured surveys
drone photography 136
drone survey *see* photogrammetric surveys
Ducarel, Andrew Coltée 143
Dunstan, Saint 184, 191
Durotriges 206, 269

Eadwold, Saint 34, 127, 133–5, 161, 184
'Ecgbert's Stone' *276*, 276, 277, 278, 280
Ecgbrihtesstan *276*, 276, 277, 278, 280
Edgar, King 125, 191
Edwards, Brian 288–9
Egan, P. 289
enclosure of the Cerne Giant 24–5, *25*
enta (giants) 128, 131
equine hill figures 152, 284–96
   case studies
      Cherhill White Horse 288–9
      Devizes White Horse 290–1, **291**
   catalogue **309–10**
   current, lost and possible examples 285–6, *285*
   fieldwork
      Devizes White Horse 291–2, *292*
      Marlborough White Horse 293, *294–5*
   origin stories 286–7
   Uffington White Horse, inspiration from 287–8, 289
   see also Uffington White Horse
erosion 51, *52–3*, 53, 97–9
estate maps *142*, 143
Eton College, Berkshire 298, 302–3
excavations
   approach and methods 55–7, *56*, 77, *78–9*
   chalk fill, dating of 53–4, 82–3, 89–90, 173
   chalk fill, origins of 89
   excavated sequence 80–1, *82–3*, **85**, *96*
   Giant 1 (chalk-cut) *82–3*, 84, 98–9
   Giant 2 (chalk-filled) 53–4, *82–3*, 84–7, **85**, *95–6*, 99
   stratigraphy 78–9, **85**
   trial trench (1945) 45, 47–8
   wooden stakes *82–3*, 87–8, **88**, *97*, 172–3
experimental archaeology 24, 199–200, *199–201*, 202
Eynsham, Oxfordshire 125, *126*

folklore 35–6, *36*
Foxe, J., *Actes and Monuments (Book of Martyrs)* 263–4
Freke, Thomas 155
Fry, Stephen 218
fyrd (shire levies) 124, 277

Gale, John 21, 40, 41–2
Gaulish warriors 265, *266*
Gelling, Margaret 282
Gentleman's Magazine 144, *144*
geoarchaeology
   discussion 97–9, 158–9
   hillwash deposits 95, *97*, *106*
   infill events 105–7, *106*
   sediment, accumulation of 53–5, *53–4*
   soil profiles 92–4, 99–103

upslope and downslope deposits 53–4, 82–3, 90–5, 95, 97
Geoffrey of Monmouth, *History of the Kings of Britain* 34, 272
geoglyphs 247
geology 11
geophysical surveys 68–77
    earth resistance survey (1979–80) 43, 68–73, *72*
    earth resistance survey (1989–95) 73, 264–5, *265*
    ground penetrating radar and resistivity surveys (2023) 73–7, *75–6*, 160
    previous work 41–4, *42–3*, 70, 160
    other sites
        Cerne Abbey 188, *189*
        Long Man of Wilmington 260–1
George III, King 255
giant-heroes 128–31, 133–4, 135
giant hill figures 243, 271–2, 281, *281*
    see also Long Man of Wilmington
giant- placenames 131–3, 211, 282
Giant's Hill 139
Glorious Revolution (1688) *see* William III, King
Godfrey, Ian 254
Gog-Magog 34, 243, 282, 320
Goscelin of Canterbury 34, 134, 161, 268, 273
'great men' of archaeology 204–5, 209–11, 214
Grimm, Samuel Hieronymous 143–4, *143*
Grinsell, Leslie 18, 19, *19*, 146, 203, 204, 205
guidebooks 206–7

Hamalainen, Pekka 249
Hardwick Old Hall, Derbyshire 243
Hardy, Jonathan 140
Hardy, Thomas 46–7, 165, 172
Harrow Way 276, 277, 278, 283
Hawkes, Jacquetta 206–7, 213
health messages **217**, *217*, 218
Helith 38, 128, 209, 210, 273
Hercules
    Anglo-Saxon associations 123, 124, 127, 128, 130–1, 135, 159–60, 280–1
    Cerne Giant identified as 205, 209, 210
    iconography 128, 212, 273
    William III and 153, *153*, 154–6, *155*
heroes, Anglo-Saxon 128–31, 133–4, 135
hill figures 297–306, 307–20
    additions to the Cerne Giant 152, 216–18, **217**, *217*
    as archaeological monuments 310–15
        archaeological research 313–15
        history and development of 311, *312*
        maintenance and renewal 312–13
        protection of 310–11, 313
        surveys as baseline records 312
    catalogue 308, **309–10**
    Chiltern hill figures 300–5, *300–2*, *304*
    distribution of 247, 297–8
    of giants 243, 258, 271–2, 281–2, *281*
    guidebooks 206–7
    meanings and purposes 161, 254, 299–300, 305–6, 307
    as monument type 307–8
    niche literature about 205–6
    research agenda 316–20
        characterisation and setting 319
        dating 319
        historical research and social history 316–17
        lost figures 320
        maintenance and renewal 320
        reinstating lost images or parts 320
        surveys and recording 317–19
    valuing of 315–16
    see also Cerne Giant; equine hill figures; geoglyphs; Long Man of Wilmington; Uffington White Horse
hillwash deposits (*see* colluvium) 95, 97, *106*
Hoade, Bill 77
Hoare family 46
Hoare, Alda 26, 46, 48, 163, 165

Hoare, Sir Henry 25–6, 46, 137, 163, 165, 177
Holles, Denzil, 1st Baron 30–1, 38, 194, 196
Holloway, William, 'Giant of Trendle Hill: A Legendary Tale' 208–9
horse cults 254–5
horse hill figures *see* equine hill figures; Uffington White Horse
horse iconography 254, 255
humorous observations 218–19
hundred meeting places 277
hundred names 278–9
Hutchins, Rev. John
    creation of the Cerne Giant 194
    description of the Cerne Giant 90, 140–1, 164
    maintenance of the Cerne Giant 26, 172, 193
    measurement of the Cerne Giant 40
    sketch of the Cerne Giant 141–2, *142*, *144*, 145
Hutton, Ronald 10, 37, 38, 157, 162, 197, 198
Hygelac, King 129–30

'Icknield' 283
Icknield Way 276, 281, 282, 305
    see also Ridgeway
iconography
    of Hercules 128, 212, 273
    of horses 254, 255
    of phalluses 269–70, *270*
    identity of the Cerne Giant 38, 128, 196, 273
    see also Eadwold, Saint; Hercules
Iglea 276
Iley Oak, Sutton Veny, Wiltshire 276–7, *276*
images of the Cerne Giant 215–21
    advertising and marketing 215–16, *216*, **217**
    global audiences 215–16
    social media 218–19, 221
    social responsibility 217–18, *217*
    see also photographs of the Cerne Giant
Iron Age coins 254
Iron Age period 206, 269

Kingston, Robert Lumley 143
Kite, Edward 290
Kitson, Peter 283
Knollys, Eardley 47, 138, 166, 173, 178

Lakota people 248–9
land snails 113–20
    discussion 119–20
    methods 57, 82–3, *112*, 114–15
    results 115–19, **116**, *118*, 162, 273
land-use history 115–19, **116**, *118*, 162
Landscape Mysteries (BBC TV) 5, 44–5, 234–5
Langtree, Gloucestershire 275–6, 277
legends 35–6, *36*, 132, 208–9, 211, 282
Legg, Rodney 18, 23, 205
Leland, John 30, 32, 35, 38, 125, 126, 162, 198, 272, 273
Lewes martyrs 263, *263*
Liber monstrorum de diversis generibus 129, 130
local engagement 151–2
Long Man of Wilmington, East Sussex 231–46
    archaeological history 234, *234*
    bricks 240–1, 242, 259–60
    chalk rubble 260
    changes to 261
    comparison with Cerne Giant 258
    conclusions 245–6
    construction of 90, 240–1, 242
    dating of 231, 260, 261, 262–3
    discussion 241–5
        dating of the figure 242, 245–6
        historical context 242–4
        instability of the slope *232*, *236*, 241
        intervisibility 243–4
        ownership 244–5

drawings 261, *262*
excavations
in 1969 *234*, 259
in 2002 234–7, *234*, *236–8*, 260
in 2004 233–4, 238, *239*, 261–2
geophysical survey 260–1
maintenance and renewal 174
OSL dating 5, 45, 235, 237, *237–8*, 242
pagan worshippers 233, *233*
Ravilious painting 47, *168*
trench outline 261–2, *262*
Lyon, Harriet 147

maintenance and renewal events
camouflaging 137–8
dating of 53–4, *82–3*, 89–90, 173
developmental sequence 105–7, *106*
excavated sequence 53–4, 80–1, *82–3*, 84–7, **85**, 94–5, *95–6*
grass and turf management 137, 138–9, 140–1
history of 25–9
problems of 311, 312
proposals for 227–8, 320
records, lack of 173–4, 311, 315–16
responsibility for 298, 312–13
trenched form 167–8, 171–2, 173
other sites
Long Man of Wilmington 174
Uffington White Horse 255, *255*, 256, 287, 298
see also condition of the Cerne Giant
maps *142*, 143
Marlborough White Horse, Wiltshire *285*, 293, *294–5*, 311, 314
Marples, Morris 35, 37, 205, 287, 305
martyrs 263, *263*
Maton, W.G. 144
maypole dancing 35, 209
meanings and purposes of hill figures 161, 254, 299–300, 305–6, 307
measured surveys
Cerne Giant *16*, 31–3, 38–41, **41**, 163–4
Marlborough White Horse 293, *295*
see also drawings; photogrammetric surveys
media coverage 151
see also radio broadcast; social media; television programmes
military assembly places *see* assembly places
military badges **310**, 314
milk bottles 150, *150*
molluscs *see* land snails
Moot Hill Piece 276
Moulsford, Oxfordshire 276
Mount Rushmore National Memorial 249
November 217, 218
mustering sites *see* assembly places
myths 35–6, *36*, 132, 208–9, 211, 282

National Trust
centennial Cerne project 4–7
guidebooks 206
maintenance and renewal events 26–8
archaeological evidence 53–4, 80–1, *82–3*, 84–7, 89–90, 94–5, *95–6*, 107
media coverage 218–19, *220*
records, lack of 173, 174, 311, 315–16
management plan (1974) 323–4, 329–30
ownership of the Cerne Giant 45–9
see also Uffington White Horse
navigation 283
Newman, Paul 205
Norden, John & John 38, 147–8, 162, 198
Norse mythology 282

Old English poems 211
see also Beowulf
Olympic torch 217–18

Orosius, *Seven Books of History Against the Pagans* 128
Osgood, Richard 218, *219*
OSL (optically stimulated luminescence) dating 5, 57–60, 108–13
background 108
concerns about 274
discussion 111–13, **111**
early deposits (Giant 1) 111–12, *112*
later deposits (post Giant 2) *112*, 113
measurement
age estimation 60
D_e and D_r values 59–60
mechanisms and principles 57–8
results 109–11, **110–11**
OSL sample 1 331–3, **332–3**, *332*
sample preparation 58–9
sampling 55–6, *56*, *82–3*, 108, **108**, *109*, *112*
other sites
Long Man of Wilmington 5, 45, 235, 237, *237–8*, 242
Uffington White Horse 5, 253
Osmington White Horse, Dorset 314

pagan monuments 212, 267–8
pagan origins of the Cerne Giant 213, 268, 269–70
pagan worshippers 233, *233*
paintings of hill figures 47, 166, *168*
Papworth, Martin 3, 4, 22, 40, 49, 65, 222–3
Payne, E.J. 305
Petrie, Sir Flinders
Cerne Giant
description 18, 26, 172
interpretation of 209, 211, 212
survey *16*, 39, 40, 46, 164
survey pegs 88
The Hill-figures of England 204
Uffington White Horse and 251
Pewsey White Horse, Wiltshire *285*, 286, 312
photogrammetric surveys
Cerne Giant 62, *63*, 64–5, *66–8*, 67–8
Devizes White Horse 291–2, *292*
photographs of the Cerne Giant 136–7
aerial
early 20th cent. *167*, 196
mid 20th cent. *167*, *171*
from the ground
early 20th cent. 138, *139*, 150
mid 20th cent. *165*
21st cent. *169*, *182*
Piggott, Stuart
dating of the Cerne Giant 209–10, 211, 252
Hercules, Cerne giant as 204–5
radio broadcast 325–8
restoration work 138
trial trench 45, 47–8
Pitman, Gerald 18–19
Pitstone Hill, Buckinghamshire 303, 304, *304*, 305
Pitt Rivers family 46
Pitt Rivers, Gen. Augustus 46, 90, 140, 204
placenames 131–3, 211, 275–7, 278–9, 280, 282
Plenderleath, William Charles 31, 164, 286–8, 290
Plymouth Hoe, Devon 282, 306
Pococke, Richard 26, 30, 140, 176
poems 208–9, 211
see also Beowulf
Pollard, Josh 254
pre-existing monuments 212
prehistoric Cerne Giant *19*, 197, 203–7, 208–13
prehistoric clubs 212
prehistoric monuments 131, 132–3, 211, 212
see also Wayland's Smithy
Priest, S.-J. 305
Protestant martyrs 242, 263, *263*
Pryce, Benjamin *142*, 143
puppets 151
purpose of chalk figures 161, 254, 299–300, 305–6, 307

radio broadcast 325–8
railings 24–5, 64, 66–7, 140
Ravilious, Eric 47, 166, 168, 172, 231
reasons for making hill figures 161, 254, 299–300, 305–6, 307
Regularis Concordia 191
research agendas
    Cerne Giant 222–8
        management issues 227–8
        movement of the giant 224–5
        proposals 225–7
        topographical and geophysical surveys 222–4, 225–6
    national agenda 316–20
        characterisation and setting 319
        dating 319
        historical research and social history 316–17
        lost figures 320
        maintenance and renewal 320
        reinstating lost images or parts 320
        surveys and recording 317–19
Ridgeway 132, 250–1, 276, *276*, 277–8, 282–3
rivers 12
Ross, Anne 205

Sackville family 244
sacred places 248–9
saints *see* Augustine; Dunstan; Eadwold
Scheduled Monument status 46, 148, 153, 313
schools 151–2, 290–1, 293, 296
scourings *see* maintenance and renewal events
sculpting *see* anthropogenic sculpting
Scutchamer Knob, Berkshire 132, *276*, 277
Second World War 137–8
sediment, accumulation of 53–5, *53–4*
Semple, Sarah 281, 283
seventeenth century 243
    see also antiquarian interest; Cromwell, Oliver
sheep 192–3
shire assembly places *see* assembly places
shire levies (fyrd) 124, 277
Shotover Hill, Oxford 282
Silley River 12
sixteenth century 243, 263–4, *263–4*, 273
    bricks 242
Sloper, Joseph Marler 290
Smart, Thomas William Wake 4, 37, 146
Smith, George 290
Smith, Ivan 136
snails *see* land snails
'Snob's Horse' *see* Devizes White Horse
social media 218–19, 221
soil micromorphology 57, 91
soil profiles *see* geoarchaeology
soil sampling 57, 91
    see also geoarchaeology; land snails
soils 11–12
South Downs 306
stakes, wooden 82–3, 87–8, **88**, 97, 172–3
Stane hundred 278–9
Stuart period 243
    see also antiquarian interest; Cromwell, Oliver
Stukeley, William 4, 30, 37, 38, 40, 142, 166, 203–5, 328
Sun Horse 254
surveys *see* auger surveys; geophysical surveys; measured surveys; photogrammetric surveys; topographic surveys
survival of beliefs 210, 211, 213
Swift, Jonathan, *Gulliver's Travels* 156
Swinbeorg 276, 278

television programmes 5, 20, 44–5, 68–9, **217**, 218, 234–5
Temple Mount, Jerusalem 248
tenth-century dating 111–12, **111**, 163, 192, 193, 194
'Tenth-Century Reformation' 124, 184, 190, 191
thirteenth-century dating **111**, 113, 163, 193, 194
Thomas, Roger 214
timeline 175–80

Toms, Phil 108, *109*, 111
topographic surveys 22, 40
    see also photogrammetric surveys
Trendle, The 279
    association with the Cerne Giant 131
    digital terrain model 65, *66*
    interpretations of 209, 210
    location 13, *13–14*, *17*, 25
    maypole and Morris dancing 35
Treves, Sir Frederick 28, 138, 139
Trial (1996) *see* Cerne Giant: an antiquity on trial
true plan image 67–8, *68*
Tudor period 243, 263–4, *263–4*, 273
    bricks 242
turf monument, Cerne Giant as 137, 138
Twitter *see* X (formerly Twitter)
Tysoe Red Horse, Warwickshire 297, 298–9, 306

Uffington White Horse, Oxfordshire 247–57
    chalk fill 90
    continuity of attention to 254–5
    dating of 197
    excavations 252, *253*
    inspiration for Wiltshire White Horses 287–8, 289
    legends 132
    location and setting *248*, 249–51, *250*, 253
    management of 252, *253*, 256
    OSL dating 5, 253
    protection of 251
    Ravilious painting 47
    restoration 48
    sacred places 248–9
    scourings and Pastimes 255, *255*, 256, 287, 298
    shrinkage 256–7, *256*
    survey (2022) 256–7, *256*
    theories about 251–2, 254

Vale, Vivian 153, 195
vegetation 323–4
Viking armies 124, 190–1
visibility of the Cerne Giant 136–41, 145

Walter of Coventry 34, 273
Wandlebury Camp, Cambridgeshire 281–2, *281*
Way, Charles 140
wayfinding 283, 305
Wayland's Smithy, Oxfordshire 131, 132
Webb, Richard 289
Weland 131–2
Wendel- placenames 282
Wessex Morris Men 35, *36*
Wessex Ridgeway 276, *276*, 277, 278
    see also Ridgeway
West Saxon kingdom 122–4, 132–3
Westbury White Horse, Wiltshire 285, *285*, 286, 296, 297, 313
white horses *see* equine hill figures
Whiteleaf Cross, Buckinghamshire 298, 300–2, *300–1*, 304, *304*
Wild Man 212
Wilfaræsdun 275, *276*, 277, 278, 280
William III, King 153–6, *153–5*
William of Malmesbury, *De Gestis Pontificorum Anglorum* 34, 125, 134, 273
Williams, Thomas 281
Wilmington Priory, East Sussex 241, 242, 244
Wiltshire's white horses *see* equine hill figures
Wise, Francis 140, 286, 297, 300, 301
Woden 129, 131
wooden clubs 212
wooden stakes 82–3, 87–8, **88**, 97, 172–3
Woodman, Richard 263, *263*
World War II 137–8

X (formerly Twitter) 218